Power Electronics and Power Systems

The Power Electronics and Power Systems Series encompasses power electronics, electric power restructuring, and holistic coverage of power systems. The Series comprises advanced textbooks, state-of-the-art titles, research monographs, professional books, and reference works related to the areas of electric power transmission and distribution, energy markets and regulation, electronic devices, electric machines and drives, computational techniques, and power converters and inverters. The Series features leading international scholars and researchers within authored books and edited compilations. All titles are peer reviewed prior to publication to ensure the highest quality content. To inquire about contributing to the Power Electronics and Power Systems Series, please contact Dr. Joe Chow, Administrative Dean of the College of Engineering and Professor of Electrical, Computer and Systems Engineering, Rensselaer Polytechnic Institute, Jonsson Engineering Center, Office 7012, 110 8th Street, Troy, NY USA, 518-276-6374, chowj@rpi.edu.

More information about this series at http://www.springer.com/series/6403

Jakob Stoustrup · Anuradha Annaswamy
Aranya Chakrabortty · Zhihua Qu
Editors

Smart Grid Control

Overview and Research Opportunities

 Springer

Editors
Jakob Stoustrup
Department of Electronic Systems
Aalborg University
Aalborg, Denmark

Anuradha Annaswamy
Department of Mechanical Engineering
MIT
Cambridge, MA, USA

Aranya Chakrabortty
Department of Electrical
    and Computer Engineering
North Carolina State University
Raleigh, NC, USA

Zhihua Qu
Department of Electrical
    and Computer Engineering
University of Central Florida
Orlando, FL, USA

ISSN 2196-3185 ISSN 2196-3193 (electronic)
Power Electronics and Power Systems
ISBN 978-3-319-98309-7 ISBN 978-3-319-98310-3 (eBook)
https://doi.org/10.1007/978-3-319-98310-3

Library of Congress Control Number: 2018950811

This Springer imprint is published by the registered company Springer Nature Switzerland AG
The registered company address is: Gewerbestrasse 11, 6330 Cham, Switzerland

Foreword I: Smart Grid Control, A Power Systems Perspective

The accelerating shift of energy supply from large central generating stations to smaller producers such as wind farms, solar PV farms, rooftop solar PVs, and energy storage systems, collectively known as distributed energy resources, has far exceeded the expectation of power system experts. Simultaneous to the steep drop in costs of renewable equipment and installations which prompted this rapid pace, intelligent and low-end sensors to measure power variables are also becoming low cost. The confluence of sensors and pervasive computer networks allows for the monitoring and feedback control of power systems at a larger geographical scale and with finer granularity. Previous unobservable remote dynamics are now visible within fractions of a second.

It has been recognized that current power system controls would not be entirely adequate to handle future smart power grid with very high penetration of renewables and long-distance transmission of such sustainable energy. System rotational inertias would be reduced such that frequency regulation would be more challenging. Renewable resources are taken as must-runs at the present time, but their variability poses additional cycling requirements from conventional generators. Allocating sufficient reserves to back up the renewables may be costly and not readily accommodated in electricity markets that were originally designed without considerations of renewable resources. Automation in power control functions also exposes its communication systems to cyber intrusion, with potentially severe consequences.

Recently, many control system researchers have taken a keen interest in examining the control issues in the future power grid and developing novel solutions. A "Smart Grid Vision" document was recently prepared by the IEEE Control System Society, outlining a number of potential control concepts and techniques that can be useful or should be explored to meet the challenges of the future power grid. This volume in the Springer Power Electronics and Power Systems Series is both an update of the earlier vision document, a necessity in this fast changing energy development environment, and an elaboration in more detail some of the areas in which controls can make contributions.

This Springer volume is fortunate to have four leading researchers on control applications in future power grid, Drs. J. Stoustrup, A. Annaswamy, A. Chakrabortty, and Z. Qu, to organize this effort. In addition to providing their own articles, they invited articles from over 20 renowned researchers, not only from control systems, power electronics, and power systems, but also from researchers who are grounded in signal processing, computer networking, optimization theory, and economics. The contributors have been asked to write provocatively and share their best ideas. The articles are divided into four topic areas, each containing a survey article, followed by in-depth discourses of more specific new results and ideas. A reader interested in future power grid control research may benefit from a careful study of one or more of these topic areas.

Joe H. Chow
Institute Professor of Engineering
Electrical, Computer, and Systems Engineering
Campus Director, NSF/DOE CURENT ERC
Rensselaer Polytechnic Institute
Editor, Springer Power Electronics and Power Systems Series

Foreword II: Smart Grids and Controls: A Global Perspective

As the world grows more interconnected, we are becoming surrounded by complex networked systems. These systems consist of numerous components interlinked in complicated webs. As a result of the number of components and their intricate interconnections, complex networked systems are extremely difficult to design, analyze, control, and protect. Despite these challenges, understanding complex networked systems is becoming critical. It is in this context that I express my gratitude to the authors and editors of this volume of exceptional work. The *Smart Grid Control: Overview and Research Opportunities*, edited by distinguished colleagues Drs. Annaswamy, Chakrabortty, Qu, and Stoustrup—with peer-reviewed articles written by superb teams of researchers and leaders in this field, is a timely and lasting contribution to the field of smart grid.

From a broader context, worldwide, the electricity infrastructure and service requirements are being dramatically changed to meet the sustainable demand of the twenty-first century. Electricity distribution is generally being challenged worldwide by growing concerns of greenhouse emissions and sustainability, aging infrastructure, and increasing demands for digital quality power. As a result of digital technology and its digitization of society, the nature of electricity generation, transmission, and distribution is undergoing a profound shift emphasized by many smart grid case studies. A fully automated electronically engaged smart grid holds the potential of doubling the consumer service reliability level and significantly improving the energy efficiency. The envisioned smart grid architecture is enabling the electric power industry globally to evolve from the traditional model relying on large centralized power plants owned by utilities to one that is much more diverse in terms of electricity generation, ownership of the assets, and integration of new distributed energy resources.

Associated with this transformation are significant challenges. The resulting system is increasingly interconnected, complex, dynamic, distributed, and nonlinear, with intra- and interconnections with human owners, operators, markets, generating units, flexible consumers, smart storage devices, and smart meters. No single entity has complete control over its operation, nor does any such entity have the ability to evaluate, monitor, and manage it in real time. Performance specifications, as in any critical infrastructures, abound in a smart grid as well. Most notable are Security, Quality, Reliability, and Availability (SQRA) of the overall system. In addition to these, a smart grid needs to have the ability to self-heal following an outage through real-time monitoring by the grid operators to the precursors or signatures of impending faults, using advanced sensor technology including Phasor Measurement Units (PMUs). This provides the potential operators to react swiftly, through rapid isolation, or by restoring balance by manipulating various field devices to respond automatically. What makes the smart grid vision especially difficult to realize is that these performance metrics are linked to multiple operational, spatial, and energy levels distributed across the entire grid.

Besides these multitudinous levels, power systems are also multi-scaled in the time domain, from nanoseconds to decades. The relative time of action for different types of events, from normal to extreme, varies depending on the nature and speed of the disturbance, and the need for coordination. The timescale of actions and operations within the power grid (often continental in scale) ranges from: microseconds to milliseconds for wave effects and fast dynamics (such as lightning or from nanoseconds to microseconds for propagation of the EMP), milliseconds for switching overvoltages, 100 ms or a few cycles for fault protection, 1–10 s for tie-line load frequency control, 10 s–1 h for economic load dispatch, 1 h to a day or longer for load management, load forecasting, and generation scheduling, and several years to a decade for new transmission or generation planning and integration. Given the above compelling drivers for smart grids, the emergence of several new stakeholders, all of whom are highly interconnected, and the fact that they have to be coordinated, at multiple timescales, it is clear that controls take a center stage in smart grids. Control systems are needed across broad temporal, geographical, and industry scales—from devices to systems, from fuel sources to consumers, from utility pricing to demand response, and so on in order to realize the complete smart grid vision.

Across the globe, the foundational and transformative role of controls and systems science has long been recognized and acknowledged in multiple ways. A more recent one is the vision document that I had the honor of coediting with Drs. Anu Annaswamy, Tariq Samad, and Chris Demarco [1] published in 2013, which outlined research opportunities and challenges that smart grids have elicited from the controls community. The second is the articulation of domains and subdomains that come together to lead to the Smart Grid Vision [2] by the IEEE Smart Grid Initiative. Started in 2009, this initiative has become the most successful cross-society endeavor, where all Smart Grid activities carried out by a total of 14 IEEE societies are showcased and disseminated through peer-reviewed webinars, tutorials, monthly newsletters, web portals as tools for collaboration, and compendium of important articles that appear in transactions and magazines of various societies—with participation of over 155,000 members from over 190 nations and territories across the globe. It should be noted that in [2], the role of controls is clearly acknowledged as a foundational support system. I am therefore delighted to see that this important volume precisely captures this key foundational area. The overall volume together with the four areas of electricity markets, wide-area control, distributed control, and cybersecurity capture the loci of controls activities to make the smart grid vision a reality. Together, we can serve this transformative vision/modernization to meet the global needs of twenty-first-century societies. The twin pillars of controls and the broader areas of systems science, two foundational areas of smart grids, enable prosperity and power progress in responsible and sustainable ways, and need your committed engagement, feedback, and support.

S. Massoud Amin
Director of the Technological Leadership Institute (TLI)
Honeywell/H.W. Sweatt Chair in Technological Leadership
Professor of Electrical and Computer Engineering
University Distinguished Teaching Professor
University of Minnesota

References

1. A.M. Annaswamy, T. Samad, M. Amin, C. De Marco (eds.), *IEEE Vision for Smart Grid Controls: 2030 and Beyond* (IEEE, 2013)
2. https://smartgrid.ieee.org/domains. Accessed 24 Feb 2018

Foreword III: Smart Grid Controls—Visions of the Future

According to the U.S. National Academy of Engineering, electrification was the greatest achievement of engineering in the twentieth century. Electrification is enabled by the electric power grid, a marvel of large-scale, spatiotemporal engineered system that operates with impressive levels of reliability, efficiency, and economy. It is among the most critical civil infrastructures at the center of our way of living.

Indeed, infrastructures are essential to civilization and society. Historically, they have defined the level of development of societies. In addition to the electric grid, water supply and distribution, roads, airports, electric grids, oil and gas pipelines, communications, hospitals, and banking are excellent examples of infrastructures. The Internet is the latest in this collection of our civilization's infrastructures.

Infrastructures result from very large public and private investments. Infrastructure decisions have long-term impacts that stretch for decades and centuries. For example, our current social structure and lifestyle has been shaped by transport system infrastructure decisions made 100 years ago.

The infusion and integration of sensors, communications, networking, computing and control into the traditional hard physical infrastructures is a major transformation whose impact will be felt for decades to come. Among other ideas, "infrastructure-as-a-service" is a key to this transformation. Smart roads, smart cars, and smart electric grids are at the forefront of this transformation as cyber-physical-social infrastructure systems.

While we cannot know the way people will live and work in 2068, we do know that there will be major changes from the way we live and work now. Therefore, the potential for flexibility inherent in algorithm and software-driven cyber-physical-social infrastructures may well turn out to be the greatest value in this transformation.

The recent hurricanes that devastated Puerto Rico, Texas, and Florida are stark reminders of the vulnerability of the critical infrastructures to natural and man-made disasters. With global warming, it is likely that such disruptive events will be more frequent and more extreme. A great promise of the smart electric grids lies in their potential to make the electricity system more resilient. That is, the electric power system can be restored to a certain minimal level of operational performance much more quickly than the current practice. Monitoring and control systems for self-healing in smart grids will be a key to this increased resilience.

Transition to a low-carbon economy is critical for mitigating global warming. For the energy sector, which constitutes 8–9% of the global economy, this requires replacing fossil fuels with renewable sources of energy such as wind and solar electricity generation. These electricity production sources are inherently variable and uncertain and present enormous obstacle to their large-scale integration into the power systems. Whereas availability of cost-effective electric energy storage would be revolutionary and therefore is the focus of large numbers of research efforts, smart grid systems will be essential to the operation of power grids with large-scale deployment of wind and solar electricity and replacement of fossil fuel based energy sources.

Infrastructure systems are not merely technological. They are deeply integrated into societal structures: homes, workplaces, public spaces and therefore in manufacturing, education, health care, entertainment, services, transport, agriculture, etc. Thus, human behavior, as individuals and in groups, is an essential driver of the behavior and performance of infrastructure systems. Smart electric grids are thus an excellent exemplar for "cyber-physical-human" or "cyber-physical-social" systems. Their analysis and design will require much greater integration of insights and knowledge from the social-behavioral-economic sciences for their analysis, design, and operation.

As the various chapters and articles in this book illustrate, control systems engineering and technology will play a central role in the realization of the benefits from investments into smart electric grids. The tutorial chapters provide a nice overview while challenge articles articulate significant challenges and opportunities. With increased uncertainty and variability, there are numerous control and decision challenges faced by market participants as well as system operators in electricity markets, for various energy and grid products and services, where advanced techniques from multistage stochastic control, estimation, prediction and optimization have great potential. With the proliferation of distributed renewable generation, storage, electric vehicles, and smart appliances along with pervasive sensing through (IoT based) sensing systems, there are very interesting and important opportunities for distributed control and optimization algorithms to extract value from these resources while supporting grid reliability and power quality. Wide-area control and monitoring will be enabled by improving communications and greater computing capability over large geographic regions. Finally, cybersecurity is very likely to remain a high priority and continuing and evolving challenge as the smart grid technologies are deployed in the field.

The electric power system is one of the largest engineered networked systems. As a result, the smart electric grid field will offer a rich set of problems and opportunities for networked control systems. Thus, there is great potential for smart electric grids to catalyze new fundamental contributions to the control systems field and contribute to its growth.

Irvine, CA, USA

Pramod P. Khargonekar
Distinguished Professor of Electrical
Engineering and Computer Science and
Vice Chancellor for Research
University of California

Preface

A smart grid is an end-to-end cyber-enabled electric power system, from fuel source, to generation, transmission, distribution, and end use, that has the potential to (i) enable integration of intermittent renewable energy sources and help decarbonize power systems, (ii) allow reliable and secure 2-way power and information flows, (iii) enable energy efficiency, effective demand management, and customer choice, (iv) provide self-healing capability from power disturbance events, and (v) operate resiliently against physical and cyber attacks. Central to the realization of all of these goals is a control-centric approach. The increased deployment of feedback and communication implies that feedback loops are being closed where they have never been closed before, across multiple temporal and spatial scales, thereby creating a gold mine of opportunities for control. Control systems are needed to facilitate decision-making under myriad uncertainties, across broad temporal, geographical, and industry scales—from devices to systems, from fuel sources to consumers, from utility pricing to demand response, and so on.

The IEEE report [1], "Vision for Smart Grid Controls: A Roadmap for 2030 and Beyond," published in 2013, provided an overview of the role of smart grid control, its loci, possible impact, and research challenges. Fifteen different control topics were identified as those where controls play a dominant part. Given the tremendous state of flux in R&D in all things Smart Grid, it is not surprising that since the publication of the IEEE report, the frontiers of research in Smart Grid in general as well as Smart Grid Control in particular have changed significantly. This volume is an effort to capture the current landscape of this high-intensity research topic, and outline the available research opportunities.

Traditional control topics in power grids were for the most part prevalent in transmission and distribution problems, and focused on transient stability and steady-state optimization. Control problems such as Automatic Generation Control, and volt-VAR control were the most common centers of research activity. The emerging picture of smart grid control is significantly different. One of the biggest drivers of a smart grid is a high penetration of renewable energy resources. A complete integration of these resources introduces a host of challenges of coordination, analytics, information processing, monitoring, optimization, estimation, protection, and resiliency. All of these challenges are control-centric in nature, and require a significantly different set of tools compared to the traditional approaches used for solving control problems in transmission and distribution. These challenges have to be addressed at all subsystems of the grid, starting from generation, through transmission and distribution, to the end user. Faster decisions need to be made in markets, with accommodations of the stochastic elements introduced due to intermittencies and uncertainties in renewables. The underlying communication topology is changing with more stakeholders entering the picture, requiring frequent and reliable communication. The grid periphery is becoming more intelligent, with opportunities to measure, monitor, process information, and communicate decisions. And decisions need to be carried out at several points of the grid, and have to be addressed at multiple timescales, all the way from planning and economic dispatch at a longer time horizon of years, months, days, and minutes to operation at the faster timescales of automatic generation control, droop control, and sub-second transient stability phenomena. At the core of all of these challenges are decision-making, information processing, modeling, optimization, and control. These problems and the underlying approaches that lead to satisfactory solutions all lie completely within the purview of the activities of the Control Systems Society.

Of these large set of problems, four broad topics are worth noting, around each one of which there has been a tremendous level of research activity, and make up this volume. These topics are electricity markets, wide-area systems, distributed control, and cyber-physical security. Markets address planning and operations issues related to economic dispatch, those in wide-area control address large-scale dynamics that arise due to spatial interconnections, those in distributed control address decision-making across the entire grid as its edge intelligence grows, and those in security address all aspects of grid security that need to be addressed as more and more portals open up in the grid to collect information and make decisions. The major R&D challenges in these four topics are outlined below, and form the subject matter for the 17 articles that follow.

1. Markets

Increasing penetration of renewables necessitates new approaches and solutions to the design of electricity markets, many of which are centered around a dynamic perspective. The volatility inherent to wind power producers (WPPs) has posed challenges to the operations of RTOs which have gradually modified their regulations as their reliance on wind power increases. The variability and uncertainty of renewable generation will substantially increase the need for operational reserves to balance supply and demand instantaneously and continuously. Under low adoption of wind power, RTOs have opted for limited regulation and control over the power output of WPPs, allowing them to inject their generation when available, and treating them as negative load. As wind volatility becomes a more significant part of the energy balance problem and causes high congestion costs and significant reliability challenges, this practice has begun to change, with a need for evaluating dynamic market mechanisms to carry out market dispatch.

Another forthcoming challenge is that the total system inertia and contingency reserve capacity decrease as non-dispatchable renewable generation displaces conventional generation. This results in the reduction in the amount of critical operating decisions that need be made from minutes to seconds or even sub-seconds. Therefore, it is becoming extremely difficult for system operators to maintain the stability and reliability of their networks. In order to facilitate the paradigm shift to achieve higher energy efficiency in the future, more flexible and fast acting resources are needed to handle the uncertainties and variabilities introduced by such uncontrollable and intermittent energy resources. Design of forward markets that help guard against risks due to large forecast errors may be needed. How storage can be introduced into the market structure so as to alleviate these forecast errors needs to be investigated.

A prevailing trend to combat the uncertainties on the generation side is to reduce uncertainty on the load side through Demand Response (DR) including methods such as direct load control and transactive control. Systems and control tools that can provide guidelines and foundations for these emerging trends are therefore imperative. An overall framework including models and methods for the quantification and realization of performance metrics such as robustness, resilience, and reliability needs to be developed. The successful demonstration projects on transactive control by the Pacific Northwest National Lab as well as the promising approaches of renewables indicate that there are a number of opportunities for the controls community to develop such a rigorous theoretical framework for integration of DR and renewables into the electricity market. Yet another challenge pertains to the setting up of a retail market, where varied issues need to be addressed including the services provided by aggregators, both of distributed generation and flexible demand, appropriate coordination that ensures economic and physical goals of the distribution grid, and accommodates demand response structures of direct load control and transactive control.

2. Distributed Control

To effectively integrate rooftop PV, storage devices, controllable loads, and other Distributed Energy Resources (DERs), their dynamic changes need to be monitored and, when possible, appropriately controlled or coordinated as much as possible. The changes of renewable generation are stochastic and may be on different timescales than other DERs, and as such the coordination of DER devices requires both spatial diversity and temporal diversity in order to reduce the spinning reserves in the overall power system. In vastly expansive distribution networks, Advanced

Metering Infrastructure (AMI), Internet of Things (IoT), and communication networks can provide local information to enable distributed optimization and controls.
Distributed optimization can maximize individual objective functions as well as
provide voltage support and other ancillary services. Distributed cooperative control can utilize all the available information to coordinate local control/optimization
actions so that a common system optimization/control can be reached. DERs may
suffer from issues of low inertia and harmonics, necessitating a systematic
deployment of distributed controls to compensate for these shortcomings. The
challenges and benefits of designing distributed controls are to take full advantage
of local information and achieve the grid-edge intelligence of addressing the distinct prosumers' interests and grid operational requirements.

3. Wide-Area Control

The US Northeast blackout of 2003, followed by the timely emergence of sophisticated GPS-synchronized digital instrumentation technologies such as Wide-Area
Measurement Systems (WAMS) led utility owners to understand how the interconnected nature of the grid topology essentially couples their controller performance with that of others, and thereby forced them to look beyond using only local
feedback and instead use wide-area measurement feedback. Some of the challenges
lie in designing suitable communication networks so as to be able to collect and
process very large volumes of real-time data produced by such thousands of PMUs.
But several other challenges correspond to control-centric challenges. For example,
the impact of the unreliable and insecure communication and computation infrastructure, especially long delays and packet loss uncertainties over wide-area networks, on the development of new WAMS applications is not well understood.
Uncontrolled delays in a network can easily destabilize distributed estimation
algorithms for wide-area oscillation monitoring using PMU data from geographically dispersed locations. Another major challenge is privacy of PMU data as utility
companies are often shy in sharing data from a large number of observable points
within their operating regions with other companies. Equally important is cybersecurity of the data as even the slightest tampering of Synchrophasors, whether
through denial-of-service attacks or data manipulation attacks, can cause catastrophical instabilities in the grid. What we need is a cyber-physical architecture that
explicitly brings out potential solutions to all of these concerns, how data from
multitudes of geographically dispersed PMUs can be shared across a large grid via a
secure communication medium for successful execution of critical transmission
system operations, how the various binding factors in this distributed communication system can pose bottlenecks, and how these bottlenecks can be mitigated to
guarantee the stability and performance of the grid.

4. Cyber-Physical Security and Control

While wide-area controls are typically implemented within SCADA, an isolated industrial control system (ICS) with dedicated communication network, a more open and network-enabled control architecture of cyber-physical-human system will become prominent due to the proliferation of PMUs, micro-PMUs, AMI and other IoT/networking technologies, to the expansion of electricity market from the bulk transmission network to distribution networks, and to distributed controls and optimization. The ever-increasing uses of information technology and communication technology make the grid vulnerable to cyber intrusions, false data attacks, and coordinated control/measurement attacks. Various scenarios such as inside attack, denial-of-service attack, switch/breaker attack, interdiction attack, data alteration, and spoofing attack have to be investigated. For each of these potential attacks, defense mechanisms such as enhanced passive/active state estimation algorithms against data attacks should be developed. A systematic design with a layered approach is needed to address monitoring and optimization/control functions at the levels of physical layer, control layer, communication layer, network layer supervisory layer, and market layer. And finally, resilient architectures such as competitive control need to be developed to ensure the overall system dynamic stability in the presence of potential attacks, especially during the period when multilevel monitoring is active and attacks are present but yet to be identified. As attack strategies evolve with more sophistication, defense mechanisms have to be more advanced. All of these challenges fall under the fourth category of cyber-physical security and control.

Aalborg, Denmark
Cambridge, USA
Raleigh, USA
Orlando, USA

Jakob Stoustrup
Anuradha Annaswamy
Aranya Chakrabortty
Zhihua Qu

Reference

1. A.M. Annaswamy, T. Samad, M. Amin, C. De Marco (eds.), *IEEE Vision for Smart Grid Controls: 2030 and Beyond* (IEEE, 2013)

Contents

Part III   Wide-Area Control Using Real-Time Data

Wide-Area Communication and Control: A Cyber-Physical
Perspective   139
Aranya Chakrabortty

Research Challenges for Design and Implementation of Wide-Area
Control   165
Aranya Chakrabortty

Signal Processing in Smart Grids: From Data to Reliable
Information   173
Meng Wang

Part I
Electricity Markets

Electricity Markets in the United States: A Brief History, Current Operations, and Trends

Thomas R. Nudell, Anuradha M. Annaswamy, Jianming Lian, Karanjit Kalsi and David D'Achiardi

Abstract The global energy landscape is witnessing a concerted effort toward grid modernization. Motivated by sustainability, skyrocketing demand for electricity, and the inability of a legacy infrastructure to accommodate distributed and intermittent resources, a cyber-physical infrastructure is emerging to embrace zero-emission energy assets such as wind and solar generation and results in a smart grid that delivers green, reliable, and affordable power. A key ingredient of this infrastructure is electricity markets, the first layer of decision-making in a smart grid. This chapter provides an overview of electricity markets which can be viewed as the backdrop for their emerging role in a modernized, cyber-enabled grid. Starting from a brief history of the electricity markets in the United States, the article proceeds to delineate the current market structure, and closes with a description of current trends and emerging directions.

1 Introduction

An electricity market enables trade of electricity between suppliers and consumers. An efficient market is one where electricity is traded at a price that minimizes the cost of generation while supplying the demand. The overall market goals are to ensure efficient pricing of electricity generation, incentivize enhanced grid services and infrastructure maintenance. The outputs of the electricity market can, therefore, be viewed as set-points for the actual units that generate or consume electricity. As electricity cannot be stored in large quantities at the current cost of energy storage, the amount of electricity generated must match the demand at every instant of time. It is, therefore, not surprising that electricity markets range over a broad timescale, from years to seconds, to accommodate planning as well as operations. Examples include markets for Forward Capacity, Energy, and Ancillary Services.

T. R. Nudell
Smart Wires Inc., Union City, CA, USA

A. M. Annaswamy (✉) · D. D'Achiardi
MIT, Cambridge, MA 02139, USA
e-mail: aanna@mit.edu

J. Lian · K. Kalsi
Pacific Northwest National Laboratory, Richland, WA 99352, USA

© Springer Nature Switzerland AG 2019
J. Stoustrup et al. (eds.), *Smart Grid Control*, Power Electronics
and Power Systems, https://doi.org/10.1007/978-3-319-98310-3_1

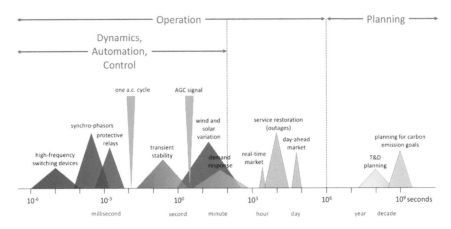

Fig. 1 Illustration of typical planning and operation market timescales (adapted from [55])

While economic theory is the underlying tool utilized in order to govern the principles of electricity markets, such a tool alone is not sufficient, as the products and services transacted in electricity markets have to interact with the physical grid and satisfy its constraints. That is, electricity markets lie in the intersection of two systems, the financial and the physical, which makes their analysis and synthesis highly challenging. What makes it even harder is the current transformation that the grid is witnessing, toward modernization, toward a cyber-enabled architecture, toward a smart grid. This transformation is, therefore, providing a cause for revisiting the electricity market structure, its mechanisms, and its overall coupling with the physical power grid.

Figure 1 shows typical timescales of commonly found markets in the US with respect to other power system planning and operation processes. Because of the multi-year lead times for building electric power plants and transmission projects, planning markets exist in many places in the US in order to ensure that the overall supply of electricity will be able to meet projected demand. Markets that govern operation, termed day ahead (DA) and real-time (RT) markets, ensure that the instantaneous supply of and demand for electric power are balanced in a least-cost manner. The DA market clears a day prior to operation for 24 hourly intervals, while the RT market clears an hour ahead of operation for 5–15 min intervals. Whether in planning or in operations, these markets operate following certain rules and guidelines, which are set by regional transmission operators (RTOs), in accordance with regulators appointed by the government.

In order to set the stage for the impact of the Smart Grid Vision on the market structure, in the following sections, this tutorial seeks to provide an overview of electricity market structure in the United States. A brief history of the electricity market is provided in Sect. 2. An overview of the market structure is delineated in Sect. 3. Some of the major changes that the smart grid paradigm has precipitated are discussed in Sect. 4.

2   A Brief History of Electricity in the US

Since the invention of electricity in the eighteenth century, the evolution of the electricity market can be organized into three parts, the War of Currents and rise of the vertically integrated firm (1880s–1930s) leading up to a viable business model for generating and delivering electricity, the regulated utility (1930–1970), and subsequent deregulation (1970–1990). Each of these parts are described in the sections below.

2.1   War of Currents and Rise of the Vertically Integrated Utility

Subsequent to the understanding of the generation of electricity, the technological battle that ensued pertains to the use of AC (championed by Nicola Tesla) versus DC (championed by Thomas Edison) for power generation and transmission. Edison's support for DC stemmed from the fact that his well-known invention of the light bulb needed a distribution network as a foundation for large-scale expansion, and he believed that low-voltage (110 V) direct current (DC) was the only safe way to distribute electric power. On December 17, 1880, he founded the Edison Illuminating Company and went on to establish the first investor-owned electric utility in 1882 at the Pearl Street Station. From the Pearl Street Station, Edison operated a low-voltage DC "microgrid", which provided 110 V DC to 59 customers in lower Manhattan in New York City [1]. A foil to this technology came from Tesla, who had initially worked for the Continental Edison company tasked with the redesign of Edison's DC generators, and came to believe that many of the DC generators' demerits could be overcome with AC-transmission. The subsequent battle of ideals, now famously dubbed as the War of Currents, would be won by Tesla, and led to a series of US patents that laid the foundation for the AC-alternative to Edison's DC system. These patents were then sold to the Westinghouse Electric Company in 1888. Its owner, George Westinghouse, took advantage of the limited transmission range of low-voltage DC-power, and expanded transmission to beyond urban centers. Subsequently, Westinghouse and his AC distribution system prevailed. The War of Currents ended when Thomas Edison, facing shrinking profits relative to his AC rivals, merged his company with a more successful AC firm, the Thomas-Houston Electric Company, to form General Electric in 1892. Battles between GE and Westinghouse continued for the next few years.

The next step in the development of modern electricity markets in the US was entrepreneurial rather than technological. This step can be attributed to Samuel Insull, who introduced a demand-adjusted billing system in which there were two tiers of prices: one for low demand times and one for high demand times. This strategy increased profits by increasing overall power consumption, allowing the continuous running of base-load plants leading to better returns. Insull's holding companies grew in value to $500 million with a capital investment of only $27 million [68]. The stock market crash of 1929 and the ensuing Great Depression, however, introduced several singularities into the picture leading to a collapse of Insull's enterprise.

The above discussions indicate that economies of scale combined with concerns over reliability led to a firm establishment of the current grid infrastructure of AC generation and transmission. Large, vertically integrated utilities that generated, transmitted, and distributed power—and which were natural monopolies—arose to capture the economies of scale. After the collapse of Insull's company, it also became clear that these natural monopolies required regulatory oversight. This, in turn, led to Congress passing the Public Utility Holding Company Act (PUHCA) in 1935, which enabled state regulation of electric utilities, and gave federal oversight responsibilities to the Securities and Exchange Commission (SEC) and the FPC.

2.2   NERC, FERC, and Deregulation

The rapid expansion of electricity demand over the next few decades led to frequent brownouts in the 1960s, culminating in a massive blackout across the eastern seaboard in 1965, led to the creation of the National Electric Reliability Council (NERC) in 1968 that subsequently became the North American Reliability Corporation [29]. NERC divided North America into several interconnected regions and oversaw these entities to fulfill its mandate of ensuring reliability of the power system.

The energy crisis in the 70s, caused in part by the oil embargo, led to a shortage of natural gas, and rising oil prices. Due to the inefficient oversight of the FPC, Congress reorganized it as the Federal Energy Regulatory Commission (FERC), an independent commission within the newly formed Department of Energy in 1977. FERC worked to develop simpler approval procedures and eliminated the direct oversight of utilities, regulating instead the transmission grid, wholesale markets, and approvals of important mergers and acquisitions in the energy sector.

As a direct response to the energy crisis, Congress enacted the Public Utility Regulatory Policies Act (PURPA) in 1978, which promoted conservation, domestic energy production, and development of efficient co-generation and non-fossil fuel resources. PURPA also opened the market to non-utility generators or independent power producers (IPP) who could produce power at a lower cost than the vertically integrated utility, in which case the utility was mandated to buy this cheaper power and pass the "avoided cost" savings to their customers. This was an important first step toward broader restructuring of the electricity industry [56].

The late 1970s and 1980s saw continued, but gradual, deregulation of the energy sector. The Energy Policy Act of 1992 gave FERC the authority to mandate that a utility provides transmission access to eligible wholesale entities, including wholesale buyers such as large industrial customers and exempt wholesale generators (merchant generators). This was an important step in the development of bulk electricity markets in the US. It is important to note that retail competition and consumer choice, are not, and never were, under the authority of FERC, rather these decisions belong to state legislatures and regulators. Finally, in the 1990s, FERC issued a series of orders that led to modern-day wholesale electricity markets.

FERC Order 888, often referred to as the "open access" rule required utilities to unbundle wholesale generation and power marketing, identified ancillary services required to operate a bulk power system. To achieve the goal of open access, five non-profit Independent System Operators (ISOs) were created, California Independent System Operator (CAISO), New York ISO (NYISO), Electric Reliability Council of Texas (ERCOT), Midcontinent Independent System Operator (MISO), and ISO New England (ISO-NE). FERC Order 889 created the Open Access Same-time Information System (OASIS), which specified standards of conduct that would allow the transmission customers described in Order 888 to have nondiscriminatory access to the transmission grid, which was ensured by wholesale electricity markets run by the ISOs. FERC Order 2000 established guidelines that a transmission entity must meet to qualify as a regional transmission operator (RTO) and required that all public utilities that own, operate, or control transmission networks must "make certain filings with respect to forming and participating in an RTO" [23]. Every US ISO is also designated as an RTO—additional, non-ISO RTOs include PJM Interconnection (PJM) and Southwest Power Pool (SPP)—whose role of RTOs is largely similar to ISOs, but with additional responsibility for the reliable operation and expansion of the transmission grid.

FERC continues to issue rulings to improve market operation and ensure that consumers receive the lowest cost for reliable electricity, notable examples being Order 745 (in 2011) and Order 825 (in 2016). These are discussed in the subsequent sections, and are related to oversight of the emerging concepts of Demand Response and Settlement Reform, respectively.

3   An Introduction to Wholesale Energy Market Operation

Every RTO in the US operates multiple wholesale electricity markets, where various products and services are bought and sold, including bulk energy, financial transmission rights, and ancillary services. In this section, we focus on wholesale *energy* markets. We start by describing market objectives, followed by an introduction to day ahead (DA) and real-time (RT) energy market operation, typical unit commitment and economic dispatch (UC and ED) problem formulation, and, finally, an overview of typical settlement rules. This section is not meant to be a comprehensive guide to market products or operation in any particular RTO, but rather an overview of the energy market operation. The goal of this section is to provide a flavor of the kinds of problems that ISOs formulate and solve today. For details of the DA and RT markets as well as markets for forward capacity and ancillary services, we refer the reader to the publicly available best practice manuals and user guides published by the each [35, 57, 58, 62, 66].

3.1 Market Objectives

Every RTO in the US operates DA and RT markets, with the primary objective of maximizing overall social welfare—i.e., maximize the sum of consumer and producer surplus by maximizing their utility functions and minimizing their cost functions. Other market objectives include providing incentives for market participants to follow commitments and dispatch instructions, transparency, maintaining system reliability, and ensuring that suppliers have an opportunity to recover their costs. We discuss a generic implementation of a fully centralized UC in the DA market to determine the generators that will run in the operating day, followed by a nodal ED in the DA and RT markets. In the ED markets, we assume energy and reserve capacity are cleared simultaneously, which is a common practice in most RTOs today. We discuss a DA market model that is settled ex-ante (i.e., before operation) on hourly intervals, and an RT market that is settled ex-post (i.e., after operation) based on 5-min intervals.

3.2 Day Ahead and Real-Time Energy Markets

The main product that DA and RT markets deal with is energy, which is specified as a power set-point over an interval, including start time and duration, at a specific location on the network. Both DA and RT markets include a security constrained[1] ED to dispatch power in the most economical way possible given the forecast and physical operating conditions. The DA energy market also requires a security constrained UC to optimally schedule generators to ensure that they will be available to provide energy and other ancillary services in real time.

3.2.1 Day Ahead Markets

For simplicity, we ignore the security constraints of DA and RT operation, we begin with the basic UC and ED problem formulations. Inputs to the UC problem include load and weather forecasts, regulation and reserve requirements, and, from each market participant, bid/offer curves, start-up, and shut-down costs, generator parameters such as ramp up/down rates, along with integer minimum up/downtime constraints. We consider full network power flow constraints in our formulation. The output of the UC problem is the set of generators that will run at each of the 24 intervals in the operating day, which is called the *day ahead operating schedule*. Figure 2 shows a very simplified operational timeline of the DA market. Table 1 introduces all notations used in this article.

With these assumptions in place, the unit commitment problem is a mixed integer (linear) program (MIP) of the following form:

---

[1] Security constraints are additional constraints that ensure line flows do not exceed specified limits following the occurrence of any one of a set of specified contingencies.

Table 1 Notation used in the market problem formulations

Symbol	Description
$\mathcal{N}$	Set of buses in the network
$\mathcal{E}$	Set of branches in the network
$\mathcal{M}$	Set of generators participating in the market
$\mathcal{L}$	Set of loads participating in the market
$i \sim n$	Generator $i \in \mathcal{M}$ adjacent to bus $n \in \mathcal{N}$
p_{it}	Power dispatch of generator $i \in \mathcal{M}$ at time t
$\overline{p}_i$	Maximum stable power output of generator $i \in \mathcal{M}$
$\underline{p}_i$	Minimum stable power output of generator $i \in \mathcal{M}$
d_{jt}	Demand of load $j \in \mathcal{L}$ at time t
$k \sim n$	Line $k \in \mathcal{E}$ incident to bus $n \in \mathcal{N}$
$\overline{d}_j$	Maximum demand of load $j \in \mathcal{L}$
$\underline{d}_j$	Minimum demand of load $j \in \mathcal{L}$
D_t	Total demand at time t
r_{it}^u	Up-reserve capacity of generator $i \in \mathcal{M}$ at time t
r_{it}^d	Down-reserve capacity of generator $i \in \mathcal{M}$ at time t
R_t^u	Total up-reserve requirement at time t
R_t^d	Total down-reserve requirement at time t
u_{it}	Commitment flag of generator $i \in \mathcal{M}$ at time t, $u_{it} \in \{0, 1\}$
a_{it}	No-load cost for generator $i \in \mathcal{M}$ at time t
b_{it}	Marginal generation cost for $i \in \mathcal{M}$ or marginal utility of consumption for load $j \in \mathcal{L}$ at time t
z_{it}^u	Start-up flag for generator $i \in \mathcal{M}$ at time t, $z_{it}^u \in \{0, 1\}$
z_{it}^d	Shut-down flag for generator $i \in \mathcal{M}$ at time t, $z_{it}^d \in \{0, 1\}$
s_{it}^u	Start-up cost for generator $i \in \mathcal{M}$ at time t
s_{it}^d	Shut-down cost for generator $i \in \mathcal{M}$ at time t
$\overline{\Delta}_i^u$	Maximum up-ramp capability for generator $i \in \mathcal{M}$
$\overline{\Delta}_i^d$	Maximum down-ramp capability for generator $i \in \mathcal{M}$
f_k	Flow of real power on branch $k \in \mathcal{E}$
$\overline{f}_k$	Maximum allowable flow on branch $k \in \mathcal{E}$
$\underline{f}_k$	Minimum allowable flow on branch $k \in \mathcal{E}$

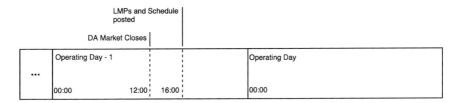

Fig. 2 Simplified DA scheduling timeline

$$\underset{\mathbf{x}}{\text{minimize}} \quad \mathbf{1}_{24}^{T}\mathbf{C}\mathbf{x}$$

$$\text{subject to} \tag{1}$$

$$\mathbf{H}\mathbf{x} - \mathbf{b} = \mathbf{0}$$

$$\mathbf{G}\mathbf{x} \leq \mathbf{0}$$

where some of the decision variables (elements in $\mathbf{x}$) are binary start-up shut-down decisions. Modern optimization software packages employ branch and bound as well as branch and cut algorithms to solve these types of problems. The UC problem can be formulated as

$$\underset{p,r,u,z}{\text{minimize}} \sum_{t=1}^{24} \sum_{i \in \mathcal{M}} \left[u_{it}a_{it} + b_{it}p_{it} + z_{it}^{u}s_{it}^{u} + z_{it}^{d}s_{it}^{d} \right]$$

subject to

$$\sum_{i \in \mathcal{M}} p_{it} = D_t = \sum_{j \in \mathcal{L}} d_{jt}$$

$$\sum_{i \in \mathcal{M}} r_{it}^{u} \geq R_t^{u}$$

$$\sum_{i \in \mathcal{M}} r_{it}^{d} \geq R_t^{d}$$

$$p_{it} + r_{it}^{u} \leq u_{it}\overline{p}_{it}$$

$$p_{it} - r_{it}^{u} \geq u_{it}\underline{p}_{it}$$

$$p_{it} - p_{i(t-1)} \leq \overline{\Delta}_i^{u}$$

$$p_{i(t-1)} - p_{it} \leq \overline{\Delta}_i^{d}$$

$$u_{it} - u_{i(t-1)} \leq z_{it}^{u}$$

$$u_{i(t-1)} - u_{it} \leq z_{it}^{d}$$

After the day ahead operating schedule has been set, the units are dispatched in one hour intervals for every hour of the operating day. Inputs to the DA Economic Dispatch (DA-ED) include weather, load, reserve requirements, bid and offer curves, commitment status, and reserve levels of each generator, in addition to the full DC-load flow network. The outputs are generator setpoints for every hour, locational marginal prices (LMPs), and reserve prices. With linearized cost curves and DC-load flow assumptions, for each $t = 1, \ldots, 24$, ED is a linear program of the following form:

$$\underset{\mathbf{x}}{\text{minimize}} \quad \mathbf{c}^{T}\mathbf{x}$$

$$\text{subject to} \tag{2}$$

$$\mathbf{H}\mathbf{x} - \mathbf{b} = \mathbf{0}$$

$$\mathbf{G}\mathbf{x} \leq \mathbf{0}$$

Specifically, the DA-ED can be written as

$$\underset{p,d,r}{\text{minimize}} \sum_{i \in \mathcal{M}} b_i\, p_i - \sum_{j \in \mathcal{L}} b_j d_j$$

subject to

$$\sum_{i \sim n} p_i - \sum_{j \sim n} d_i + \sum_{k \sim n} f_k = 0, \quad \forall n \in \mathcal{N}$$

$$f_k \leq \overline{f}_k$$

$$f_k \geq \underline{f}_k$$

$$\sum_{i \in \mathcal{M}} r_i^u + \sum_{j \in \mathcal{L}} r_j^u \geq R^u \qquad\qquad (3)$$

$$\sum_{i \in \mathcal{M}} r_i^d + \sum_{j \in \mathcal{L}} r_j^d \geq R^d$$

$$p_i + r_i^u \leq \overline{p}_i$$

$$p_i - r_i^d \geq \underline{p}_i$$

$$d_j + r_j^u \leq \overline{d}_j$$

$$d_j - r_j^d \geq \underline{d}_j$$

The DA-ED results in a binding agreement to buy, sell, or reserve the cleared energy, defined as the specified power setpoint over the one hour interval, at the LMP. The LMP, often denoted by λ_n, represents the cost to serve the next increment of load at node $n \in \mathcal{N}$. Mathematically, the λ_n is the sum of the Lagrange multiplier of the power balance constraints (sometimes called "system lambda" or energy price) and the Lagrange multipliers of the flow constraints (congestion component) described in (3).

$$\lambda_n = \lambda_{energy} + \sum_k \mu_{n,k} \qquad\qquad (4)$$

where λ_{energy} denotes the energy component of the LMP and $\mu_{n,k}$ denotes the congestion component of the LMP. The energy component, λ_{energy}, is the shadow price of the energy balance constraint in (3). Note that all real-world markets include a loss component of LMP—or at least a modification of the energy component to account for losses—which has been ignored for simplicity. The congestion component, $\mu_{n,k}$ is the shadow price of the flow constraints for branch k weighted by the impact it has on node n. The dispatch schedule and LMPs are binding, which results in a single settlement for each market participant each time they are dispatched.

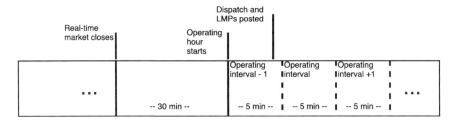

Fig. 3 Simplified RT market operation timeline

3.2.2   Real-Time Market

During the operating day, the DA-ED is augmented by the RT-ED according to the physical operation of the transmission network. Inputs to the RT market include the DA-ED as initial conditions, real-time topology, and network flows from a state estimator, in addition to the inputs used in the DA-ED. Today, it is a common practice for the RTO to run the markets on an operating hour schedule, where bids and offers are collected for an entire hour at once. The market will close at least 30 min before the operating hour begins, which implies that all inputs from market participants must be collected before this time. Throughout the operating hour, the ED is solved, typically every 5 min, to determine the dispatch and LMP, which are used in a second settlement. Figure 3 shows a typical timeline for real-time market operation and dispatch of the second operating interval in the operating hour. This timeline corresponds to current operation at ISO-NE in the Northeastern United States, where certain parameters to the DA bids and offers may be submitted no later than 30 min before the start of the next operating hour to be used as inputs to the real-time market. Other RTOs may differ in the specific timing, but for the most part operate in a similar manner.

3.3   Regulation Markets

In addition to energy and reserves, most RTOs also operate markets for a number of different ancillary services, the most prominent of which is the regulation market. Regulation actually consists of two separate products—capacity and service. Regulation capacity is measured in MW over a specified interval, and ensures a certain amount of room is available for the operator to deviate from the real-time market dispatch. Regulation service, measured in MW/min, is necessary for the RTO to be able to instantaneously match supply and load. Typically, the regulation service market will operate subsequently to the DA and RT energy and reserve markets, while regulation capacity is co-optimized with energy and reserves.

In order to participate in the regulation service market, resources must be able to receive the automatic generation control (AGC) dispatch signal, which is sent every 2–4 s. Additionally, they must be able to respond to AGC signal, and demonstrate a minimum performance standard. The market clears with the least-cost set of resources needed to meet the regulation service requirement.

While the market is cleared on a least-cost basis, the RTO will dispatch resources in a way that maximizes system performance. In other words, once the resources have been cleared in the market, they are not dispatched economically. The disconnection between reliability requirements and economic operation is an opportunity for new market design and/or new products.

3.4 Settlement Rules

Settlement rules precisely specify how market participants pay or get paid for the energy or services they provide. Thus, these rules are a key component of market design and the ability of a market to achieve its objectives.

The settlement rules take into account generator schedules, dispatch orders, actual produced or consumed energy and services, and the cleared price for energy and services. However, settlements are not simply equal to the price times the quantity of power delivered at a specific time and location. As settlements often occur ex-post, they may be based on an average price over a period of time or an average quantity over a given geographical region. Settlement rules may also include other components to provide proper incentives for market participants to continue participating in the market and to follow the operational dispatches. While the intention of the market is to provide an incentive for participants to continue providing a good or service while maximizing overall social welfare, poorly implemented settlement rules may undermine this objective.

In regulation markets, it is common to provide market participants with a regulation *performance* score, a number between 0 and 1, that measures how well the dispatch signal was followed. The settlements are then made based on the cleared regulation price (or hourly average regulation service price) multiplied by the performance score. This type of settlement incentivizes market participants to closely follow the dispatch signal, which is essential for regulation services. In addition to a performance score, regulation settlements include a make-whole component, which ensures that the generator is always compensated for the costs it incurs to provide regulation service. For example, in ISO-NE the regulation, settlement includes *payments* for any market participant that provides regulation service or capacity and includes *charges* for any market participant with an obligation to serve load. The net settlement is computed as

$$S_{net} = R_{service} + R_{capacity} + R_{mwp} - \left(R_{charge} + R_{mwc} \right). \qquad (5)$$

The regulation service payment $R_{service}$ is based on the cleared service price, the service provided and the performance score. The regulation capacity payment $R_{capacity}$ is based on the cleared capacity price and regulation capacity—which are co-optimized in the energy market—weighted by the performance score. The regulation charge R_{charge} and regulation make-whole charge R_{mwc} are owed by each market participant based on their relative real-time load obligation. This ensures that generators can be fairly compensated for providing regulation services, particularly in times of large load deviations. We refer the reader to [34] for more details on settlement rules and [25] for a market design that leads to reduced make-whole payments by the ISO.

Up to this point, we have summarized the history, development, and operation of wholesale electricity markets in the United States. The next section looks at trends in electricity markets today which will drive their future development, including distributed generation, demand-side management, direct load control, and transactive control.

4   Current Trends in the Electricity Market

Environmental concerns and economic and political requirements [3], have put pressure on the electric power industry to significantly increase electricity generation and to search for new sources of energy. Renewable resources not only provide the capability of reduced CO_2 emissions, but also have a low if not near-zero marginal cost of energy.

The volatility inherent to wind power producers (WPPs) has posed challenges to the operations of RTOs, which have gradually modified their regulations as their reliance on wind power increases. The variability and uncertainty of renewable generation will substantially increase the need for operational reserves to balance supply and demand instantaneously and continuously [32, 50, 65]. Under low adoption of wind power, RTOs have opted for limited regulation and control over the power output of WPPs, allowing them to inject their generation when available, and treating them as negative load. As wind volatility becomes a more significant part of the energy balance problem and causes high congestion costs and significant reliability challenges, this practice has begun to change, with RTOs opting for additional market mechanisms such as market dispatch and penalties for unmet commitments in energy markets.

Another forthcoming challenge is the total system inertia and contingency reserve capacity decrease as non-dispatchable renewable generation displaces conventional generation. This results in the reduction in the amount of critical operating decisions that need be made from minutes to seconds or even sub-seconds. Therefore, it is becoming extremely difficult for system operators to maintain the stability and reliability of their networks. In order to facilitate the paradigm shift to achieve higher energy efficiency in the future, more flexible and fast-acting resources are needed to handle the uncertainties and variabilities introduced by such uncontrollable and intermittent energy resources. A prevailing trend to combat the uncertainties on the generation side, is to reduce uncertainty on the load side through demand-side management, direct load control, and transactive control. These emerging trends and associated challenges are discussed in the subsequent sections.

4.1 Dispatchable Wind Power

The first large-scale WPPs were integrated to the California electric grid during the 1980s, motivated by the Public Utility Regulatory Policies Act of 1978 which required companies to purchase a certain amount of renewable energy [20]. Given the comparatively small dependence on these power plants and the intermittency associated with their wind resource, limited control over wind generation was initially required throughout the different RTOs. Wind power plants were treated as negative loads, that is, their instantaneous power output was always purchased at the market-clearing price. This meant that the volatility of their generation was largely absorbed by other power plants, mimicking variations in electric demand [9].

Between the 1980s and the early 2000s, the average capital expense in wind generation dropped by close to 65% while average capacity factors (a proxy for performance) improved by over 20%, even when curtailments due to congestion and grid reliability concerns are included [44]. Decreasing costs and increasing efficiencies prompted energy developers to invest in WPPs. In states with significant wind resources in rural areas, such as Texas, investments in transmission lines were made to bring power from windy regions to load centers. ERCOTs Competitive Renewable Energy Zone expansion invested $7 billion between 2008 and 2013, increasing the capacity of the West–East Corridor by 18.5 GW [45]. Such investments have allowed for an integration of 74 GW of wind power capacity to the US electric grid by 2015, primarily in the Midwest, Texas, and California [20].

Ancillary markets, such as balance reserves, and relatively fast-clearing real-time energy markets have enabled the integration of WPPs to higher fractions of total generation through the use of fast-ramping, low relative efficiency, natural gas-fired power plants [7, 9, 45]. RTOs with larger fractions of wind integration, such as MISO and CAISO, have run into issues in treating wind as a negative load, leading to additional technology implementation and control. Wind constitutes approximately 7% of the generation mix in MISO and CAISO, with peaks well over 50% renewable energy [10, 26]. These RTOs have resorted to economic dispatch systems, where wind is curtailed when additional generation poses a threat to transmission lines or the energy balance of the region. In California, approximately 1% of potential renewable generation is curtailed due to operational concerns [10].

Even in RTO operating regions where wind is dispatched by a market, costs associated with wind volatility are socialized among generators and consumers, as most integration mechanisms focus on internalizing congestion-related operational stresses under high generation but fall short of forcing WPPs to internalize costs for low production. However, wind power producers are on track to face significant penalties when energy commitments are not met. ISO-NE is an example of an operating region where the RTO is requiring wind to bid in day-ahead energy markets for planning and capacity commitment firming (implemented technology for remote dispatch by mid-2016, requiring Day-Ahead bidding by mid-2019) [33]. By charging penalties when generation commitments are not met, RTOs can pass on costs related to balance reserve requirements and fast-ramping of other power plants to

WPPs. The transition from negative load to economic dispatch and firming of generation commitments lead wind power producers to internalize costs associated with wind resource volatility and more accurately reflect the cost of their generation in the market.

4.2 Demand-Side Management

Current power grid operation predominantly relies on scheduling and regulating generation resources to supply electric loads and balance load changes. Due to inherent limitations of most conventional generators in providing fast-ramping capacities, the power grid solely based on supply-side control will not be able to support the large-scale integration of renewable energy. Alternatively, in addition to generators, electric loads can be used to balance between supply and demand. This practice is often referred to as the demand-side control or demand response (DR).

Traditionally, electric loads were considered to be passive and non-dispatchable elements of the power grid. However, various grid services that were traditionally delivered by generators only [12] can now be provided by a collection of electric loads through proper coordination and control with required speed, accuracy, and magnitude. Popular load types used for DR are thermostatically controlled loads (TCLs), including residential air conditioners, water heaters, and refrigerators, deferrable loads such as dryers and electric vehicles, and commercial HVAC (heating, ventilation, and air-conditioning) systems. Due to the large population size and fast aggregated ramping rate of these electric loads, DR has an enormous potential to reliably and economically offset the dynamic variability introduced by renewable generation.

Besides the emergence of DR, another growing trend in the power grid is the integration of distributed energy resources such as distributed generator and energy storage in power distribution systems. These distributed energy resources (DERs) are small and highly flexible compared with conventional generators. If appropriately coordinated and controlled, DERs and DR can collectively become a valuable system asset playing an increasingly important role in the future smart grid. Their seamless integration into power distribution systems will lead to efficient grid operation and high renewable penetration without compromising the stability and reliability of the power grid.

4.3 Direct Load Control

The coordination and control of electric loads to provide various grid services have been extensively studied in the literature. Direct load control (DLC) is one of the most popular demand response approaches. It allows electric loads to be remotely controlled by an aggregator (for example, utility company) based on prior mutual financial agreements, referred to as contracts. Traditional DR programs use DLC to deliver services such as peak shaving and load shifting [16, 19, 43]. The latest development in this area focuses on modeling and control of electric loads such as TLCs [6, 11, 37, 41, 54, 76], plug-in electric vehicles [48, 67], and data center servers [15, 47] to provide various grid services including frequency regulation and load following. In addition, there are also efforts on the design of financial contracts between the aggregator and individual loads under DLC. The essential step in the design of DLC is the development of an aggregated model that can accurately capture the collective dynamics of the load population.

Existing works on aggregated modeling have focused mainly on air conditioners and water heaters [5, 11, 18, 49]. The key idea of this approach is to characterize the evolution of the temperature density for the load population. Several first-principle-based approaches such as deterministic fluid dynamics approach [22] and stochastic differential equation approach [51] were proposed, which lead to a Fokker–Planck type of partial differential equation (PDE). In [11], the analytical solution to this PDE was derived in a much simplified setting, and provided useful insights into the transient dynamics. Besides those first-principle-based approaches, Markov-chain-based approaches were also studied in [39, 49, 53], where state transition probabilities between discrete temperature bins were derived based on simplified first-order models or directly from the simulated training data. However, both first-principle- and Markov-chain-based approaches are subject to several limitations for practical applications. First of all, most of the approaches model individual loads using a first-order differential equation, yet such a model is insufficient to capture the dynamics of TCLs that have large heat capacities. For example, in the case of air conditioners, it is essential to model the dynamics of both air and mass temperature dynamics because the house mass is so large that it will significantly affect the transients of air temperature. Second, homogeneity is a common assumption in many aggregated models, which does not hold in general. It is important to consider the diversity in load parameters in order to generate realistic aggregated responses [5, 39, 75]. The methods in [39, 49, 61] considered the heterogeneity in some parameters but still assume the homogeneity for the rest of parameters. Finally, the existing aggregated models usually allow the existence of short cycling for individual TCLs. These models cannot be directly applied to air conditioners for which there exist protection schemes that prevent the device from the short cycling. Hence, a new Markov-chain-based aggregated modeling that accounts for second-order equivalent thermal parameter models [72] of individual air conditioners was proposed in [13, 76, 77] to systematically address all the above issues.

Several non-density-based methods have also been proposed in [36, 37, 42], whose the main objective is to represent the aggregated dynamics using simple linear state-space or transfer function models. Compared with aggregated modeling, the design of aggregated controller is relatively simpler. With a good aggregated model of the load population, many well-established control methods such as Model Predictive Control [39], Lyapunov-based control [5], or simple inverse control [53] can be directly applied to regulate the aggregated power response so that it matches the given reference signal.

The most important advantage of DLC is that it can achieve reliable and accurate aggregated load response. Its implementation in practice, however, has been challenged often due to privacy and security concerns. This is because most of the models require information about the state of the end-use appliance owned by customers in order to design control strategies. On the other hand, there are also concerns that DLC signals could be disruptive to local constraints and inevitably result in adverse effects such as response fatigue [27]. Another important paradigm for demand response as an alternative to DLC is price responsive control (PRC), which sends price signals to customers so that they can individually and voluntarily manage their local demand. Unfortunately, under existing PRC schemes, it is difficult to achieve an acceptable level of predictable and reliable aggregate load response. *Transactive control* is a more comprehensive approach compared to PRC, and addresses the reliability concerns while maintaining the privacy and security advantages that PRC has over DLC. This is the focus of the following section.

4.4 Transactive Control

The most common examples of PRC in place today include time of use (TOU) pricing, critical peak pricing (CPP), and real-time pricing (RTP) [2, 8, 14, 30]. There have been many demonstration projects [22] to validate the performance of PRC in terms of payment reduction, load shifting, and power shaving. However, the existing approaches either directly pass the wholesale energy price to customers or modify the wholesale price in a heuristic way. Therefore, it is very difficult for PRC to achieve predictable and reliable aggregated load response that is essential in various demand response applications.

Transactive control, which is sometimes referred to as market-based control, has been proposed as an alternative to PRC that can integrate DERs and DR in power distribution systems and then into the transmission system to realize the transactive operation for the entire power grid. It shares the same advantage of PRC in preserving customer privacy by using internal price as the control signal. However, the internal price is systematically designed according to specific control objectives, which can be dramatically different from the wholesale price (see, for example, [17, 46]). Hence, transactive control shares the advantage of DLC, and avoids the shortcomings of PRC, in having a more predictable and reliable aggregated load response. Because transactive control borrows ideas from microeconomics [52] into the controller design, it is amenable to problems where self-interested customers are involved [21, 63]. Furthermore, it can also greatly facilitate the coordination and control between DERs and electric loads in the future distribution systems [73].

The idea behind transactive control is actually not new, and can be traced to concepts outlined in [64]. These concepts also recognize that different regions are structured in a variety of ways that cover wholesale power markets, electricity delivery markets, retail markets, and vertically integrated service provider markets. Transactive approaches appear to have the potential to be incorporated into many different structures and mechanisms that allow them to coexist with present operational approaches.

Current research activities on transactive control have mainly focused on innovating end-user loads with enhanced intelligence and launching field demonstrations involving various parties such as grid operators, energy supply companies, vendors, and regulators. In the United States, three major field demonstration projects were executed under the support from the U.S. Department of Energy to illustrate and prove the technology feasibility of transactive control in practice. Each of these three projects is summarized below, details of which can be found in [40].

4.4.1 Olympic Peninsula Demonstration

The Olympic Peninsula Demonstration (2006–2007) [24, 28] was the first proof-of-concept demonstration project on transactive control. This demonstration was located on the Olympic Peninsula of Washington State. The Penninsula is served by a capacity-constrained, radial transmission system connection to the Pacific Northwest power grid. The area had been experiencing a significant population growth and it had already projected by that time that power transmission capacity in the region may be inadequate to supply–demand during extremely cold winter conditions. The objective of this project was to evaluate practical and economical alternatives to new transmission and distribution construction by coordinating distributed energy resources and electric loads for congestion management. It used a 5-min double-auction market to coordinate five 40 HP water pumps distributed between two municipal water-pumping stations, two distributed diesel generators (175 and 600 kW), and electric water and space heating loads of 112 residential houses. This demonstration established the viability of transactive control in achieving multiple objectives such as peak load reduction and energy cost saving.

4.4.2 AEP gridSMART® Demonstration

The AEP gridSMART® Demonstration (2010–2014) [70, 71] was built upon the technology implemented in the Olympic Peninsula Demonstration and market-based incentive signals. This project used a 5-min double-auction market again to coordinate residential and control air conditioners on each of four distribution feeders for congestion management. However, it introduced an additional real-time pricing (RTP) component by incorporating PJMs 5-min wholesale locational marginal price (LMP). The overview of the RTP system design in this demonstration is illustrated in Fig. 4.

The AEP gridSMART® demonstration had three specific objectives [69]. The first is to build a transactive control platform to demonstrate the capability of responsive end-user loads in providing benefits to the utility and the consumer. The second was to actively educate consumers in innovative business models that encourage flexibility in energy use in return for reward and energy saving. The third is to record

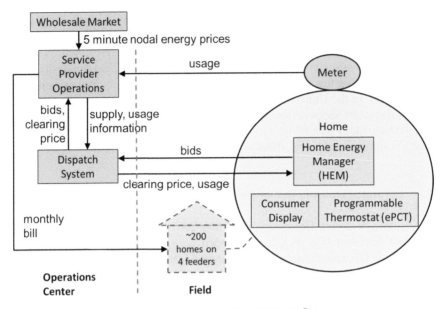

Fig. 4 Overview of the RTP system in the AEP gridSMART® Demonstration (reproduced from [40] ©2016IEEE and used with permission)

the system operation to study the technology performance and also the consumer behaviors under varying operating conditions.

4.4.3   Pacific Northwest Smart Grid Demonstration

The Pacific Northwest Smart Grid Demonstration (PNWSGD) (2010–2015) [31, 60] was a unique demonstration of unprecedented geographic breadth across five Pacific Northwest states—Idaho, Montana, Oregon, Washington, and Wyoming as shown in Fig. 5. There were 55 unique instantiations of distinct smart grid systems demonstrated at the project site. The local objectives for these systems included improved efficiency and reliability, energy conservation, and demand responsiveness. In this demonstration project, a new transactive approach was deployed to coordinate distributed energy resources and address regional objectives including the mitigation of renewable energy intermittency and the flattering of system load. Unlike the one based on the double-auction market, this approach is based on peer-to-peer negotiations as illustrated in Fig. 6. The major objective of this demonstration was to establish a more efficient and effective power grid that can simultaneously reduce fossil fuel consumption and CO_2 emissions, improve system stability and reliability, increase renewable penetration, and provide greater flexibility for customers.

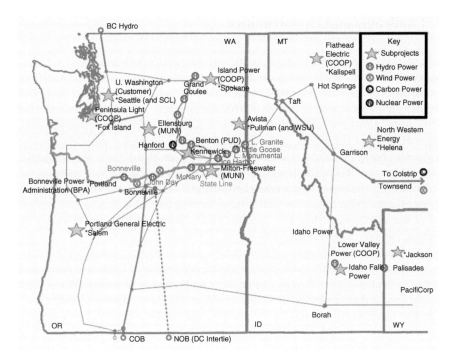

Fig. 5 The geographical region, participants and major generation and transmission of the PNWSGD (reproduced from [40] ©2016IEEE and used with permission)

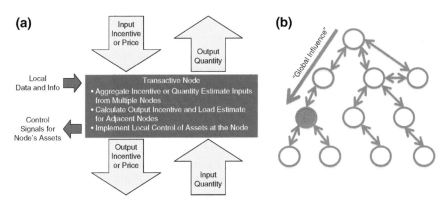

Fig. 6 Overview of the transactive approach deployed in the PNWSGD (reproduced from [40] ©2016IEEE and used with permission)

5   Challenges and Opportunities

In the sections above, we have attempted to introduce the reader to electricity markets, the first building block that lays the foundation for a reliable and affordable electricity infrastructure. With a focus on the United States, we have outlined a brief history of electricity markets, current operation of wholesale markets, and emerging trends. Due to the overall trend toward deregulation, the structure of markets covering planning and operations of the power grid, and the emphasis on DERs and DRs, similar changes are being investigated in electricity markets across the globe, though details of the workings of RTOs, wholesale markets, and retail markets differ. In addition, the evolution of and emerging topics in markets are tightly interwoven with technological advances in the cyber aspects as well as other technological domains such as storage and power electronics. Here too, specific trends and topics that are dominant in different parts of the world have differed.

A clear message that is apparent from the discussions above is that changes in market structures are needed because of the growing penetration of DERs and because of the high potential of DRs. Systems and control tools that can provide guidelines and foundations for these emerging trends are therefore imperative. An overall framework including models and methods for the quantification and realization of performance metrics such as robustness, resilience, and reliability need to be developed. The successful demonstration projects on transactive control as well as the promising approaches of DERs indicate that there are a number of opportunities for the controls community to develop such a rigorous theoretical framework for integration of DR and DERs into the electricity market. In the three challenge articles that follow, some of the opportunities and forward-looking directions are discussed. These span theoretical issues such as non-convexities and multiple timescales, practical challenges related to hedging in future markets, setting up retail markets, integration of multiple demand response units, and a redesign of markets and control to have better ancillary services. A brief summary of each of these articles follows. Several other directions on developing a dynamic framework with hierarchical, co-optimization, passivity, and game theory based components are currently being pursued to develop new solutions and architectures in electricity markets (see for example [4, 25, 38, 59, 74]) and are not included in this volume.

The paper *Some Emerging challenges in Electricity Markets* authored by S. Bose and S.H. Low focuses on five different challenges that are precipitated due to the growing penetration of renewables. The first concerns a fundamental theoretical issue of how non-convexities in constraints and feasibility sets need to be addressed, and how pricing these non-convexities may be used as incentives for introducing corrective action. The second is the need to understand how forward markets can be designed so as to guard against risks against large forecast errors with a large penetration of renewables-based generation. The third is a challenge associated with integrating storage devices with a possible approach that views storage devices as entities similar to those in a transmission infrastructure that helps manage congestion. How the hedging of associated price variations needs to be carried out is addressed.

The fourth challenge pertains to strategic market players and the use of game theory for analyzing strategic interactions. The final challenge pertains to the setting up of a retail market, where varied issues need to be addressed including the services provided by aggregators, both of distributed generation and flexible demand, appropriate coordination that ensures economic and physical goals of the distribution grid, and varied demand response structures such as direct load control and Transactive control.

The paper *Incentivizing Market and Control for Ancillary Services in Dynamic Power Grids* authored by K. Uchida, K. Hirata, and Y. Wasa addresses a redesign of the energy management system in an electricity market, with the goal of improving the quality of ancillary services. With increasing renewables, a high- speed market-clearing structure may be called for that ensures all private information of market players and reliable performance of the related frequency response. Limitations and fundamental challenges related to market design and the necessary incentives, when it comes to the integration of the requisite economic models and dynamic characteristics of the power grid are discussed. A specific model-based method is outlined as a possible approach for overcoming these challenges.

The paper *Long term challenges for future electricity markets in the presence of Distributed Energy Resources* authored by S. Muhanji, A. Muzhikyan, and A.M. Farid outlines three main challenges including (i) a simultaneous management of the technical and economic performance of the electricity grid (ii) spanning multiple timescales during operation, and (iii) enabling multiple demand-side resources. Reducing day-ahead and real-time market time steps so as to reduce load following, ramping, and regulation reserve requirements, development of new control architectures that are able to respond quickly to real-time changes in grid operations, and market structures that enable the participation of and proper compensation for such services are stressed in (i). Multi-timescale approaches so as to reduce the impact of net load variability and forecast error away from scheduled set points and related perspectives are stressed in (ii). Design and analysis of transactive control of demand-side management so as to lead to an appropriate balance of physical as well as economic signals is underscored in (iii).

Acknowledgements This work was supported in part by the NSF Award no. EFRI-1441301.

References

1. A brief history of con edison: 'electricity'. Electronic (1998), https://www.coned.com. Accessed 2 Dec 2016
2. H. Allcott, Rethinking real-time electricity pricing. Resour. Energy Econ. **33**(4), 820–842 (2011)
3. M. Amin, A. Annaswamy, A. Cadena, D. Callaway, E. Camacho, M. Caramanis, J. Chow, D. Dotta, A. Farid, P. Flikkema et al., *IEEE Vision for Smart Grid Controls: 2030 and Beyond* (2013), pp. 1–168

4. A. Annaswamy, A. Malekpour, B. Baros, Emerging research topics in control for a smart infrastructure. J. Ann. Rev. Control (2016)
5. S. Banash, H. Fathy, Modeling and control insights into demand-side energy management through setpoint control of thermostatic loads, in *Proceedings of the 2011 American Control Conference*, San Francisco, CA (2011)
6. S. Bashash, H.K. Fathy, Modeling and control of aggregate air conditioning loads for robust renewable power management. IEEE Trans. Control Syst. Technol. **21**(4), 1318–1327 (2013). https://doi.org/10.1109/TCST.2012.2204261
7. E. Bitar, K. Poolla, P. Khargonekar, R. Rajagopal, P. Varaiya, F. Wu, Selling random wind, in *2012 45th Hawaii International Conference on System Science (HICSS)* (IEEE, 2012), pp. 1931–1937
8. S. Borenstein, M. Jaske, A. Ros, Dynamic pricing, advanced metering, and demand response in electricity markets. J. Am. Chem. Soc. **128**(12), 4136–4145 (2002)
9. P. Brown, Us renewable electricity: how does wind generation impact competitive power markets. Congr. Res. Serv. **7** (2012)
10. CAISO: Impacts of renewable energy on grid operations (2017)
11. D.S. Callaway, Tapping the energy storage potential in electric loads to deliver load following and regulation with application to wind energy. Energy Convers. Manag. **50**(5), 1389–1400 (2009)
12. D.S. Callaway, I.A. Hiskens, Achieving controllability of elctric loads. Proc. IEEE **99**(1), 184–199 (2011)
13. C.Y. Chang, W. Zhang, J. Lian, K. Kalsi, Modeling and control of aggregated air conditioning loads under realistic conditions, in *Proceedings of the 2013 IEEE PES Innovative Smart Grid Technologies Conference (ISGT)* (2013), pp. 1–6. https://doi.org/10.1109/ISGT.2013.6497895
14. H. Chao, Price-responsive demand management for a smart grid world. Electr. J. **23**(1), 7–20 (2010)
15. H. Chen, A.K. Coskun, M.C. Caramanis, Real-time power control of data centers for providing regulation service, in *Proceedings of the 52nd IEEE Conference on Decision and Control* (2013), pp. 4314–4321. https://doi.org/10.1109/CDC.2013.6760553
16. J. Chen, F. Lee, A. Breipohl, R. Adapa, Scheduling direct load contorl to minimize system operation cost. IEEE Trans. Power Syst. **10**(4), 1994–2001 (1995)
17. L. Chen, N. Li, S.H. Low, J.C. Doyle, Two market models for demand response in power networks, in *Proceedings of the 2010 First IEEE International Conference on Smart Grid Communications*, Gaithersburg, MD (2010). https://doi.org/10.1109/SMARTGRID.2010.5622076
18. C.Y. Chong, A.S. Debs, Statistical synthesis of power system functional load models, in *Proceedings of the 1979 18th IEEE Conference on Decision and Control Including the Symposium on Adaptive Processes* (vol. 2) (1979), pp. 264–269
19. W.C. Chu, B.K. Chen, C.K. Fu, IEEE Trans. Power Appar. Syst. Scheduling of direct load control to minimize load reduction for a utility suffering from generation shortage **8**(4), 1525–1530 (1993)
20. Department of Energy U.D., History of U.S. Wind Energy (2017), https://energy.gov/eere/wind/history-us-wind-energy
21. M. Fahrioglu, F.L. Alvarado, Designing incentive compatible contracts for effective demand management. IEEE Trans. Power Syst. **15**(4), 1255–1260 (2000). https://doi.org/10.1109/59.898098
22. A. Faruqui, S. Sergici, A. Sharif, The impact of informational feedback on energy consumption: a survey of the experimental evidence. Energy **35**(4), 1598–1608 (2010)
23. FERC: FERC Order 2000 (1999)
24. J.C. Fuller, K.P. Schneider, D. Chassin, Analysis of residential demand response and double-auction markets, in *Proceedings of the 2011 IEEE Power and Energy Society General Meeting*, Detroit, MI (2011), pp. 1–7. https://doi.org/10.1109/PES.2011.6039827
25. M. Garcia, T. Nudell, A. Annaswamy, A dynamic regulation market mechanism for improved financial settlements in wholesale electricity markets, in *Proceedings of the 2017 American Control Conference*, Seattle (2017)

26. E. Gimon, R. Orvis, S. Aggarwal, Renewables curtailment: what we can learn from grid operations in california and the midwest (2015), https://www.greentechmedia.com/articles/read/renewables-curtailment-in-california-and-the-midwest-what-can-we-learn-from#gs.bkwQBJc

27. C. Goldman, N. Hopper, R. Bharvirkar, B. Neenam, P. Cappers, Estimating demand response market potential among large commercial and industrial customer: a scoping study. Technical report (Lawrence Berkeley Nat. Lab., Berkeley, CA, 2007)

28. D. J. Hammerstrom et al.: Pacific Northwest GridWise^{TM} testbed demonstration projects. Part I: Olympic Peninsula project. Tech. Rep. PNNL-17167 (Pacific Northwest National Laboratory, 2007)

29. R. Hirsh, *Power Loss: The Origins of Deregulation and Restructuring in the American Electric Utility Industry* (The MIT Press, 2002)

30. W.W. Hogan, Demand response compensation, net benefits and cost allocation: comments. Electr. J. **23**(9), 19–24 (2010)

31. P. Huang, J. Kalagnanam, R. Natarajan, D. Hammerstrom, R. Melton, M. Sharma, R. Ambrosio, Analytics and transactive control design for the Pacific Northwest smart grid demonstration project, in *Proceedings of the 2010 First IEEE International Conference on Smart Grid Communications*, Gaithersburg, MD (2010), pp. 449–454. https://doi.org/10.1109/SMARTGRID.2010.5622083

32. Integration of renewable resources, Operational requirements and generation fleet capability at 20% RPS, Technical report (California Independent System Operator (CAISO), 2010)

33. ISO-NE: Do not exceed dispatch (2017), https://www.iso-ne.com/participate/support/customer-readiness-outlook/do-not-exceed-dispatch

34. ISO New England: ISO New England Inc., Transmission, Markets, and Service Tariff, Section III: Market Rule 1 (2017), https://www.iso-ne.com/participate/rules-procedures/tariff/market-rule-1

35. ISO New England: ISO New England Manual for Market Operations: Manual M-11 (2017), https://www.iso-ne.com/participate/rules-procedures/manuals

36. K. Kalsi, F. Chassin, D. Chassin, Aggregated modeling of thermostatic loads in demand response: a systems and control perspective, in *Proceedings of the 50th IEEE Conference on Decision and Control*, Orlando, FL (2011)

37. K. Kalsi, M. Elizondo, J. Fuller, S. Lu, D. Chassin, Development and validation of aggregated models for thermostatic controlled loads with demand response, in *Proceedings of the 45th Hawaii International Conference on System Sciences*, Maui, HI (2012)

38. A. Kiani, A.M. Annaswamy, A dynamic mechanism for wholesale energy market: stability and robustness. IEEE Trans. Smart Grid, **5**(6), 2877–2888 (2014)

39. S. Koch, J. Mathieu, D. Callaway, Modeling and control of aggregated heterogeneous thermostatically controlled loads for ancillary services, in *Proceedings of the 17th Power System Computation Conference*, Stockholm, Sweden (2011)

40. K. Kok, S. Widergren, A society of devices: integrating intelligent distributed resources with transactive energy. IEEE Power Energy Mag. **14**(3), 34–45 (2016). https://doi.org/10.1109/MPE.2016.2524962

41. J. Kondoh, N. Lu, D.J. Hammerstrom, An evaluation of the water heater load potential for providing regulation service. IEEE Trans. Power Syst. **26**(3), 1309–1316 (2011). https://doi.org/10.1109/TPWRS.2010.2090909

42. S. Kundu, N. Sinitsyn, S. Backhaus, I. Hiskens, Modeling and control of thermostatically controlled loads, in *Proceedings of the 17th Power Systems Computation Conference*, Stokholm, Sweden (2011)

43. C. Kurucz, D. Brandt, S. Sim, A linear programming model for reducing system peak through customer load control programs. IEEE Trans. Power Syst. **11**(4), 1817–1824 (1996)

44. E. Lantz, M. Hand, R. Wiser, Past and future cost of wind energy: Preprint. Technical report (National Renewable Energy Laboratory (NREL), Golden, CO, 2012)

45. W. Lasher, The competitive renewable energy zones process (2014), https://energy.gov/sites/prod/files/2014/08/f18/c_lasher_qer_santafe_presentation.pdf

46. N. Li, L. Chen, S.H. Low, Optimal demand response based on utility maximization in power networks, in *Proceedings of the 2011 IEEE Power and Energy Society General Meeting*, Detroit, MI (2011), pp. 1–8. https://doi.org/10.1109/PES.2011.6039082

47. S. Li, M. Brocanelli, W. Zhang, X. Wang, Integrated power management of data centers and electric vehicles for energy and regulation market participation. IEEE Trans. Smart Grid **5**(5), 2283–2294 (2014). https://doi.org/10.1109/TSG.2014.2321519

48. J. Liu, S. Li, W. Zhang, J.L. Mathieu, G. Rizzoni, Planning and control of electric vehicles using dynamic energy capacity models, in *Proceedings of the 52nd IEEE Conference on Decision and Control* (2013), pp. 379–384. https://doi.org/10.1109/CDC.2013.6759911

49. N. Lu, D.P. Chassin, S.E. Widergren, Modeling uncertainties in aggregated thermostatically controlled loads using a state queueing model. IEEE Trans. Power Syst. **20**(2), 725–733 (2005). https://doi.org/10.1109/TPWRS.2005.846072

50. Y. Makarov, C. Loutan, J. Ma, P. de Mello, Operational impacts of wind generation on california power systems. IEEE Trans. Power Syst. **24**(2), 1039–1050 (2009)

51. R. Malhame, C.Y. Chong, Electric load model synthesis by diffusion approximation of a high-order hybrid-state stochastic system. IEEE Trans. Autom. Control **30**(9), 854–860 (1985)

52. A. Mas-Colell, M.D. Whinston, J.R. Green, *Microeconomic Theory* (Oxford University Press, New York, NY, 1995)

53. J.L. Mathieu, D.S. Callaway, State estimation and control of heterogeneous thermostatically controlled loads for load following, in *Proceedings of the 45th Hawaii International Conference on System Sciences*, Maui, HI (2012), pp. 2002–2011

54. J.L. Mathieu, S. Koch, D.S. Callaway, State estimation and control of electric loads to manage real-time energy imbalance. IEEE Trans. Power Syst. **28**(1), 430–440 (2013). https://doi.org/10.1109/TPWRS.2012.2204074

55. A. Meier, Integration of renewable generation in California: coordination challenges in time and space, in *11th International Conference on Electrical Power Quality and Utilisation* (2011), pp. 1–6. https://doi.org/10.1109/EPQU.2011.6128888

56. R.J. Michaels, Electricity and its regulation, in *The Concise Encyclopedia of Economics*. Library of Economics and Liberty, http://www.econlib.org/library/Enc/ElectricityandItsRegulation.html

57. Midcontinent ISO: Energy and Operating Reserve Markets Business Practices Manual (2016)

58. New York ISO: Market Participants Users Guide (2016)

59. Y. Okawa, T. Namerikawa, Z. Qu, Passivity-based stability analysis of dynamic electricity pricing with power flow, in *Proceedings of the 2017 Conference on Decision and Control*, Melbourne, Australia (2017)

60. Pacific Northwest Smart Grid Demonstration project technology performance report. Volume 1: Technology performance. Technical Report, PNWD-4438 (Battelle Memorial Institute, 2015)

61. C. Perfumo, E. Kofman, J.H. Braslavsky, J.K. Ward, Load management: model-based control of aggregate power for populations of thermostatically controlled load. Energy Convers. Manag. **55**, 36–48 (2012)

62. PJM: PJM Manual 11: Energy & Ancillary Services Market Operation (2016)

63. P. Samadi, H. Mohsenian-Rad, R. Schober, V.W.S. Wong, Advanced demand side management for the future smart grid using mechanism design. IEEE Trans. Smart Grid **3**(3), 1170–1180 (2012). https://doi.org/10.1109/TSG.2012.2203341

64. F. Schweppe, M. Caramanis, R. Tabors, R. Spot Pricing of Electricity, Bohn, Kluwer Academic Publishers (1988)

65. J. Smith, M. Milligan, E. DeMeo, B. Parsons, Utility wind integration and operating impact state of the art. IEEE Trans. Power Syst. **22**(3), 900–908 (2007)

66. Southwest Power Pool: Market Protocols: SPP Integrated Marketplace (2017)

67. S. Vandael, B. Claessens, M. Hommelberg, T. Holvoet, G. Deconinck, A scalable three-step approach for demand side management of plug-in hybrid vehicles. IEEE Trans. Smart Grid **4**(2), 720–728 (2013). https://doi.org/10.1109/TSG.2012.2213847

68. J.F. Wasik, *The Merchant of Power: Sam Insull, Thomas Edison, and the Creation of the Modern Metropolis* (St. Martin's Press, 2006)

69. S. Widergren, C.M. Marinovici, J.C. Fuller, S. Krishnappa, D.P. Chassin, A. Somani, Customer engagement in AEP gridsmart residential transactive system, in *Proceedings of the 2014 Transactive Energy Framework Workshop Proceedings* (Folsom, CA, 2014)
70. S.E. Widergren, J.C. Fuller, M.C. Marinovici, A. Somani, Residential transactive control demonstration, in *Proceedings of the 2014 Innovative Smart Grid Technologies Conference*, Washington, DC (2014). https://doi.org/10.1109/ISGT.2014.6816405
71. S. E. Widergren et al., AEP Ohio gridSMART® demonstration project: real-time pricing demonstration analysis. Technical report, PNNL-23192 (Pacific Northwest National Laboratory, 2014)
72. N.W. Wilson, B.S. Wagner, W.G. Colborne, Equivalent thermal parameters for an occupied gas-heated house. ASHRAE Trans. **91**(CONF-850606-) (1985)
73. D. Wu, J. Lian, Y. Sun, T. Yang, J. Hansen, Hierarchical control framework for integrated coordination between DERs and demand response. Electr. Power Syst Res, **150**, 45–54, September (2017)
74. A. Zeinalzadeh, N. Gomes de Aguiar, S. Baros, A.M. Annaswamy, I. Chakraborty, V. Gupta, Using natural gas reserves to mitigate intermittence of renewables in the day ahead market, in *Proceedings of the 2017 Conference on Decision and Control*. Melbourne, Australia (2017)
75. W. Zhang, K. Kalsi, J. Fuller, M. Elizondo, D. Chassin, Aggregate model for heterogeneous thermostatically controlled loads with demand response, in *Proceedings of the 2012 IEEE Power and Energy Society General Meeting*, San Diego, CA (2012)
76. W. Zhang, J. Lian, C.Y. Chang, K. Kalsi, Aggregated modeling and control of air conditioning loads for demand response. IEEE Trans. Power Syst. **28**(4), 4655–4664 (2013)
77. W. Zhang, J. Lian, C.Y. Chang, K. Kalsi, Y. Sun, Reduced order modeling of aggregated thermostatic loads with demand response, in *Proceedings of the 51st IEEE Conference on Decision and Control*, Maui, HI (2012)

Some Emerging Challenges in Electricity Markets

Subhonmesh Bose and Steven H. Low

Energy deregulation in the 90s led to the development of power markets in the United States. The Public Utilities Regulatory Policies Act (PURPA) in 1978 laid down the early foundations of deregulation. Subsequent legislations included the Energy Policy Act of 1992 (EPAct92) and FERC Order No. 888 in 1996. They established the rules to "remove impediments to competition in the wholesale bulk power marketplace" by promoting "non-discriminatory transmission services" [1]. These legislations led to the development of two different market architectures in different parts of the US. In one, utility companies established a bilateral market to transact with independent power producers and/or other utilities. In others, a third-party nonprofit facilitator—an Independent System Operator (ISO) or a Regional Transmission Organization (RTO)—was established to mediate between the buyers and the sellers of power at the wholesale level. Our discussion in this article will primarily revolve around the latter.

Heated debates have continued to shape the rules of engagement in wholesale electricity markets over the past two decades. As a result, markets in various parts of the US today do share commonalities, but the specific rules are often diverse. And now, the electricity markets are undergoing a rapid transformation in an effort to accommodate a deepening penetration of variable (and often distributed) renewable supply. Renewable resources like wind and solar energy can reduce carbon emissions and ensure a sustainable energy future. Their power output, however, is *uncertain* (difficult to forecast), *intermittent* (exhibit large fluctuations), and *largely uncontrollable* (not dispatchable on command). Besides profoundly affecting the control

S. Bose (✉)
Department of Electrical and Computer Engineering, University of Illinois
at Urbana–Champaign, Urbana 61801, IL, USA
e-mail: boses@illinois.edu

S. H. Low
Departments of Computing and Mathematical Sciences & Electrical Engineering,
California Institute of Technology, Pasadena 91125, CA, USA
e-mail: slow@caltech.edu

© Springer Nature Switzerland AG 2019
J. Stoustrup et al. (eds.), *Smart Grid Control*, Power Electronics
and Power Systems, https://doi.org/10.1007/978-3-319-98310-3_2

paradigm of the power systems, such resources have challenged the current market designs. Not only are they transforming the wholesale markets, the design of new retail markets is also under consideration.

Academics, industry, and policymakers have amassed a significant literature on electricity markets. In this article, we do not attempt to provide a comprehensive overview of that literature, but focus on a few research directions that we deem important in response to the rise of renewable supply and distributed energy resources (DERs) in the grid. Our discussion largely pertains to the electricity markets in the US. We begin by providing a quick primer on the principles underlying the wholesale markets that are managed by ISOs/RTOs. We then outline some research challenges that must be tackled to extend these principles to the emerging energy landscape.

1 Wholesale Electricity Market Primer

Unlike traditional marketplaces, the physics of the underlying grid impacts the outcomes of the trade in electricity markets. The ISOs/RTOs implement a market clearing mechanism to facilitate this trade within the confines of the power network. In such a centrally managed networked marketplace, what prices are deemed meaningful? Schweppe et al. in [51] provide an answer to that question based on competitive equilibrium theory. Their proposal, the locational marginal pricing (LMP) scheme, has become a ubiquitous feature of today's wholesale electricity markets. We now describe an idealized model of such a wholesale market and explain the locational marginal prices. Our description below does not include various practical considerations. Section 2 alludes to some of these considerations for which the stylized model presented here serves as a prelude.

Consider a power network with n nodes, where each node i has a dispatchable generator. Let this generator incur a cost of $c_i(g_i)$ for supplying $g_i \in \mathcal{G}_i$, where $c_i : \mathcal{G}_i \to \mathbb{R}$ is a smooth increasing strictly convex function ($\mathbb{R}$ denotes the set of real numbers). The convex set $\mathcal{G}_i$ encodes the possible power outputs of that generator. Let node i also have a known inelastic demand $d_i \geq 0$. In the sequel, we use boldface symbols to distinguish vectors and matrices, and use the notation $\mathbf{v}$ to represent a column vector of the form $(v_1, \ldots, v_n)^\mathsf{T}$. Consider the following optimization problem that seeks a socially efficient generation schedule that minimizes the total cost.

$$\underset{\mathbf{g} \in \mathbb{R}^n}{\text{minimize}} \quad \sum_{i=1}^{n} c_i(g_i), \tag{1a}$$

$$\text{subject to} \quad \mathbf{A}(\mathbf{g} - \mathbf{d}) \leq \mathbf{b}, \tag{1b}$$

$$\mathbf{1}^\mathsf{T}(\mathbf{g} - \mathbf{d}) = 0, \tag{1c}$$

$$g_i \in \mathcal{G}_i, \ i = 1, \ldots, n, \tag{1d}$$

where $\mathbf{1}$ denotes a vector of all ones, and $\mathbf{A} \in \mathbb{R}^{2m \times n}$ in (1b) maps the vector of nodal power injections $\mathbf{g} - \mathbf{d}$ to the vector of directed power flows on m transmission lines using a linearized power flow model (the so-called DC approximation). This widely used approximation uses fixed voltage magnitudes, ignores transmission line losses, and assumes the differences between voltage phase angles across transmission lines to be small. See [52] for more details. With these approximations, (1b) says that the power flows over the transmission lines respect their power transfer capacities $\mathbf{b} \in \mathbb{R}^{2m}$, (1c) says that demand and supply of power must balance over the entire network, and (1d) models the supply constraints on each generator.

Let $\boldsymbol{\mu} \in \mathbb{R}^{2m}_+$ and $\gamma \in \mathbb{R}$ be the Lagrange multipliers associated with (1b) and (1c), respectively. A partial Lagrangian for (1) is then given by

$$\mathscr{L}(\mathbf{g}, \boldsymbol{\mu}, \gamma) := \sum_{i=1}^{n} c_i(g_i) + (\mathbf{A}^\mathsf{T}\boldsymbol{\mu} - \gamma\mathbf{1})^\mathsf{T}(\mathbf{g} - \mathbf{d}) - \boldsymbol{\mu}^\mathsf{T}\mathbf{b}.$$

Let $(\mathbf{g}^*, \boldsymbol{\mu}^*, \gamma^*)$ be a primal–dual optimal solution for (1). With the optimal Lagrange multipliers $\boldsymbol{\mu}^*$ and γ^*, define the following vector of *nodal prices* or *locational marginal prices* LMPs:

$$\boldsymbol{\lambda}^* := \gamma^*\mathbf{1} - \mathbf{A}^\mathsf{T}\boldsymbol{\mu}^* \in \mathbb{R}^n. \tag{2}$$

Schweppe et al. [51] advocates the use of $\boldsymbol{\lambda}^*$ as the nodal energy prices. That is, generator (load) at bus i is paid (pays) λ_i^* for each unit of energy she generates (consumes). This pricing scheme is justified by the fact that $(\mathbf{g}^*, \boldsymbol{\lambda}^*)$ constitutes an *efficient competitive equilibrium*, i.e., $(\mathbf{g}^*, \boldsymbol{\lambda}^*)$ satisfies the following conditions.

1. *Incentive compatibility.* Suppose generator i is price-taking in that she believes that her supply offer cannot meaningfully alter the price the ISO pays her at. Given the nodal price λ_i^*, the generator will then produce the ISO's prescribed quantity g_i^* in an effort to maximize her own profit, i.e.,

$$g_i^* = \operatorname*{argmax}_{g_i \in \mathscr{G}_i} \ \lambda_i^* g_i - c_i(g_i).$$

2. *Market clearing.* The total power production matches the total demand.
3. *Efficient dispatch.* Under the assumption that generator i is price-taking, her supply offer will essentially be the parametric maximizer of $pg_i - c_i(g_i)$ over $\mathscr{G}_i$ as a function of the price p. The ISO can then infer the cost curve c_i (up to an additive constant) required in (1) from the submitted supply offer, and hence, is able to find the socially efficient generation schedule through this market mechanism.

It is straightforward to draw the above conclusions using the Karush–Kuhn–Tucker optimality conditions for (1). LMPs are marginal in the sense that $\boldsymbol{\lambda}^*$ measures the sensitivity of the aggregate optimal cost to the vector of nodal demands, i.e.,

$$\lambda^* = \nabla_{\mathbf{d}} \left[\sum_{i=1}^{n} c_i(g_i^*) \right].$$

Interestingly, if the transmission constraints are not binding at the optimal solution, then $\mu^* = \mathbf{0}$ and $\lambda_i^* = \gamma^*$ for each $i = 1, \dots, n$. That is, LMPs become spatially homogenous without network congestion. With congestion, these prices can vary with location. Perhaps counterintuitive, a nodal price λ_i^* can be positive or negative, and hence, sometimes a load can be *paid* to consume. LMPs and their properties have been discussed in great detail in many papers over the last two decades. We refer the interested readers to [31, 59] for more insights. Keeping the above market model in mind, we now turn toward the emerging market challenges in accommodating variable renewable supply and DERs.

2 Future Market Challenges

We discuss five challenges and opportunities in electricity markets that we see as becoming increasingly important moving forward.

2.1 *Pricing Nonconvexities*

Competitive equilibrium provides an elegant theoretical support for the LMP-based market mechanism. Real power markets, however, introduce *nonconvexities* to the optimization problem in (1). Besides compounding the computational difficulties in solving the resulting market clearing problem, it is generally difficult to derive a pricing strategy that enjoys the same economic rationale as the one above. Pricing such nonconvexities has given rise to its own literature, e.g., see [21, 44, 49], and the references therein.

Nonconvexity arises in various different ways. For example, unit commitment decisions introduce integrality constraints into the market clearing problem.[1] Slow-ramping dispatchable generators typically require forward planning. At the forward stage—a day or a few hours ahead of the time of power delivery—one has to decide when such generators will come online to produce power over a time horizon. It is straightforward to extend (1) to include ON/OFF status of the generators as additional decision variables (with associated costs and constraints) in a multiperiod formulation. These decisions are binary in nature, and render the market clearing problem nonconvex. Among others, convex hull pricing has been suggested to design compensation schemes for a market clearing problem with integrality constraints [21, 50].

---

[1] Modeling transmission switching capabilities also gives rise to integrality constraints [43].

The power flow model in (1) is another source of nonconvexity. While the DC approximation defines a convex set of feasible power injections, more accurate nonlinear power flow equations define a nonconvex feasible set. The challenge of solving such an optimization problem over a power network is well documented. The literature on this so-called *AC optimal power flow problem* is extensive; see [10, 11, 14, 17, 18, 24, 33–36] for a sequence of surveys. Again, the derivation of meaningful prices to accompany such a dispatch can be challenging. A generator can also provide voltage and reactive power support in addition to supplying real power. An LMP-based market mechanism, as described, does not pay for such services.

The challenges in pricing nonconvexities are well known. Various "side payments" and "uplift payments" have been designed on top of LMPs to pay for them. Integration of variable renewable energy makes it imperative to investigate some of them more closely in the near future. To provide an example, many have identified the threats to voltage stability from pockets of large-scale photovoltaic installations [15]. To mitigate such risks, one can design a market to incentivize corrective actions. One might conceive a compensation scheme in such a market to accompany the solution of an optimal power flow problem similar to (1), but with nonlinear AC power flow equations—a nonconvex market clearing problem.

2.2 Hedging via Forward Markets

Variability characteristics of renewable resources such as wind and solar pose a fundamental challenge to their integration into the grid. Operational protocols and market structures need adjustments to respond to that challenge. To illustrate that need, notice that errors in day-ahead aggregate load forecasts for the bulk power system are roughly 1–3%. The same errors in forecasts of regional wind energy output are 6–8% in comparison [29]. Uncertainty in net demand (nominal demand less wind power output) will therefore significantly change for wind penetration beyond 15–20% of average load.[2] Scheduling decisions, whether taken a day or a few hours in advance of power delivery, have to account for that uncertainty.

The fleet of dispatchable generators cannot always respond to variations in net demand in realtime. Cheap baseload generators often have limited ramping capabilities and long start-up/shut-down processes. Consequently, one has to plan forward when and for how long such units will operate over a time horizon.

One way to schedule generators under uncertainty is the so-called *certainty equivalent* approach. This approach to unit commitment and economic dispatch replaces all uncertain parameters at the forward stage by their certainty surrogates. That is, the forward dispatch is computed based on a point forecast of uncertain demand and supply. Deviations from said forecasts are balanced in realtime. To explain it mathematically, inherit the notation in Sect. 1 with the addition that the available capacity of production in realtime is random, given by $\mathscr{G}_i(\omega)$. Here, ω encodes a *scenario*, a

---

[2]The exact penetration level beyond which the uncertainty is significant will no doubt depend on the statistics of the wind and the nature of the power system.

collection of which is given by Ω. Dispatchable generators will typically have identical $\mathscr{G}_i$ across scenarios. For simplicity, ignore the forward (binary) commitment decisions, but rather suppose that the forward schedule defines a *set-point* g_i^0 for generator at bus i. Then, its ramping capability is modeled as $|g_i(\omega) - g_i^0| \leq R_i$. Slow-ramping baseload generators have a smaller R compared to fast-ramping peaker power plants. Then, a certainty equivalent approach for forward scheduling can assume the following form:

$$\underset{\mathbf{g}^0 \in \mathbb{R}^n}{\text{minimize}} \quad \sum_{i=1}^{n} c_i(g_i^0), \tag{3}$$

$$\text{subject to} \quad \mathbf{A}(\mathbf{g}^0 - \mathbf{d}) \leq \mathbf{b},$$

$$\mathbf{1}^{\mathsf{T}}(\mathbf{g}^0 - \mathbf{d}) = 0,$$

$$g_i^0 \in \hat{\mathscr{G}}_i, \ i = 1, \ldots, n.$$

Here, $\hat{\mathscr{G}}_i$ is a certainty surrogate for the random production capacity. If $\mathscr{G}_i(\omega) = [0, \bar{g}_i(\omega)]$, then an example of such a surrogate is $\hat{\mathscr{G}}_i := [0, \mathbb{E}[\bar{g}_i(\omega)]]$, where $\mathbb{E}$ stands for expectation. One can price the outcome of Eq. (3) along similar lines as in Sect. 1. When scenario ω is realized, the realtime dispatch is computed by minimizing $\sum_{i=1}^{n} c_i(g_i(\omega))$—possibly with an added ramping cost for deviating from the setpoint—such that $\mathbf{g}(\omega)$ balances $\mathbf{d}$ across the network, and it satisfies $|g_i(\omega) - g_i^0| \leq R_i$, and $g_i(\omega) \in \mathscr{G}_i(\omega)$ at each bus i. Again, the realtime prices can be computed along the same lines as in Sect. 1. The realtime settlement, however, only pays for deviations of $\mathbf{g}$ from $\mathbf{g}^0$ at the realtime price. The forward schedule so computed and priced is agnostic to realtime balancing costs, making this approach inefficient. This inefficiency has long been recognized. To circumvent it, one can alternately make use of a stochastic formulation to schedule forward. Within our idealized market model, it can be written as

$$\text{minimize} \quad \sum_{i=1}^{n} \mathbb{E}[c_i(g_i(\boldsymbol{\omega}))], \tag{4a}$$

$$\text{subject to} \quad \mathbf{A}(\mathbf{g}(\boldsymbol{\omega}) - \mathbf{d}) \leq \mathbf{b}, \tag{4b}$$

$$\mathbf{1}^{\mathsf{T}}(\mathbf{g}(\boldsymbol{\omega}) - \mathbf{d}) = 0, \tag{4c}$$

$$|g_i(\boldsymbol{\omega}) - g_i^0| \leq R_i, \tag{4d}$$

$$g_i(\boldsymbol{\omega}) \in \mathscr{G}_i(\boldsymbol{\omega}), \tag{4e}$$

$$i = 1, \ldots, n, \ \boldsymbol{\omega} \in \boldsymbol{\Omega}$$

over the variables $\mathbf{g}^0 \in \mathbb{R}^n$, and $\mathbf{g} : \Omega \to \mathbb{R}^n$. The above stochastic approach is designed to maximize efficiency, but makes it difficult to define a meaningful forward settlement. To that end, authors in [45] introduce an additional constraint in (4) that balances demand with generator set points. Then, they define a forward payment based on the optimal Lagrange multiplier for that constraint. While it suggests an interesting workaround, this approach has its limitations. Notice that the forward set points for peaker power plants and renewable generators lack any physical meaning. As a result, the rationale behind their forward payment scheme becomes unclear. In addition, such stochastic economic dispatch-based forward settlement can be revenue inadequate in certain scenarios. That is, the system operator can end up cash negative after settling the payments of market participants. The difficulty in defining forward settlements for a stochastic economic dispatch model is chronicled in [7, 37]. They also propose alternate settlement schemes; each has its own pros and cons. In parallel with the stochastic paradigm, different dispatch procedures based on adaptive robust optimization have also been proposed. For example, [32] advocate the dynamic construction of look-ahead uncertainty sets, based on which one can correct past commitment or dispatch decisions with revised information about impending uncertainty closer to the time of power delivery. Designing an accompanying payment mechanism remains an interesting direction for future research.

In general, when forecast errors in demand and supply are relatively small, one can generally expect a forward nominal dispatch to be close to the physical dispatch in realtime. The resulting compensation then becomes meaningful, and the forward settlement will typically comprise a significant portion of the total payment to the market participants. With deep penetration of renewable supply, the forecasts will be less accurate. A forward market that compensates for a nominal dispatch will therefore prove less effective for participants to mitigate financial risks. The design of alternate and practically implementable forward market mechanism is thus an important research direction. Extending that logic further, the growing penetration of renewable supply will make the market participants' payments more volatile. Therefore, the participants will have to bear larger financial risks on a daily basis. Such risks can adversely affect the competitiveness of the market in the long run. New financial instruments are needed to counter such risks. To illustrate, notice that peaker power plants are often necessary to ensure that supply can meet demand even with the least available renewable supply. However, they may not always be required to produce, if enough renewable supply is realized. Short-term "capacity" payments via *flexible ramping products* have been adopted in some markets to address this question, e.g., in the markets managed by California ISO and Midcontinent ISO. Such contracts allow flexible generators to earn money when they remain available but are not called upon to supply power. Toward the same goal of risk mitigation, some have also suggested using swing options and call options, e.g., in [2, 26]. Additional research is necessary to judge the applicability of such ideas within practical market settings and to carefully characterize their impacts on market competitiveness.

Financial instruments will no doubt help make the energy marketplace attractive for both renewable power producers and flexible dispatchable generators. One must, however, analyze and monitor their possible interactions. As a cautionary tale, an example in [28] points out how a firm can strategically utilize their financial position in one instrument to favorably influence the reward from another.

2.3   Integrating Storage Technologies

Energy storage devices are expected to play an important role in integrating renewable supply. One way to integrate them at scale is to treat them like any other distributed energy resource operated by a storage owner-operator. Another way is to treat them as part of the infrastructure like transmission lines where their operation and compensation are determined by the market outcome. We illustrate the second approach that is proposed in [38, 55].

We start by explaining financial instruments that help hedge against *spatial* variations of LMPs when transmission lines are congested. Then, we illustrate how the same principle carries over to the case of hedging against *temporal* price variations using energy storage. Recall our definition of the locational marginal prices λ^* in (2). A generator at node i is paid $\lambda_i^* g_i^*$, while a demander at the same node pays $\lambda_i^* d_i$. After settling the payments with the market participants, the system operator is left with a *merchandising surplus* of

$$\text{MS} := \sum_{i=1}^{n} \lambda_i^* (d_i - g_i^*) = - \sum_{i=1}^{n} \lambda_i^* q_i^*, \tag{5}$$

where $\mathbf{q}^* := \mathbf{g}^* - \mathbf{d}$ is the vector of nodal power injections. If the transmission lines are not congested at the optimal dispatch, then $\text{MS} = 0$. This follows from (1c) and the fact that all nodal prices become equal. When there is congestion, one can prove that $\text{MS} \geq 0$ at an efficient competitive equilibrium. Stated differently, the ISO never runs cash negative upon settling payments with market participants.

An ISO is a nonprofit market facilitator and is not entitled to keep the merchandising surplus. It reallocates a positive MS (that arises due to congestion) through *financial transmission rights* (FTR) [23, 46]. A holder of FTR, e.g., a grid operator or one who has bought these rights on a secondary market, will receive a fraction of the MS that the ISO collects in clearing the spot market. Formally, a point-to-point FTR is identified by a power $p_{ij} \in \mathbb{R}_+$ with the interpretation that the holder grants the injection of p_{ij} amount of power into the grid at bus i and a withdrawal of the same amount from bus j. The FTR holder is entitled to receive a *rent* (or a liability) equal to $(\lambda_i - \lambda_j) p_{ij}$. It can be shown that if a collection of FTR positions, together with the physical power flows over the grid, is feasible (i.e., the injections satisfy (1b)–(1d)), then the FTR rent is no greater than MS. That is, the *simultaneous feasibility* ensures that the ISO has sufficient revenue to cover the issued FTRs [23, 59]. Alternately, flow-based congestion rents have also been proposed (defined in terms of μ and the line flows, instead of λ and the bus injections); they are called flowgate rights (FGRs) [12, 13]. FTRs and FGRs are convenient instruments for hedging against spatial price variations, and also provide market signals to incentivize investment in transmission capacity.

An interesting proposal is presented in [38, 55] to integrate storage devices into the market framework for hedging temporal price variations using instruments similar to FTRs/FGRs. Consider a time horizon $t = 0, 1, \ldots, T - 1$. Let the corresponding generations and (possibly time-varying) transmission line capacities be given by

$\mathbf{q}(t) := \mathbf{g}(t) - \mathbf{d}(t) \in \mathbb{R}^n$ and $\mathbf{b}(t) \in \mathbb{R}^m$ at each time t, respectively. Suppose there is a storage device with capacity $s_i \geq 0$ at each bus i. Denote by $u_i(t)$, the power extracted from the device at bus i and time t. The convention is such that $u_i(t) < 0$ denotes charging that device. Assuming zero initial state, the state of charge at time t of the device at bus i depends on past inputs $u_i(\tau), \tau = 0, \ldots, t-1$. The state of charge over the time horizon can then be expressed as $\mathbf{L}\mathbf{u}_i$ for a suitably defined matrix $\mathbf{L}$ and $\mathbf{u}_i := (u_i(0), \ldots, u_i(T-1))^\mathsf{T}$. We must have $\mathbf{0} \leq \mathbf{L}\mathbf{u}_i \leq s_i\mathbf{1}$ for each $i = 1, \ldots, n$ to ensure that the state of charge respects the storage capacity constraints. Here, $\mathbf{1}$ denotes a vector of all ones. Using the notation introduced, the problem in (1) can be extended to multiple periods as

$$\text{minimize} \quad \sum_{t=0}^{T-1}\sum_{i=1}^{n} c_i(g_i(t)), \tag{6a}$$

$$\text{subject to} \quad \mathbf{q}(t) = \mathbf{g}(t) - \mathbf{d}(t), \tag{6b}$$

$$\mathbf{A}(\mathbf{q}(t) + \mathbf{u}(t)) \leq \mathbf{b}, \tag{6c}$$

$$\mathbf{1}^\mathsf{T}(\mathbf{q}(t) + \mathbf{u}(t)) = 0, \tag{6d}$$

$$\mathbf{0} \leq \mathbf{L}\mathbf{u}_i \leq s_i\mathbf{1}, \tag{6e}$$

$$g_i(t) \in \mathcal{G}_i, \tag{6f}$$

$$i = 1, \ldots, n, \ t = 0, \ldots, T-1$$

over the variables $\mathbf{g}(t) \in \mathbb{R}^n, \mathbf{q}(t) \in \mathbb{R}^n$ for each t and $\mathbf{u}_i \in \mathbb{R}^T$ for each i. As in Sect. 1, let $\boldsymbol{\mu}(t), \gamma(t)$ denote the Lagrange multipliers associated with (6c), (6d), respectively, and $\lambda(t)$ be the associated LMPs at time t. Let $\underline{\boldsymbol{v}}_i, \overline{\boldsymbol{v}}_i$ denote the Lagrange multipliers associated with the lower and upper bounds in (6e), respectively. From hereon, let $(\mathbf{g}(t), \mathbf{q}(t), \mathbf{u}_i, \boldsymbol{\mu}(t), \gamma(t), \underline{\boldsymbol{v}}_i, \overline{\boldsymbol{v}}_i)$ denote the variables at a primal–dual optimum of (6).

Defining the power flow on the line from bus i to bus j at time t as $f_{ij}(t)$, the power balance at bus i yields $q_i(t) + u_i(t) = \sum_j f_{ij}(t)$. That relation allows us to write the merchandising surplus defined in (5) as a sum of two terms—the transmission congestion surplus (TCS) and the storage congestion surplus (SCS).

$$-\underbrace{\sum_{t=0}^{T-1}\sum_{i=1}^{n}\lambda_i(t)q_i(t)}_{\text{MS}} = \underbrace{\frac{1}{2}\sum_{t=0}^{T-1}\sum_{i,j=1}^{n}(\lambda_i(t)-\lambda_j(t))f_{ij}(t)}_{\text{TCS}} + \underbrace{\sum_{t=0}^{T-1}\sum_{i=1}^{n}\lambda_i(t)u_i(t)}_{\text{SCS}} \tag{7}$$

As proven in [38], TCS is always sufficient to cover the rent from FTRs defined above. Identify *financial storage right* (FSR) as a power profile $\boldsymbol{v}_i \in \mathbb{R}^T$ with the interpretation that the holder of an FSR $\boldsymbol{v}_i$ grants the withdrawal of power $v_i(t)$ from storage device i at time t and will receive a rent (or liability) equal to $\sum_t \lambda_i(t)v_i(t)$. FSR is an instrument similar in spirit to FTR; the latter allows its holder to hedge against spatial price variations, and the former does the same for temporal ones. FSR entitles its holder to inter-temporal arbitrage gains that a storage device generates under socially optimal operation. FSRs also provide market signals to incentivize investment in storage capacities.

An alternative to FSRs is *energy capacity rights* (ECRs) as described in [55]. Identify an ECR with $e_i \in \mathbb{R}_+^T$ that the holder grants the storage of $e_i(t)$ amount of energy at the device at bus i and time t, for which she receives a rent equal to $\sum_t \overline{v}_i(t)e_i(t)$. Recall that $\overline{v}_i(t)$ is the shadow price associated with the storage capacity constraint at bus i. Much like an FGR, an ECR is a convenient instrument for compensating property rights to specific energy devices. A key result of [38, 55] is the revenue adequacy of these compensation schemes. That is, the ISO remains cash positive after settling the payments of the loads, generators, FTR/FGR holders, and FSR/ECR holders as long as the physical power flows are simultaneously feasible with those implied by the FTR/FGR and the FSR/ECR positions. See [38] for more details.

Several questions need to be addressed to make these proposed mechanisms viable in practice. For example, an ISO needs to check a certain simultaneous feasibility condition to issue such rights through an auction to ensure its revenue adequacy. That condition, however, involves future dispatch decisions, unbeknownst to the ISO at the time of that auction. A causal enforcement of such a feasibility condition then becomes impossible. An ISO will possibly run periodic auctions for such rights based on forecasts of future operations. It remains to be seen as to how often the ISO runs cash negative with such a mechanism, given that the realized demand/supply conditions can substantially deviate from their forecasts with deepening penetration of variable renewable energy. One also has to analyze how FSRs/ECRs will affect storage participation in ancillary service markets, in which storage is expected to play a significant role.

2.4 Mitigating Strategic Interactions

The locational marginal pricing mechanism is derived under the premise that each generator is price-taking. Under this assumption, they will reveal their true marginal costs in their supply offers. In reality, market participants can be strategic in that they can partly anticipate the effect of their bid or offer on the resulting prices that in turn define their payments. Game theory has been widely used to analyze such strategic interactions. Electricity markets are particularly challenging to analyze, however. Since power is procured over multiple timescales, it requires one to model a multi-settlement market. Further, market participants typically engage repeatedly in the energy auctions, and when they do, they may not be privy to a complete network model or the cost/utility structures of other market participants. Also, each market participant only observes their locational marginal prices from the last auction and not the offers/bids from other market participants. Ideally, one would need to analyze the electricity market as a *dynamic game with incomplete and imperfect information*. Unless one resorts to simulations (e.g., using an agent-based paradigm [57]), deriving structural insights from such formulations becomes untenable without making substantial simplifications.

In the interest of mathematical tractability, many have studied a static game of complete information. Within that category, some have studied a *supply function* competition model on a copperplate power system (ignoring the underlying power network). In that model, each generator submits a supply offer that represents how much power that generator is willing to supply as a function of the price it is paid at. This competition model mirrors current practice in many wholesale markets. Equilibrium analysis is still quite challenging, given that the strategy set of each player is a function space. One often restricts attention to a parameterized class of supply offers to further simplify the analysis. For example, authors in [25] present a scalar parameterized supply function game that exhibits bounded efficiency loss. A capacitated version of the same is derived in [62]. And authors in [30] provide an elegant characterization of the equilibria for a networked variant of the model in [62]. The networked case with affine supply functions is studied in [54], where the authors report a lack of equilibrium. See [3, 4, 19, 20, 27, 47, 48, 60, 61] for other references on this competition model.

Another line of work simplifies the analysis of electricity markets to a networked Cournot model, where each generator is strategic and competes via quantity offers. See [5, 9, 22, 41, 63] and references therein for examples. Such models generalize the classical (non-networked) Cournot competition. While electricity markets do not employ Cournot-type quantity offers in their mechanisms, conclusions on price movements from Cournot markets have been known to correlate well with that from supply function models [58]. And they are substantially easier to analyze than their supply function counterparts.

What is the ultimate goal of such game theoretic models? First, one can characterize how the incentives are aligned for specific market participants to act strategically. While legal barriers exist to limit strategic behavior, such models can identify potential transgressors.[3] Second, such models can inform market design. They can provide insights into how the system operator can alter its market clearing procedure, knowing that market participants can be strategic, e.g., [9]. This is precisely the realm of mechanism design theory. However, adopting popular mechanisms such as the Vickerey–Clarkes–Groves (VCG) auctions to clear electricity markets will entail a substantial modification of the current market structure. For example, nodally uniform linear pricing based compensation schemes have to be sacrificed in favor of one with nonlinear pricing. Such tectonic shift in market structure is perhaps less realistic. Therefore, an interesting research challenge is to identify a market mechanism that maintains the simplicity of a parametric offer/bid-based mechanism and a nodally uniform linear pricing scheme, but limits the impacts of strategic interaction.

Strategic interaction in wholesale markets has largely been studied for a power network with a collection of dispatchable generators. The case with variable renewable supply will define an interesting and important direction for research in the coming years.

---

[3]Structural market power indices attempt to reveal the same without an explicit game theoretic analysis; see [8, 56].

2.5 Retail Markets for DERs

The low-voltage distribution grid is at the cusp of a historic transformation. This part of the power system has traditionally comprised a collection of consumers with largely inelastic demands. Such consumers are now rapidly becoming *prosumers* who own devices capable of generation, storage, and active demand side management. Generation—e.g., from rooftop photovoltaic panels—can be stochastic in nature. Storage can encompass at-home batteries or electric vehicles. Thermostatically controlled loads as well as other smart appliances can alter their power draw on demand. Such DERs are individually too small to provide meaningful services to other customers within a distribution grid and even less so to the bulk power system. They can be impactful when aggregated, however. The active management of a collection of DERs can transform the distribution grid from being a passive demander to a dynamic participant of the energy ecosystem.

Concomitant to the operational challenges that arise in DER management is the design of a market structure to organize these services. Distribution utility companies or other retail aggregators can assume the role of a DER coordinator. Question arises as to who will manage a market for DERs or DER aggregators. Consider the example of New York state's Reforming the Energy Vision (NY-REV) initiative. Their Public Service Commission has approved the state's utility companies to become *distribution system platform (DSP) providers*. Such a market platform will ideally provide alternate revenue streams for the utilities that have traditionally relied on returns from grid investments. Many have advocated the creation of an independent entity called *distribution system operator* (DSO) to provide such platform services for DER management. Another model for DER participation is that defined by California ISO where DER owners with sufficiently large assets or aggregations of them can offer demand-response services directly in the wholesale market. Designing the rules of engagement for this emerging retail marketplace will be an important area of research. The participants are the customers who own and operate the resources, the retail aggregators who interface between the customers and a market platform, and the utility company that maintains the distribution grid and may additionally manage a retail market or play in it.

What services can retail aggregators provide from coordinating a collection of DERs? DERs can transform the distribution grid into a vibrant energy economy that is flexible and able to respond to the needs of both the distribution and the transmission grids. For example, they can collectively reduce the total energy demanded over a time horizon when needed. They can provide flexible capacity to act as reserves with the ability to compensate for variability in supply from renewable resources in the bulk power system. They can also track regulation signals to balance the second-by-second fluctuations of demand and supply in the transmission network. Furthermore, some DERs can inject or extract reactive power. That in turn can provide voltage support in the low and medium voltage distribution grids. The authors in [42] and a sequence of followup papers argue that distribution network costs currently account for up to 35% of the total electricity cost. This cost can be significantly reduced by utilizing DERs for such grid services.

For retail market design, the authors of [42] among others advocate the use of distribution LMPs, emulating and extending its wholesale counterpart to the distribution grids. That proposal breaks away from the widely used practice of fixed, average-cost based, and possibly tiered retail pricing schemes administered by the utility companies. The hope is to replace that scheme with one that reflects the locational value of a DER to the engineering and economic needs of the grid, as recognized in [16]. Identifying the right engineering architecture (including power electronics based hardware and communication technology) and algorithms (including ones to solve large-scale optimal power flow problems with nonlinear AC power flow models under uncertainty) for DER coordination will define important directions for research. The initiative from New York and California among others indicate that the power industry has recognized the value of DERs and the possible creation of retail markets to enable their participation. Such markets may very well become a reality within the next decade or so. Its successful implementation will complete the deregulation process of the power industry that began in the 1990s with the unbundling of the generation sector at the bulk power level.

How can retail aggregators control and coordinate a collection of spatially distributed resources? Two different paradigms for coordination have been explored. One paradigm is that of *direct load control*, wherein a retail aggregator wields some level of control on the resources themselves. For example, a retail aggregator can have the ability to turn an air conditioner on or off for a customer who signs up for direct load control. The other paradigm is that of *transactive control*, where the retail aggregator posts an incentive for a certain service, and the customer responds to that incentive. Transactive control can rely on a human in the loop, or it can be automated. For example, an energy management system for a house can be programmed to accept a request to turn an air conditioner off, if the monetary reward for that action exceeds a certain threshold.

The compensation scheme for customers can vary widely as well. Some have advocated aggregators to pay for the ability to interrupt the service to a customer. The payment then depends on the frequency of allowable interruptions [53]. Some pay for the flexibility to defer demand, and the payment is either deadline-differentiated [6] or duration-differentiated [40]. Some compensation schemes have been proposed with specific types of distributed resources in mind such as electric vehicles or in-home batteries. A major challenge in this line of inquiry is the difficulty for retail aggregators to accurately measure the response of customers to their commands. For example, if a retail aggregator compensates for temporary reduction in power consumption, they need to establish a baseline demand in order to calculate that reduction. One can surmise that customers can game the baseline measurement. See [39] for a recently proposed mechanism to elicit truthful baselines from customers.

3   Conclusion

A sound market design for electricity is crucial to reap the benefits of technological advances in the power grid. The electricity industry in the US and beyond is on an aggressive path toward the adoption of renewable generation and distributed energy resources. How do we integrate these technologies in system operations and market mechanisms? In this manuscript, we have discussed five topics surrounding this transformation: pricing nonconvexities, forward market designs, markets for energy storage, understanding strategic interactions, and retail market design. For each topic, we have summarized some known results and pointed to some questions for future research. The list of selected topics, references, and potential research directions are not exhaustive. We do not cover, for example, market designs for long-run resource adequacy or reliability in the face of rapid adoption of renewable power and retirements of coal/nuclear power plants. We also do not discuss markets for operating reserves and regulation products. Such topics are left for similar future endeavors.

Acknowledgements We thank Prof. Alejandro D. Domínguez-García, Khaled Alshehri, and Mariola Ndrio at UIUC and Prof. Eilyan Bitar at Cornell for helpful discussions.

References

1. FERC Order No. 888. *Promoting Wholesale Competition Through Open Access Nondiscriminatory Transmission Services by Public Utilities: Recovery of Stranded Costs by Public Utilities and Transmitting Utilities* (1996)
2. K. Alshehri, S. Bose, and T. Başar. Cash-settled options for electricity markets under uncertainty, in *Proceedings of the International Federation of Automatic Control Wold Congress* (2017)
3. R. Baldick, R. Grant, E. Kahn, Theory and application of linear supply function equilibrium in electricity markets. J. Regulat. Econom. **25**(2), 143–167 (2004)
4. R. Baldick, W.W. Hogan, et al. *Capacity constrained supply function equilibrium models of electricity markets: Stability, non-decreasing constraints, and function space iterations* (University of California Energy Institute, 2001)
5. J. Barquín, M. Vázquez, *Cournot equilibrium in power networks* (Universidad Pontificia Comillas, Madrid, Instituto de Investigación Tecnológica, 2005)
6. E. Bitar, Y. Xu, Deadline differentiated pricing of deferrable electric loads. IEEE Trans. Smart Grid **8**(1), 13–25 (2017)
7. S. Bose, On the design of wholesale electricity markets under uncertainty, in *Proceedings of the 53rd Annual Allerton Conference on Communication, Control, and Computing* (IEEE, 2015) pp. 203–210
8. S. Bose, C. Wu, Y. Xu, A. Wierman, H. Mohsenian-Rad, A unifying market power measure for deregulated transmission-constrained electricity markets. IEEE Trans. Power Syst. **30**(5), 2338–2348 (2015)
9. D. Cai, S. Bose, and A. Wierman, On the role of a market maker in networked Cournot competition. arXiv:1701.08896 (2017)
10. M.B. Cain, R.P. O'Neill, and A. Castillo, History of Optimal Power Flow and Formulations (Federal Energy Regulatory Commission, 2012) pp. 1–36
11. F. Capitanescu, Critical review of recent advances and further developments needed in AC optimal power flow. Elect. Power Syst. Res. **136**, 57–68 (2016)

12. H.P. Chao, S. Peck, A market mechanism for electric power transmission. J. Regulat. Econom. **10**, 25–59 (1996)
13. H.P. Chao, S. Peck, S. Oren, R. Wilson, Flow-based transmission rights and congestion management. Electric. J. **13**(8), 38–58 (2000)
14. H.W. Dommel, W.F. Tinney, Optimal power flow solutions. IEEE Trans. Power Apparatus Syst. **10**, 1866–1876 (1968)
15. S. Eftekharnejad, V. Vittal, G.T. Heydt, B. Keel, J. Loehr, Impact of increased penetration of photovoltaic generation on power systems. IEEE Trans. Power Syst. **28**(2), 893–901 (2013)
16. S. Fine, P. De Martini, S. Succar, M. Robison, *The value in distributed energy: It's all about location, location, location* (2014)
17. S. Frank, I. Steponavice, S. Rebennack, Optimal power flow: a bibliographic survey i. Energy Syst. **3**(3), 221–258 (2012)
18. S. Frank, I. Steponavice, S. Rebennack, Optimal power flow: a bibliographic survey ii. Energy Syst. **3**(3), 259–289 (2012)
19. R.J. Green. Increasing competition in the British electricity spot market. J. Indust. Econom. 205–216 (1996)
20. R.J. Green, D.M. Newbery. Competition in the British electricity spot market. J. Political Econom. pages 929–953 (1992)
21. P.R. Gribik, W.W. Hogan, S.L. Pope, Market-clearing electricity prices and energy uplift (Cambridge, MA, 2007)
22. B.F. Hobbs, Linear complementarity models of Nash-Cournot competition in bilateral and POOLCO power markets. IEEE Trans. Power Syst. **16**(2), 194–202 (2001)
23. W.W. Hogan, Contract networks for electric power transmission. J. Regulat. Econom. **4**(3), 211–242 (1992)
24. M. Huneault, F. Galiana, A survey of the optimal power flow literature. IEEE Trans. Power Syst. **6**(2), 762–770 (1991)
25. R. Johari, J.N. Tsitsiklis, Parameterized supply function bidding: equilibrium and efficiency. Operat. Res. **59**(5), 1079–1089 (2011)
26. J. Keppo, Pricing of electricity swing options. J. Derivat. **11**(3), 26–43 (2004)
27. P. Klemperer, M. Meyer, Supply function equilibria in oligopoly under uncertainty. Econometrica 1243–1277 (1989)
28. S.D. Ledgerwood, J.P. Pfeifenberger, Using virtual bids to manipulate the value of financial transmission rights. Electr. J. **26**(9), 9–25 (2013)
29. D. Lew, M. Milligan, G. Jordan, R. Piwko, The value of wind power forecasting, in *Proceedings of the 91st AMS Annual Meeting and the 2nd Conference on Weather, Climate, and the New Energy Economy, Washington DC* (2011)
30. W. Lin, E. Bita, A structural characterization of market power in power markets. arXiv:1709.09302 (2017)
31. E. Litvinov, Design and operation of the locational marginal prices-based electricity markets. IET Generat. Trans. Distrib. **4**(2), 315–323 (2010)
32. A. Lorca, X.A. Sun, Adaptive robust optimization with dynamic uncertainty sets for multiperiod economic dispatch under significant wind. IEEE Trans. Power Syst. **30**(4), 1702–1713 (2015)
33. S.H. Low, Convex relaxation of optimal power flow, I: formulations and relaxations. IEEE Trans. Control Netw. Syst. **1**(1), 15–27 (2014)
34. S.H. Low, Convex relaxation of optimal power flow, II: exactness. IEEE Trans. Control Netw. Syst. **1**(2), 177–189 (2014)
35. J.A. Momoh, R. Adapa, M. El-Hawary, A review of selected optimal power f literature to 1993. i. nonlinear and quadratic programming approaches. IEEE Trans. Power Syst. **14**(1), 96–104 (1999)
36. J.A. Momoh, M. El-Hawary, R. Adapa, A review of selected optimal power flow literature to 1993. ii. newton, linear programming and interior point methods. IEEE Trans. Power Syst. **14**(1), 105–111 (1999)

37. J.M. Morales, M. Zugno, S. Pineda, P. Pinson, Electricity market clearing with improved scheduling of stochastic production. Eur. J. Operat. Res. **235**(3), 765–774 (2014)
38. D. Munoz-Alvarez, E. Bitar, Financial storage rights in electric power networks. J. Regulat. Econom. **1**, 1–23 (2017)
39. D. Muthirayan, D. Kalathil, K. Poolla, P. Varaiya, Mechanism design for demand response programs. arXiv:1712.07742 (2017)
40. A. Nayyar, M. Negrete-Pincetic, K. Poolla, P. Varaiya, Duration-differentiated services in electricity. arXiv:1404.1112 (2014)
41. K. Neuhoff, J. Barquin, M.G. Boots, A. Ehrenmann, B.F. Hobbs, F.A. Rijkers, M. Vazquez, Network-constrained cournot models of liberalized electricity markets: the devil is in the details. Energy Econom. **27**(3), 495–525 (2005)
42. E. Ntakou, M. Caramanis, Price discovery in dynamic power markets with low-voltage distribution-network participants, in *Proceedings of the IEEE PES T&D Conference and Exposition* (2014)
43. R.P. O'Neill, K.W. Hedman, E.A. Krall, A. Papavasiliou, S.S. Oren, Economic analysis of the N-1 reliable unit commitment and transmission switching problem using duality concepts. Energy Syst. **1**(2), 165–195 (2010)
44. R.P. O'Neill, P.M. Sotkiewicz, B.F. Hobbs, M.H. Rothkopf, W.R. Stewart, Efficient market-clearing prices in markets with nonconvexities. Eur. J. Operation. Res. **164**(1), 269–285 (2005)
45. G. Pritchard, G. Zakeri, and A. Philpott.A single-settlement, energy-only electric power market forunpredictable and intermittent participants. *Operations Research*, 58(4-part-2):1210–1219, 2010
46. J. Rosellón, T.K. (eds.). *Financial Transmission Rights: Analysis, Experiences and Prospects*, vol. 7. Springer. Lecture Notes in Energy (2013)
47. A. Rudkevich, Supply function equilibrium in power markets: learning all the way. TCA Technical Paper (1999), pp. 1299–1702)
48. A. Rudkevich, On the supply function equilibrium and its applications in electricity markets. Decision Support Syst. **40**(3), 409–425 (2005)
49. C. Ruiz, A.J. Conejo, S.A. Gabriel, Pricing non-convexities in an electricity pool. IEEE Trans. Power Syst. **27**(3), 1334–1342 (2012)
50. D.A. Schiro, T. Zheng, F. Zhao, E. Litvinov, Convex hull pricing in electricity markets: formulation, analysis, and implementation challenges. IEEE Trans. Power Syst. **31**(5), 4068–4075 (2016)
51. F. Schweppe, M. Craramanis, R. Tabors, R. Bohn, *Spot Pricing of Electricity* (1988)
52. B. Stott, J. Jardim, O. Alsaç, DC power flow revisited. IEEE Trans. Power Syst. **24**(3), 1290–1300 (2009)
53. C.-W. Tan, P.P. Varaiya, Interruptible electric power service contracts. J. Econom. Dynam. Control **17**(3), 495–517 (1993)
54. W. Tang, R. Jain, Game-theoretic analysis of the nodal pricing mechanism for electricity markets, in *Proceedings of the 52nd Annual Conference on Decision and Control (CDC)* (IEEE, 2013), pp. 562–567
55. J.A. Taylor, Financial storage right. IEEE Trans. Power Syst. **30**(2), 997–1005 (2015)
56. P. Twomey, R. J. Green, K. Neuhoff, D. Newbery, A review of the monitoring of market power: The possible roles of TSOs in monitoring for market power issues in congested transmission systems (The Cambridge-MIT Institute working paper, 2006)
57. A. Weidlich, D. Veit, A critical survey of agent-based wholesale electricity market models. Energy Econom. **30**(4), 1728–1759 (2008)
58. B. Willems, I. Rumiantseva, H. Weigt, Cournot versus supply functions: what does the data tell us? Energy Econom. **31**(1), 38–47 (2009)
59. F. Wu, P. Varaiya, P. Spiller, S. Oren, Folk theorems on transmission access: proofs and counterexamples. J. Regulat. Econom. **10**(1), 5–23 (1996)
60. L. Xu, R. Baldick, Transmission-constrained residual demand derivative in electricity markets. IEEE Trans. Power Syst. **22**(4), 1563–1573 (2007)

61. L. Xu, Y. Yu. Transmission constrained linear supply function equilibrium in power markets: method and example, in *Proceedings of the International Conference on Power System Technology (Powercon 2002)*, volume 3, pages 1349–1354. IEEE, 2002
62. Y. Xu, N. Li, S.H. Low, Demand response with capacity constrained supply function bidding. IEEE Trans. Power Syst. **31**(2), 1377–1394 (2016)
63. J. Yao, B. Willems, S. S. Oren, I. Adler, Cournot equilibrium in price-capped two-settlement electricity markets, in *Proceedings of the 38th Annual Hawaii International Conference on System Sciences (HICSS '05)*, pp. 58 (2005)

Incentivizing Market and Control for Ancillary Services in Dynamic Power Grids

Kenko Uchida, Kenji Hirata and Yasuaki Wasa

Abstract We discuss an incentivizing market and model-based approach to design the energy management and control systems, which realize high-quality ancillary services in dynamic power grids. Under the electricity liberalization, such incentivizing market should secure a high-speed market-clearing by using the market players' private information well. Inspired by contract theory in microeconomics field, we propose a novel design method of such incentivizing market on the basis of integration of the economic model and the dynamic grid model. We first outline our contract and model-based method to design the incentivizing market and clarify the basic properties of the designed market. We then discuss possibilities, limitation, and fundamental challenges in the direction of our approach and general market-based approaches.

1 Introduction

Achieving a quality assurance of electric energy, called the ancillary service, is a key target of next-generation energy management and control systems for dynamic electric smart grids where electricity liberalization is fully enforced and renewable energy is highly penetrated [1]. Frequency, voltage, and power controls, which are typical contents of the ancillary service, have been technical requirements for the electric energy supplier (e.g., see [2, 3]). Since the electricity liberalization starts, such ancillary control services have been investigated and realized in competitive

K. Uchida (✉) · Y. Wasa
Department of Electrical Engineering and Bioscience, Waseda University,
Tokyo 169-8555, Japan
e-mail: kuchida@waseda.jp

Y. Wasa
e-mail: wasa@aoni.waseda.jp

K. Hirata
Department of Mechanical Engineering, Nagaoka University of Technology,
Niigata 940-2188, Japan
e-mail: hirata@nagaokaut.ac.jp

© Springer Nature Switzerland AG 2019
J. Stoustrup et al. (eds.), *Smart Grid Control*, Power Electronics
and Power Systems, https://doi.org/10.1007/978-3-319-98310-3_3

electricity markets; early market designs consider ancillary services as constraint conditions of the optimal energy dispatch and provide simultaneously the energy dispatch and the ancillary services in a single market [4–8]; the subsequent developments [9–11], focusing on the differences not only in the transaction process of energy and ancillary services but also in the required transaction response time, have proposed ancillary service markets, which are closely interacted with energy markets but external to energy markets, in order to implement ancillary services by transacting typically spinning reserves and regulation reserves; the more recent works [12, 13] have proposed a market model implementing the frequency response ancillary service in the primary control level and pointed out the importance of incentives in ancillary market designs. In view these, future energy management and control systems should include the ancillary service markets with some incentive mechanisms, as core elements, which provide high-quality and fast-response control services to the extent of the primary level. Moreover, if we need ancillary control services of transient state, ancillary service markets should include physical models of dynamic power grids. In this article, we propose an incentivizing model and market-based approach to design the energy management and control systems which realize high-quality ancillary services in such dynamic power grids; especially, we develop a design method of such incentivizing market based on the contract-based integration of the economic model and the dynamic grid model. Commenting on the possibilities and limitation of our approach, we discuss some challenges and significant research issues in the direction of our approach and general market-based approaches.

Our approach is developed under the assumption that an energy dispatch scheduling on a future time interval has been finished in a spot energy market at the tertiary control level [14, 15], e.g., for one-hour future interval and each agent has a linearized model of his/her own system along the scheduled trajectory over the future time interval. For this linear time-varying model, we formulate a design problem of real-time regulation markets, i.e., ancillary service markets, at the secondary and primary control levels [14, 15]. Participants in the dynamic electric grid are consumers, suppliers, or prosumers, called agents, who control their own physical systems selfishly according to their own criteria, and independent public commission, called utility, who integrates economically all the controls of agents into a high-quality power demand and supply. In the integration, a market mechanism is adopted inevitably in order to secure selfish behaviors of agents in the electricity liberalization; that is, each agent bids his/her certain quantity in response to a market-clearing price, while utility (auctioneer) clears the market based on the bidding and decides the price with the high speed for regulation at the secondary and primary levels.

The market model in our approach is characterized by two terminologies: "private information" and "incentivizing market". A conventional market-clearing process based on an iterative exchange of price and quantity (private information) between utility and agents, called the tâtonnement model, does not need rigorous agents' models, but does not generally converges to an equilibrium. Moreover, even if it converges, the tâtonnement model takes generally a long time to converge to the equilibrium without agents' model information [16]. To overcome these issues, we propose a novel noniterative/one-shot market model, in which a market planner first designs contract-based incentives for the agents to report their private information (including their own model information) spontaneously to the utility so that the utility can make a high-speed market-clearing in the incentivizing market. This model needs incentivizing rewards, and the optimization process based on the rewards can be recognized as an intermediate model (the second best model) between two extremal models, namely the tâtonnement model and the so-called supply/demand function equilibrium model (the first best model) which uses for free all agents' rigorous models, i.e., agents' private information. On the basis of our incentivizing market, we discuss the relationships of our incentivizing mechanism with the Lagrange multiplier based integration/decomposition mechanism and the mechanism design.

This article has been organized as follows: In Sect. 2, we introduce a dynamic power grid model and a model-based incentivizing market model. We then outline a contract-based approach to the design of incentivizing markets. In Sect. 3, focusing on the relationship between the private information and the incentives, we discuss possibilities and limitation of our approach through some typical scenarios. From a systems and control perspective, we also provide some research directions on model-based and market-based approaches while taking into account the results of the proposed models and scenarios. In Sect. 4, we conclude our discussion.

2   Grid Model and Incentivizing Market Model

Two Layers Market In this paper, we consider the two-level architecture with the two layers market, spot energy market, and real-time regulation market (see Fig. 1). The well-known temporally separated architecture [14, 15] motivated by the conventional power system control is divided into the primary control level (voltage and frequency stabilization), the secondary control level (quasi-stationary power imbalance control), and the tertiary control level (economic dispatch). The two layers market reorganizes the conventional three-level architecture according to the functions of the markets. Our approach is developed under the assumption that an energy dispatch scheduling on a future time interval (shaded blue in Fig. 1) has been finished in a spot energy market (at the tertiary control level), and each agent has a linearized model of his/her own system along the scheduled trajectory over the future time interval (shaded red in Fig. 1). For this linear time-varying model, we formulate a design problem of energy management and control systems to realize ancillary services based on a real-time regulation market (at the secondary and primary control levels). The combination of the physical models is essentially the same as [14].

Linearized Grid Model Let us first consider the linearized time-varying model used in the ancillary market. This paper considers one of the standard grid models, e.g., the average system frequency model [17], as a generic model of high-speed response for ancillary service control problems with two area power networks and with two kinds of players: Utility and Agents. Here, we present a linearized model of each player's own system along the scheduled trajectory over a future time interval during

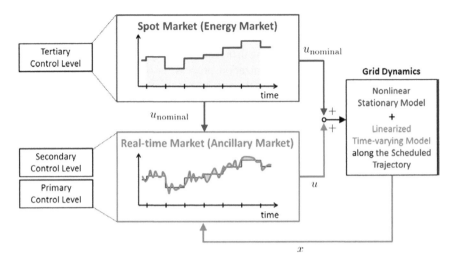

Fig. 1 Two layers market

when an energy dispatch scheduling has been finished in a spot energy market (at the tertiary control level). Of course, it is straightforward to extend the problem to the arbitrary number of agents.

The utility dynamics, which describes the deviation of the power and/or frequency balance and other deviations from physical constraints as well, obeys the following equation:

$$\frac{dx_{0t}}{dt} = f_0(t, x_{1t}, x_{2t}, \omega_t) = A_{01}(t)x_{1t} + A_{02}(t)x_{2t} + D_0(t)\omega_t, \quad t_0 \le t \le t_f, \quad (1)$$

and is evaluated by the utility's revenue functional:

$$J_0(t, x; u) = \mathbf{E}_{t,x}\left[\varphi_0(t_f, x_{t_f}) + \int_t^{t_f} l_0(\tau, x_\tau, u_\tau)d\tau\right], \quad (2)$$

where $x = (x_0^\top, x_1^\top, x_2^\top)^\top$ is the collection of the states of the utility dynamics and the two agents' dynamics and $u = (u_1^\top, u_2^\top)^\top$ is the local control inputs, respectively; ω_t is the disturbance modeled by a white Gaussian random process with zero-mean and unity-variance defined on $[t_0, t_f]$; $\mathbf{E}_{t,x}$ indicates an expectation given initial data (t, x); we use an abbreviation like $x_{0t} = x_0(t)$, $x_t = x(t)$. The agent's dynamics obeys the following equation:

$$\frac{dx_{it}}{dt} = f_i(t, x_{it}, u_{it}, \omega_t) = A_i(t)x_{it} + B_i(t)u_{it} + D_i(t)\omega_t, \quad t_0 \le t \le t_f, \quad i = 1, 2,$$
$$(3)$$

and is evaluated by the agent's revenue functional:

$$J_i(t, x; u) = \mathbf{E}_{t,x}\left[\varphi_i(t_f, x_{it_f}) + \int_t^{t_f} l_i(\tau, x_{i\tau}, u_{i\tau})d\tau\right], \quad i = 1, 2, \quad (4)$$

where x_i is the state of agent i indicating typically the deviation of power generation or consumption from the scheduled trajectory; u_i is the control input of agent i compensating the deviation. An admissible control of agent i is a state feedback $u_{it} = u_i(t, x)$, which assures the existence of the unique state trajectory $x_t = (x_{0t}^\top, x_{1t}^\top, x_{2t}^\top)^\top$, $t_0 \leq t \leq t_f$ of the grid model defined by the Eqs. (1) and (3). We formulated the dynamic grid model with the evaluation functionals (2) and (4) defined on the future time interval from the current time t to the final time t_f, which follows from the time-consistency property of dynamic programming; in other words, our evaluation functionals are of model-predictive type.

We will develop our discussion under the assumption that the standard regularity conditions hold on the mathematical formulas, e.g., (1)–(4), and will not refer to such technical conditions on each formula, since the objective of this section is just to outline the incentivizing market and its contract-based design by using these formulas. For readers who are interested in a mathematically rigorous treatment, please see our companion paper [18].

Incentivizing Market Model Now, we present a contract-based approach to the incentivizing market design, which reformulates the conventional contract problems [19–24] adapted to the market mechanism from the systems and control perspective. To describe our market model, we need to specify the participant's private information. The private information of agent i consists of the model information $\Xi_i = (f_i, \varphi_i, l_i)$ and the online information $Z_{it} \subset \{x_{it}^{t_f}, u_{it}^{t_f}\}, i = 1, 2$, where $x_{it}^{t_f}$ and $u_{it}^{t_f}$ denote the histories of the state $x_{i\tau}, t \leq \tau \leq t_f$ and the control $u_{i\tau}, t \leq \tau \leq t_f$, respectively.

To incentivize agent's behavior in market model, we (or a market planner) use a reward (salary) functional of the following form. The reward functional:

$$W_i^w(t, x_t^{t_f}; u) = w_{if}(t_f, x_{t_f}) + w_{i0}(t, x) + \int_t^{t_f} w_{i1}(\tau, x_\tau)d\tau$$
$$+ \int_t^{t_f} w_{i2}(\tau, x_\tau)\frac{dx_\tau}{d\tau}d\tau, t_0 \leq t \leq t_f, \quad i = 1, 2, \quad (5)$$

are defined along with the trajectory $x_\tau = (x_{0\tau}^\top, x_{1\tau}^\top, x_{2\tau}^\top)^\top$, $t \leq \tau \leq t_f$ given by a control $u = (u_1^\top, u_2^\top)^\top$, where $w = (w_1, w_2)$ and $w_i = (w_{if}, w_{i0}, w_{i1}, w_{i2}), i = 1, 2$. An admissible parameter of the reward functional w, called the reward parameter, is in the same class as for the revenue functionals (2) and (4). We use the notation W_i^w so as to emphasize the dependence of W_i on the choice of the parameter w. In order to make these reward functionals play a role in the market model, we express the reward parameter w with another parameter h, called the price, such that $w = w(h)$. The price h will be decided by the utility in the market.

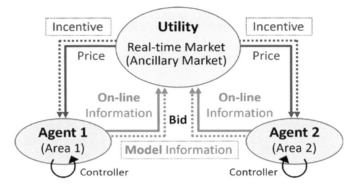

Fig. 2 Incentivizing market

The reward functionals together with the utility's revenue functional and the agent's revenue functional define the social welfare functional as

$$I^{w(h)}(t, x; u) = J_0(t, x; u) - \mathbf{E}_{t,x} \left[\sum_{i=1}^{2} W_i^{w(h)}(t, x_t^{t_f}; u) \right], \tag{6}$$

and the agent's profit functional as

$$I_i^{w(h)}(t, x; u) = J_i(t, x; u) + \mathbf{E}_{t,x} \left[W_i^{w(h)}(t, x_t^{t_f}; u) \right], \quad i = 1, 2. \tag{7}$$

A market planner designs a market mechanism with incentivizing structures (Fig. 2) and makes auction rules as well, based on the evaluation functionals and the grid model information introduced so far; the auction is performed in the following five steps:

Step 1   Utility announces the auction system, and agents decide participation.
Step 2   Agent offers his/her bid based on his/her own private information.
Step 3   Based on agents' bids, price is determined so as to maximize social welfare.
Step 4   Agent decides his/her control to maximize his/her own profit based on price.
Step 5   Utility pay rewards to agents.

Note that Steps 2, 3, and 4 will be performed continuously over a finite time interval.

Reward Design and One-shot Pricing To complete our market model, we need to fix a concrete procedure of agents' bidding by specifying agents' private information to be bidden in the market model; that is, *each agent's model information $\Xi_i = (f_i, \varphi_i, l_i)$ is sent a priori to utility, and each agent's online information to be bidden is just the current state, i.e., $Z_{it} = x_{it}$, which means that utility cannot access control input u_i.* Then, the design problem of our market is reduced to finding the reward parameter $w = (w_1, w_2)$, as a function of the price, i.e., $w = w(h)$, such that the social welfare functional is maximized by the reward parameter and the agents' control inputs each of which maximizes his/her own profit and guarantees a satisfactory level of the profit, at a market-clearing price. Such reward parameters, controls and market-clearing prices are given as solutions of the following dynamic optimization problem:

$$\max_{u,h} I^{w(h)}(t, x; u)$$

subject to
Constraint 1 (Incentive compatibility constraint):

$$I_i^{w(h)}(t, x; u) = \max_{v_i} I_i^{w(h)}(t, x; v_i, u_{-i}), \quad (i, -i) = (1, 2), (2, 1),$$

Constraint 2 (Individual rationality constraint):

$$I_i^{w(h)}(t, x; u) \geq k_{i0}(t, x), \quad i = 1, 2,$$

where k_{i0} is in the same class as for the parameter w_{i0} in the reward functional (5). Constraint 1 claims that the reward incentivizes each agent's behavior to adopt the optimal control that maximizes his/her own profit, so that the agents' controls constitute a Nash equilibrium. This implies that, even if the control profiles are not bidden, the utility can reconstruct them based on the bidden model information. On the other hand, Constraint 2 assures a prescribed level of each agent's profit that incentivizes the agent to participate in the market with his/her model information to be sent to the utility a priori. Now, if the agents' bidding is done, the utility can decide a market-clearing price immediately by carrying out the above optimization, and send it to each agent together with the reward payment in real-time. Thus, we obtain a non-iterative/one-shot incentivizing market model.

The above dynamic optimization for market-clearing has an overlapped structure of a dynamic game and an optimal control problem. We developed an approach to this optimization based on the dynamic programming (see [18] for details); first, we parameterize the reward parameter and the agents' controls with the price, and then convert our complex problem to the single optimal control problem in which the social welfare functional is maximized by the price; we also discuss the case that the optimal price is given as the gradient of the value function with respect to the current state, called the shadow price in economics literature.

3 Fundamental Challenges

On the basis of a genetic model suggested from the average system frequency model [17], we have shown an incentivizing market and model-based approach to design the energy management and control systems which realize ancillary services in dynamic power grids. The key issue of the approach is to incentivize the agents (areas) to open their private information that includes model information, which is essential to realize our model-based scheme. In this section, first, following our incentivizing mechanism briefly through two typical scenarios, we discuss the limitation of the mechanism in collecting the private information and possibilities of revisions/reformations as well. Next, toward realization of our approach, we discuss basic technical issues when the utility gathers the model information and also when the utility and the agents process online information. Finally, we provide an outline of several fundamental research directions to realize the designed markets in actual smart grids from the point of view of systems and control.

Incentive Design and Private Information The incentive design in our market model depends on the setting of the utility's and the agents' revenue functionals. Here, focusing on some special cases of the model, we discuss the roles of the revenues in incentivizing the agents to open or report their private information to the utility.

Consider the case that the utility's revenue functional is given as the sum of the original utility's revenue functional, which is assumed not to depend directly on the agents' control such that $l_0 = l_0(t, x)$, and the agents' revenue functionals. Assume further that the payment of the rewards for the agents is not liquidated in the social welfare, i.e., the utility's revenue is identical to the social welfare and each agent asks a zero level profit, i.e., $k_{i0}(t, x) \equiv 0$. In this case, the social welfare maximization is done by the so-called dual decomposition of optimal control problems and the market-clearing price is given as the shadow price, i.e., the value of the "Lagrange multiplier", of the social welfare maximization [18]. From the viewpoint of the incentive design, this design provides each agent with a zero level of the inventive to his/her participation in the market. If all the agents are not satisfied with the zero level, our model-based market mechanism does not work and requires additional incentives or legal forces for participations of strategic agents. Even in such a simple problem setting, we can see that it is not trivial issues to implement an optimal solution considering some economic constraints.

The reward design discussed so far incentivizes the agents to constitute a Nash equilibrium and to participate in the market if the profit level is over his/her expectation. However, these are assured under the tacit assumption that the agents' private information consisting of the model data and the online data is truthfully sent and bidden; if an agent fictitiously bids his/her private information, for example, the Nash equilibrium shifts or disappears; the "mechanism design" [25–28] provides a solution in such case by using additional incentives. Consider the same setting as above where the social welfare functional does not include the budget for the payment of the agents' rewards, and, on the other hand, let agent 1's (2's) profit functional have an additional reward functional of the form $J_0(t, x; u) + J_2(t, x; u)$ $(J_0(t, x; u) + J_1(t, x; u))$. Then, the additional reward provides the utility and all the agents with the same revenue, so that the optimal price from the viewpoint of the social welfare is optimal for all the agents. Therefore, if an agent sends or bids fictitiously his/her private information to the market, the agent obtains a price which is not optimal for his/her own profit. This incentivizing scheme corresponds to the

Groves mechanism [25, 26] in mechanism design literature. Finally, we should note that the additional rewards will be paid from the social welfare budget. More general incentivizing schemes of this type, especially, the schemes realizing the budget balance are sought [28].

Model Information The most significant issue of the model-based approach is to clearly identify not only the dynamic physical models but also the economics models, i.e., $\Xi_i = (f_i, \varphi_i, l_i)$ in sophisticated future smart girds installing diverse energy management systems (EMSs) and state-of-the-art grid control mechanisms. There are generally a variety of the market participants; large conventional generators, xEMSs, aggregators and prosumers combining loads and small-scale renewables, risk-sensitive utility and agents, and players with different market power. It is strongly required to enrich such reliable mathematical models in order to stabilize the grid systems and reduce financial risks. The dynamics consisting of mainly mechanical systems can be estimated by using system identification techniques developed in the system and control field. It is also important to improve the predicting accuracy of the dynamical behavior of consumers through behavioral economic analysis and data-based analysis with environmental information systems. In the model-based approach, specifically to our approach and generally, the compression of the model information is another important issue for the future, although there have been already the trials using randomized models [27, 29], a model reduction method [30].

On-line Information Reducing the online information helps privacy protection and reduction of communication loads. One of the options is to use the output feedback strategy with Kalman filter [31] and distributed/decentralized approaches based on dual decomposition and control methodologies in multi-agent systems. The most crucial issue of the approaches is that it is a very long time to converge at an appropriate equilibrium. Actually, it is necessary to appropriately determine the following items according to circumstances; the compression of the model information and the online information, the system performance, the controller complexity and the computation time to converge at an equilibrium. For instance, in case of LQG power networks, we can obtain an optimal solution analytically [27, 28, 31]. When we use the fast regulation market with nonlinear models and state constraints to require the guarantee of the computability in real time, it is valid to use the continuation and generalized minimum residual (C/GMRES) method presented in [32, 33]. If the revenue functional is approximately composed by the combination of specific basis functions, e.g., linear polynomials, step functions, and piecewise linear functions, it is expected to shorten the computation time by reporting the basis functions as the model information and only the coefficients of the basis functions as the online information. The learning in transition is one of the essential research topics and has been encouraged in systems and control field.

Market Structure, Performance, and Evaluation In the presence of the aforementioned varied participants, it is important to theoretically reveal the performance of the electricity markets such as budget loss and the efficiency (Price of Anarchy), and an influence for physical state constraints and financial limitations. To enhance the reliability of the markets in smart grids, it is also needed to prepare a legal framework promoting the truth-telling mechanism and the crackdown on a malicious report of not only the agents but also the utility. As illustrated in Fig. 1, there are not only the ancillary markets including the incentivizing markets but also energy markets. Ultimately, it is expected to organize a widespread timescale electricity-related market layer from seconds to decades, similarly to the financial markets. Development of software platform and benchmark models integrating the above complex and multiple time-layers models becomes powerful in order to make an opportunity to test and compare novel control mechanisms and to predict some trends in the near future. Such system integration based on mathematical models is to enable the quantitative evaluation of multidisciplinary cost based on system and control theory, engineering and micro-/macroeconomics required at each timescale without field experiments. Through the platform, it is also expected to fill the gap between the fundamental theory based on the systems and control approaches and the well-elaborated practice to make policy recommendations.

4 Conclusions

We have developed a model-based approach to the incentivizing market design for realizing the ancillary services in dynamic smart grids and discussed significant research challenges in the direction of our approach and general market-based approaches. The target of our market design is to provide all the participants with the transparent transactions that assure a satisfactory level from both economic and technical viewpoints for realizing ancillary services; as a promising approach, we have reformulated the conventional contract problems in economics literature and proposed a new contract problem adapted to the model-based market design on the dynamic grid; from the discussion so far, we can point out that the essential roles in this research direction should be played by systems and control, dynamic team/games, multi-agent/distributed decision-making, and so on. We can also see that many challenges are waiting for people from systems and control community to join the research on the topics discussed.

Acknowledgements This material is based upon work supported by the Japan Science and Technology Agency, CREST under Grant Number JPMJCR15K2.

References

1. M. Amin, A.M. Annaswamy, C.L. DeMarco, T. Samad, *IEEE Vision for Smart Grid Controls: 2030 and Beyond* (IEEE Press, 2013)
2. M.D. Ilic, S.X. Liu, *Hierarchical Power Systems Control—Its Value in a Changing Industry* (Springer, 1996)

3. Y.G. Rebours, D.S. Kirschen, M. Trotignon, S. Rossignol, A survey of frequency and voltage control ancillary services–Part I: Technical features. IEEE Trans. Power Syst. **22**(1), 350–357 (2007)
4. A.J. Wood, B.F. Wollenberg, *Power Generator Operation and Control* (Wiley, 1996)
5. J.W. O'Sullivan, M.J. O'Malley, Economic dispatch of a small utility with a frequency based reserve policy. IEEE Trans. Power Syst. **16**(3), 1648–1653 (1996)
6. S. Hao, A. Papalexopoulos, Reactive power pricing and management. IEEE Trans. Power Syst. **12**(1), 95–104 (1997)
7. T. Wu, M. Rothleder, Z. Alaywan, A.D. Papalexopoulos, Pricing energy and ancillary services in integrated market systems by an optimal power flow. IEEE Trans. Power Syst. **19**(1), 339–347 (2004)
8. J.F. Restrepo, F.D. Galiana, Unit commitment with primary frequency regulation constraints. IEEE Trans. Power Syst. **20**(4), 1836–1842 (2005)
9. M.A.B. Zammit, D.J. Hill, R.J. Kaye, Designing ancillary services markets for power system security. IEEE Trans. Power Syst. **15**(2), 675–680 (2000)
10. S.S. Oren, Design of ancillary service markets, in *Proceedings of the 34th Hawaii International Conference on System Sciences* (2001), pp. 1–9
11. Y. Rebours, D. Kirschen, M. Trotignon, Fundamental design issues in markets for ancillary services. Electr. J. **20**(6), 26–34 (2007)
12. E. Ela, V. Gevorgian, A. Tuohy, B. Kirby, M. Milligan, M. O'Malley, Market designs for the primary frequency response ancillary service–Part I: Motivation and design. IEEE Trans. Power Syst. **29**(1), 421–431 (2014)
13. E. Ela, V. Gevorgian, A. Tuohy, B. Kirby, M. Milligan, M. O'Malley, Market designs for the primary frequency response ancillary service–Part II: Case studies. IEEE Trans. Power Syst. **29**(1), 432–440 (2014)
14. M.D. Ilic, Toward a unified modeling and control for sustainable and resilient electric energy systems. Found. Trends Electric Energy Syst. **1**(1–2), 1–141 (2016)
15. A. Kiani, A. Annaswamy, T. Samad, A hierachical transactive control architecture for renewables integration in smart grids: analytical modeling and stability. IEEE Trans. Smart Grid **5**(4), 2054–2065 (2014)
16. K. Hirata, J.P. Hespanha, K. Uchida, Real-time pricing leading to optimal operation under distributed decision makings, in *Proceedings of the 2014 American Control Conference* (2014), pp. 1925–1932
17. A.W. Berger, F.C. Schweppe, Real time pricing to assist in load frequency control. IEEE Trans. Power Syst. **4**(3), 920–926 (1989)
18. Y. Wasa, K. Hirata, K. Uchida, Contract theory approach to incentivizing market and control design (2017), arXiv:1709.09318
19. P. Bolton, M. Dewatripont, *Contract Theory* (The MIT Press, 2005)
20. B. Holmstrom, P. Milgrom, Aggregation and linearity in the provision of intertemporal incentives. Econometrica **55**(2), 303–328 (1987)
21. H. Schattler, J. Sung, The first-order approach to the continuous time principal-agent problem with exponential utility. J. Economic Theory **61**, 331–371 (1993)
22. H.K. Koo, G. Shim, J. Sung, Optimal multi-agent performance measures for team contracts. Math. Finance **18**(4), 649–667 (2008)
23. Y. Sannikov, Contracts: The theory of dynamic principal-agent relationships and the continuous-time approach, in *Advances in Economics and Econometrics, 10th World Congress of the Econometric Society*, ed. by D. Acemoglu, M. Arellano, E. Dekel (Cambridge University Press, 2013)
24. J. Cvitanic, J. Zhang, *Contract Theory in Continuous-Time Models* (Springer, 2013)
25. D. Fudenberg, J. Tirole, *Game Theory* (The MIT Press, 1991)
26. M.O. Jackson, Mechanism theory, in *Encyclopedia of Life Support Systems*, ed. by U. Derigs (EOLSS Publishers, 2003)
27. Y. Okajima, T. Murao, K. Hirata, K. Uchida, A dynamic mechanism for LQG power networks with random type parameters and pricing delay, in *Proceedings of the 52nd IEEE Conference on Decision and Control* (2013), pp. 2384–2390

28. T. Murao, Y. Okajima, K. Hirata, K. Uchida, Dynamic balanced integration mechanism for LQG power networks with independent types, in *Proceedings of the 53rd IEEE Conference on Decision and Control* (2014), pp. 1395–1402

29. F. Farokhi, K.H. Johansson, Optimal control design under limited model information for discrete-time linear systems with stochastically-varying parameters. IEEE Trans. Autom. Control **60**(3), 684–699 (2015)

30. T. Murao, K. Hirata, K. Uchida, An approximate dynamic Integration mechanism for LQ power networks with multi-time scale structures, in *Proceedings of the 2016 European Control Conference* (2016), pp. 202–209

31. S. Matsui, T. Murao, K. Hirata, K. Uchida, A dynamic output integration mechanism for LQG power networks with random type parameters, in *Proceedings of the Asian Control Conference* (2015), pp. 1–6

32. T. Ohtsuka, A continuation/GMRES method for fast computation of nonlinear receding horizon control. Automatica **40**(4), 563–574 (2004)

33. T. Ohtsuka, A tutorial on C/GMRES and automatic code generation for nonlinear receding horizon control, in *Proceedings of the 2015 European Control Conference* (2015), pp. 73–86

Long-Term Challenges for Future Electricity Markets with Distributed Energy Resources

Steffi Muhanji, Aramazd Muzhikyan and Amro M. Farid

Abstract Recently, the academic and industrial literature has arrived at a consensus in which the electric grid evolves to a more intelligent, responsive, dynamic, flexible, and adaptive system. This evolution is caused by several drivers including decarbonization, electrified transportation, deregulation, growing electricity demand, and active consumer participation. Many of these changes will occur at the periphery of the grid, in the radial distribution system and its potentially billions of demand-side resources. Such spatially distributed energy resources naturally require equally distributed control and electricity market design approaches to enable an increasingly active "smart grid." In that regard, this chapter serves to highlight lessons recently learned from the literature and point to three open long-term challenges facing future design of electricity markets. They are (1) simultaneously manage the technical and economic performance of the electricity grid; (2) span multiple operations timescales, and (3) enable active demand-side resources. For each challenge, some recent contributions are highlighted and promising directions for future work are identified.

Keywords Smart grid controls · Variable energy resources · Energy storage
Demand-side resources · Electric microgrids · Power systems stakeholders
Electricity market structures

Mathematics Subject Classification (2010) MSC code1 · MSC code2 · More

S. Muhanji (✉) · A. Muzhikyan · A. M. Farid
The Laboratory for Intelligent Integrated Networks for Engineering Systems (LIINES),
Thayer School of Engineering, Dartmouth College, 14 Engineering Drive, Hanover,
NH 03755, USA
e-mail: steffi.o.muhanji.th@dartmouth.edu

A. Muzhikyan
e-mail: aramazd.muzhikyan.th@dartmouth.edu

A. M. Farid
e-mail: amfarid@dartmouth.edu

© Springer Nature Switzerland AG 2019
J. Stoustrup et al. (eds.), *Smart Grid Control*, Power Electronics
and Power Systems, https://doi.org/10.1007/978-3-319-98310-3_4

1 Introduction

Traditional power systems were built upon the assumption that generation was controlled by a few centralized generation facilities that were designed to serve fairly passive loads [1, 2]. This assumption has since controlled the structure of the physical power grid, power systems economics as well as regulatory measures. However, several drivers have emerged to challenge this assumption.

1.1 Power Grid Evolution Drivers

The first of these drivers is decarbonization. With rising concern about CO_2 emissions, many nations have taken major steps to lower their greenhouse gas (GHG) emissions. More specifically, the European Union has vowed to reduce their GHG emissions to 40% of 1990 levels by 2030 [3, 4] and increase their renewable energy portfolio by at least 27% in 2030 [5]. Also, the Paris Agreement signatories have set national goals to combat climate change within their own capabilities [6, 7].The renewable portfolio standard (RPS) and the mandatory green power option (MGPO) policies have been implemented in many US states to encourage renewable energy generation [8]. For example, the California renewable portfolio standard (RPS) set out to increase the percentage of renewables in the state of California to 33% by 2020 [9].

The second driver is rising electricity demand, especially in developing countries. Studies have shown that electricity demand in developing countries will continue to increase steadily by about 4% each year between 2000 and 2030, approximately tripling in that time [10–12]. In order to minimize the need for more generation capacity and its associated investment cost, techniques such as peak shaving and demand-side management are imperative [13–15].

The third driver of electrified transportation also supports decarbonization efforts. Electric vehicles offer higher well-to-wheel efficiencies and have zero operational emissions if charged using renewable energy sources [16–18]. However, studies have shown that given the temporal and spatial uncertainty of electric vehicles, a large number of plug-in electric vehicles (PEVs) in one region can potentially affect different aspects of power system operations, including balancing performance, line congestion, and system voltages. The grid must, therefore, evolve to accommodate charging schedules and energy needs of PEVs [19–22].

Fourth, deregulation of power markets promises greater social welfare, reduced electricity prices, and improved quality of service. Traditionally, power systems have consisted of vertically integrated utilities, from generation to transmission to distribution, each having monopolies over their own geographical region [23, 24]. However, as demand for electricity increased and consumption patterns became more variable, a general interest in reducing reliance on regulation and enhancing market forces to guide investments and operations have developed [24]. In time, this vertically aligned chain became more unbundled to allow for diversified and competitive wholesale prices [24–28]. As the electric power grid continues to evolve, deregulated electricity markets must continue to develop down into the distribution system so as to support these objectives.

Lastly, deregulation measures and the rise of smart grid technologies have empowered consumers to take an active role in managing electricity consumption patterns [15, 29]. Empowered consumers cause both physical and economic changes to the electricity grid [13, 29, 30]. As a result, demand becomes more controllable and capable of responding to dynamic prices and reliability signals. Demand-side management (DSM) programs offer several opportunities. These include active balancing operations in the presence of stochastic renewable energy resources, and load shifting so as to reduce new generation capacity requirements and increase the utilization of existing facilities [31]. In spite of their potential benefits, many questions remain as to how DSM programs will be implemented to realize these gains [32].

1.2 Contribution

These five drivers cause an evolution of the grid so as to become more intelligent, responsive, dynamic, flexible, and adaptive. Many of these changes will occur at the grid periphery with the integration of spatially distributed energy resources, namely, distributed generation (e.g., solar PV and small-scale wind turbines, and run-of-river hydro turbines) and demand-side resources. These in turn will necessitate their associated distributed control techniques. This work adopts the terms distributed, decentralized, and centralized control as described by Farina and Trecate [33]. In that regard, this chapter serves to highlight lessons recently learned from the literature. A central theme in these lessons is the need for holistic approaches that integrate multiple layers of control so as to achieve both technical and economic objectives [32]. The chapter also points to several open long-term challenges which require resolution to support distributed energy resources.

1.3 Outline

To that effect, the rest of the chapter is structured in three open challenges facing design of electricity markets. Section 2 discusses the need to simultaneously balance the technical and economic performance of the electric grid. Section 3 recognizes that control actions span multiple operation timescales and asserts the need for holistic assessment methods to capture potential inter-timescale coupling. Section 4 argues for active participation of demand-side resources. The chapter is brought to a conclusion in Sect. 5.

2 Challenge I: Simultaneously Manage Technical and Economic Performance

The evolution of the electricity grid will simultaneously impact its technical and economic performance [32] in large part due to the integration of variable energy resources (VERs) and demand-side resources (DSRs). Figure 1 presents this argument succinctly. The horizontal axis represents the (physical) generation and demand value chain that is connected through transmission and distribution networks. A second axis recognizes that these resources can be either stochastic or dispatchable. Finally, the vertical axis views the power grid cyber-physically with multiple layers of control decisions, automation, and information technologies. Together, this system must achieve both technical and economic control objectives. The technical side includes balancing operations, line congestion prevention, and voltage control, while the economic control weighs the investment and operating cost of integrated technologies against their impact on system performance. Thus, each newly added technology should provide measurable improvement to the holistic cost and technical performance. As such, grid control decisions must be assessed holistically to account for the techno-economic trade-offs of its associated layers.

Most academic literature on the control of the electricity grid has primarily studied a single resource layer such as variable energy [35–37], energy storage [38–40] or demand-side resources [13–15]. These studies have also focused on a single layer of power system balancing operations, such as security-constrained unit commitment (SCUC) or security-constrained economic dispatch (SCED), thus ignoring potential cost benefits of ancillary services which are drivers of overall system performance [32]. Additionally, some of these studies have been conducted on specific case studies, making generalization to other cases difficult [41–43]. Many integration studies

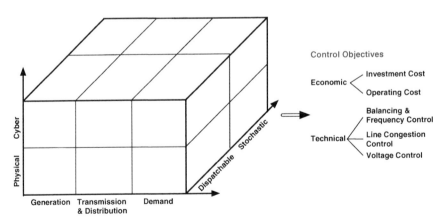

Fig. 1 The power grid is taken as a cyber-physical system composed of an energy value chain with dispatchable and stochastic elements that must fulfill certain technical and economic objectives. Reproduced from [32] ©Elsevier *Apr 1, 2016*, used with permission

ignore the cost of additional measurement and control technologies. Similarly, various grid codes impose regulations on renewable energy integration without providing a cost rationalization. Furthermore, most studies have been limited to statistical analyses that are yet to be validated by simulations. These statistical analyses are based upon either the net load variability or its forecast error [44, 45] despite recent closed-form analytical derivations showing the dependence on both factors [46]. Lastly, many of the grid control assumptions are based on the experience of system operators. This experience, albeit practically useful, is not guaranteed to remain valid as the grid evolves [34, 47]. Overall, these studies indicate a lack of holistic assessment methods that are necessary to successfully capture the techno-economic benefits of control decisions.

Recent works have proposed the concept of an integrated power grid enterprise control as a means of creating techno-economic synergies and studying their trade-offs [34, 47–51]. Originally, the concept of enterprise control [52, 53] was developed in the manufacturing sector out of the need for greater agility [54, 55] and flexibility [56–58] in response to increased competition, mass-customization, and short product life cycles. Its essence is a single simulation that includes the physical production system connected to multiple layers of control, operations, and management at their associated timescales. Over time, a number of integrated enterprise system architectures [59, 60] were developed coalescing in the current ISA-S95 standard [53, 61]. Analogously, recent work on power grids has been proposed to update operation control center architectures [62] and integrate the associated communication architectures [63]. The recent NIST interoperability initiatives further demonstrate the trend toward integrated and holistic approaches to power grid operation [64]. Other works have also proposed decentralized approaches to generation control by combining two or more market layers to achieve economic equilibria [65–67]. One such work presents a distributed optimization-based controller that combines automatic generation control (AGC) layer with the economic dispatch (ED) to achieve economic efficiency in real-time market operations [67]. These initiatives form the foundation for further and more advanced holistic control of the grid [68–73].

In power systems, enterprise control is achieved by creating a single simulation that ties the physical power grid to several layers of control and optimization so as to study the technical and economic performance simultaneously [38, 74–80]. The enterprise control model described fully in [34] holistically addresses three control layers: resource scheduling in the form of a security-constrained unit commitment (SCUC), balancing actions in the form of a security-constrained economic dispatch (SCED) and operator manual actions, and a regulation service in the form of AGC. The enterprise control diagram is shown in Fig. 2, where each consecutive layer operates at a smaller timescale, reducing the imbalances with each layer of control. This model has been used to explore the effects of timescale coupling and net load variability on balancing performance and system costs. The results show that reducing day-ahead and real-time market time steps can potentially reduce load following, ramping, and regulation reserve requirements [34], which will significantly reduce the overall system cost. Additionally, the model in [34, 47] was used to conduct a series of steady-state simulations to study the impact of integrating variable energy,

64 S. Muhanji et al.

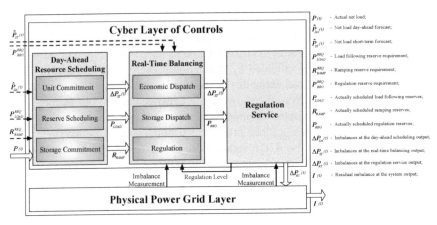

Fig. 2 A conceptual model of the power grid enterprise control simulator (adapted from [34] ©IEEE *Apr 1, 2015*, used with permission)

energy storage, and demand-side resources on power system imbalances [46–48, 75, 77, 78, 81–83].

VER volatility has increased the urgency in securing resources to provide ancillary services and ensuring proper compensation for such services. To that end, recent works have explored various ways of engaging distributed energy resources and deferrable loads in the provision of ancillary services [84, 85]. The former introduces the concept of intelligent decentralized control architecture which takes advantage of the flexibility of loads to provide ancillary services during peak hours, VER volatility, or various contingencies. Unlike other approaches, this work introduces intelligent deferrable loads that employ randomization and localized decision-making to minimize communication congestion. The control protocol minimizes information exchange between loads and balancing authorities by allowing local control loops at the load level. This architecture helps address the privacy concerns and communication constraints that arise from automatic control of loads used in the provision of ancillary services [84]. The work in [85] proposes a real-time charging and discharging controller for electric vehicles that permits tracking of the AGC signal while exploring the effects of look-ahead through model-predictive control (MPC). These two frameworks recognize the need to engage demand-side resources in market operations. It is evident that new control architectures that are able to respond quickly to real-time changes in grid operations as well as promote autonomous and decentralized decision-making must be advanced. Naturally, market structures that would enable participation of and proper compensation for such services are necessary.

Perhaps, one of the greatest challenges in the techno-economic assessment of power systems with large quantities of variable energy, energy storage, and demand-side resources is the quantitative determination of operating reserves. Power system energy resources are fundamentally constrained resources. Therefore, the degree to which they can provide spare capacity of various types is integral to their ability to respond to net load variability and forecast error away from scheduled set points. Such spare capacity has real economic value. And so for decades electricity markets have incentivized generators to provide several types of operating reserves, be they in normal or contingency operation [86]. Consequently, the focus of most renewable (i.e., variable) energy integration studies has been on estimating the required quantities of operating reserves as the grid's energy portfolio changes [32, 87–89]. The challenge here is that the taxonomy and definition of operating reserves from one power system geography to the next varies [86]. Furthermore, this taxonomy and definition is often different from the methodological foundations found in the literature [86]. There are even significant differences in the definitions found within the literature itself [86, 90–92]. Nevertheless, the literature is converging toward a consensus view that variable energy integration requires the assessment of three types of normal operating reserves: load following, ramping, and regulation [86]. Recently, Muzhikyan et al. have shown closed-form analytical derivations of the required quantities of all three types of operating reserves [46]. This work recognizes that the required quantities of operating reserves depend on endogenous characteristics of the electricity market design as well as exogenous temporal and spatial characteristics of the net load [46, 49]. This work may prove fundamental as the methodologies of renewable energy integration studies advance to account for more holistic aspects of the grid's techno-economic operation.

As the power grid continues to evolve in the coming years, it is essential that its evolution continues to be assessed techno-economically. While the above works have developed holistic assessment methodologies for today's power systems, new technologies be they physical energy resources or control technologies will continue to be introduced. In essence, the integration of each new technology should be assessed for its overall technical and economic impact. Furthermore, these integration decisions will need to be rigorously framed so as to meet these mixed objectives and their associated trade-offs. In many cases, the technical integration question will have to be considered in the context of an evolving control architecture and stakeholder jurisdictions.

3   Challenge II: Span Multiple Operations Timescales

As illustrated in Fig. 3, power system control phenomena overlap in timescales. Traditionally, power systems literature have broken these phenomena into a hierarchical control structure, namely, primary, secondary, and tertiary control. Primary control (10–0.1 Hz) performs dynamic stability analyses and generator output adjustments by implementation of automatic generator control (AGC) and automatic voltage regulators (AVR) [93, 94]. Secondary control acts in the minutes timescale and provides set points for automatic control actions for primary control. It also involves operator manual actions to ensure secure and stable performance as fast as possible. Tertiary control, which happens in tens of minutes to hours timescale, performs economic optimization to minimize the cost of generation to meet demand subject to generator

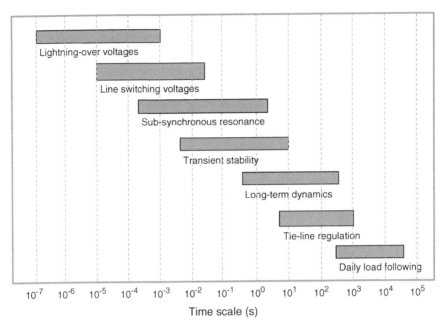

Fig. 3 Timescales of physical power grid dynamics. Reproduced from [32] ©Elsevier *Apr 1, 2016*, used with permission

capacity and line limits [93, 94]. In the past, these control actions have been studied separately under the assumption that they are independent because of their distinct timescales [32].

However, a study of the load power spectrum, shown in Fig. 4, exhibits variations across a wide range of frequencies. Similarly, multi-timescale dynamics are observed in the solar photovoltaic [95] and wind [96] power spectra shown in Figs. 5 and 6. The Federal Energy Regulatory Commission (FERC) has responded to these findings by reducing the minimum time requirement for economic dispatch to 15 min [97]. Several independent system operators (ISO) have further reduced their dispatch time to only 5 min. A recent study has shown that due to VER integration, the frequency of manual operator actions with regard to curtailment has increased significantly [37]. Furthermore, it has been shown that the probability of infeasible real-time dispatches is likely to increase in the absence of exact profile distributions for stochastic resources [98]. In summary, the integration of VER introduces dynamics at all control timescales and consequently challenges the separation of primary, secondary, and tertiary control phenomena.

Academic studies have illustrated the impacts of cross-timescale variability on power system balance and operating cost [38, 74, 77–80]. Lately, optimization-based approaches that seek to capture the timescale coupling of primary, secondary, and tertiary control of power networks with controllable loads have been introduced [99–102]. In these approaches [99–102], decoupling is achieved through decentralized

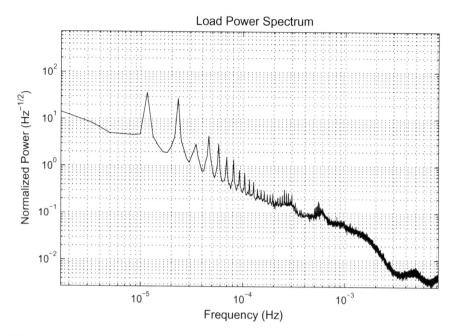

Fig. 4 Normalized power spectrum of the daily load (data from Bonneville Power Administration). Reproduced from [32] ©Elsevier *Apr 1, 2016*, used with permission

and distributed controllers, and a steady-state equilibrium of the system is illustrated. The enterprise control model presented in [34] integrates primary, secondary, and tertiary control layers into a holistic dynamic simulation to capture the inter-timescale coupling within these three layers. The simulations in [34] reveal the power grid's cross-timescale dynamic behavior.

Results from [47] demonstrate that system imbalances are significantly reduced when the timescale of the real-time market is reduced from 60 to 15 min. Additionally, the overall load following and ramping reserve requirements are decreased as shown in Figs. 7 and 8. A study of the relative merits of energy storage reserves in the balancing operations and resource layer of control shows that energy storage is effective at balancing high net load variability and small day-ahead market time step [48]. Figure 9 shows that integrating storage reduces the overall system imbalances and the amount of load following reserve requirements. Figure 10 illustrates that the system with a higher normalized variability and greater penetration of renewables will experience greater system imbalances [47]. An enterprise control model demonstrates the timescale coupling of various power system phenomena and asserts the benefits of cross-layer coupling in the holistic assessment of techno-economic trade-offs.

Multi-timescale dynamics that are introduced by VERs and DSRs imply multilayer control approaches. The challenge with a multilayer approach is that each layer of control affects the overall life-cycle properties of the system. In this context, the

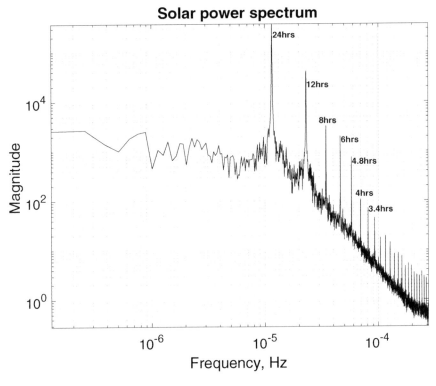

Fig. 5 Typical power spectrum of a solar PV panel

dispatchability, flexibility, stability, forecastability, and resilience of the power system would need to be studied from a multilayer and not just a single layer perspective [32]. This opens up a plethora of practical questions for the emerging theory of hybrid dynamic systems [103]. The formal analysis of such systems would provide direct guidance as the power grid continues to evolve with new control architectures.

4   Challenge III: Enable Active Demand-Side Resources

As mentioned in the introduction, the electricity grid has traditionally operated under the paradigm that generation exists to follow the exogenous variability in consumer demand [2]. This has had a significant impact on the design of grid infrastructure in that generation capacity must be sized for peak demand irrespective of how infrequently that capacity is required over the course of the year [14]. Distributed generation and demand-side resources (DSRs), as actively controlled energy resources, have the potential to reduce the need for generation capacity expansion. Their presence, however, causes the potential for upstream flows from the power grid

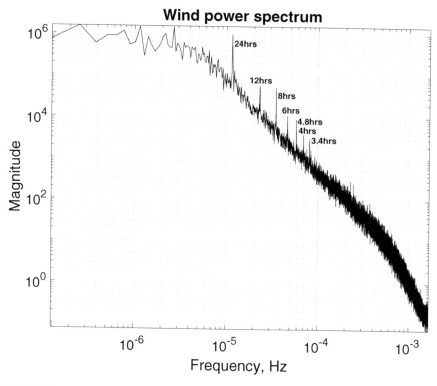

Fig. 6 Typical power spectrum of a wind turbine

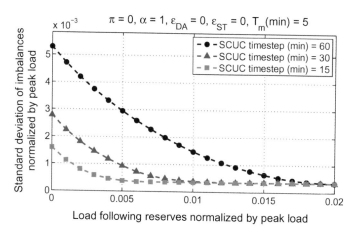

Fig. 7 The impacts of normalized load following reserves and day-ahead market time step on the normalized standard deviation of power system imbalances. Reproduced from [47] ©IEEE *Apr 1, 2015*, used with permission

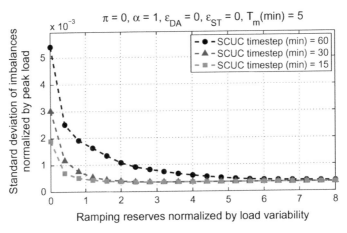

Fig. 8 The impacts of normalized ramping reserves and day-ahead market time step on the normalized standard deviation of power system imbalances. Reproduced from [47] ©IEEE *Apr 1, 2015*, used with permission

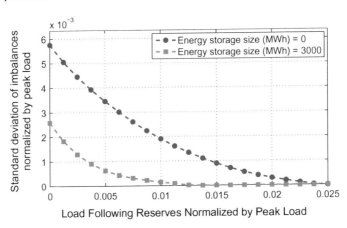

Fig. 9 The relative trade-offs of utilizing normalized load following reserves versus energy storage on the normalized standard deviation of imbalances. Reproduced from [48] ©Elsevier *Jan 1, 2016*, used with permission

periphery toward the centralized transmission system. This possibility violates another long-held assumption in the power grid where the transmission system is organized in a meshed fashion while the distribution system is organized in a radial fashion allowing power to flow outward in one direction [1, 2]. Instead, distributed generation and DSRs are set to challenge this structural assumption requiring a meshed topology on the demand-side too [40].

Similarly, power systems economics in the distribution system have been structured such that electricity prices paid by consumers are independent of system conditions [23, 26]. Those consumers that connect directly to transmission system have

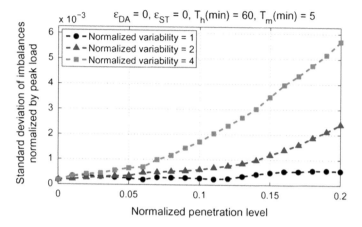

Fig. 10 Impact of VER variability on power system imbalances. Reproduced from [47] ©IEEE *Apr 1, 2015*, used with permission

been wholesale market price-takers up until only recently. Consequently, radical changes in consumer demand that result in more expensive generation do not affect the prices paid by consumers [23, 25]. Furthermore, system operators have traditionally had minimal control over the load size, often resorting to blunt solutions such as emergency load shedding, and blackouts in the most extreme situations [104]. However, as the new smart grid infrastructure is deployed, demand-side resources will play a significant role in ensuring grid stability. Consumer participation favors load flexibility and peak shifting, hence promoting grid reliability. Sensors, communication systems, automated metering, intelligent devices, and ad specialized processors have the potential to activate demand-side resources to participate in the electric system techno-economic decision-making [15]. Such technologies promote consumer participation, exploit renewable energy resources, and achieve energy savings [15].

Coordinated control of the demand side is also key to the successful integration of VERs. As shown in Fig. 11, the introduction of variable renewable energy resources erodes the dispatchability of the grid introduced by thermal power generation. DSM restores the grid's dispatchability, thereby enhancing reliability and flexibility amidst the increased stochasticity of the generation fleet [32]. In such a case, DSR can be used to reduce demand when solar PV and wind generation unexpectedly drops, meet the associated ramp profile, and even act as an ancillary service that responds to short-term frequency and voltage deviations.

DSM programs take several forms but have the common feature of market-based price signals that aim to reduce electricity consumption. DSM programs include energy efficiency, demand response (DR) [40, 105, 106], and load management programs [107, 108]. Load management programs are designed to reduce consumption or shift it to off-peak hours. Peak shifting is accomplished through real-time pricing schemes, whereby the energy price grows with the aggregated load for a given period [109]. Real-time pricing motivates consumers to purchase power during off-

Future:	Generation/Supply	Load/Demand
Well-Controlled & Dispatchable	Thermal Units: ↓ (Unsustainable cost & emissions)	Demand Side Management: (Requires new control and market design) ↑
Stochastic/ Forecasted	Renewable Energy Sources: (Can cause unmanaged grid imbalances) ↑	Conventional Loads: ↓ (Growing & needs curtailment)

Fig. 11 Demand and generation portfolio of the future grid. Reproduced from [32] ©Elsevier *Apr 1, 2016*, used with permission

peak times in order to reduce their overall energy cost [110]. The concept of real-time pricing (RTP) is, however, still very much under development. Social questions in relation to equity and access need to be considered and compensation mechanisms must address consumers with distributed generation and/or energy storage [40, 106]. Another approach to load management is direct load management (DLC). DLC is based on an agreement between utilities and consumers, whereby consumers agree to let utilities remotely control the energy consumption of some of their appliances such as lighting and thermal comfort equipment [15]. Concerns about consumer privacy have, however, resulted in less participation in DLC programs [105]. Various methods such as dynamic programming [111], fuzzy logic [112], game-theoretic [105, 110], and binary particle swarm approaches [107] have been proposed for DLC and RTP programs.

More recently, the focus in literature has shifted toward studying the impact of the dynamics introduced by shifting loads, fuel price volatility, and stochastic generation on electricity prices and market stability [65, 66, 73, 113–121]. The concept of dynamic real-time markets (DRMs) refers to market structures that are set up so as to enable active VER and DR participation and coordination in real time or near real time. In this market model, demand-side participants are price-setters rather than price-takers. To ensure real-time or near real-time coordination, extensive, flexible, and distributed communication channels capable of handling the large amounts of data generated and provide feedback in real time are imperative. DRM approaches tend to be geared toward the overall stability of the wholesale electricity markets [114, 116, 117] and enhancing the social welfare [73, 119, 121]. While some focus solely on a single layer such as regulation [66, 119], a few DRM techniques combine multiple layers of real-time market control [73]. It is, however, important to note that a significant number of these approaches have neglected to define the communication layer or rather assumed a perfect communication network [65, 66, 113–121]. This results in algorithms that fail to acknowledge communication challenges such as latency [73] that affect the resiliency of DRM structures. Naturally, this emerging diversity of DSM approaches needs to be rigorously assessed, be it techno-economically as in Challenge I, or across multi-timescales in Challenge II.

Considerable attention has been given to DSM programs in the context of load scheduling in the day-ahead market or load shifting in the real-time energy markets. In the electric power industry, these programs are implemented through optimization algorithms that aim to minimize the overall generation cost given capacity and ramping constraints [122–124]. Demand units are represented in the wholesale energy market through curtailment service providers (CSP) who bid through independent system providers (ISO) or reliability transmission organizations (RTO) [125]. The CSP has an estimated baseline consumption—consumption without demand response—from which load reductions can be measured. Load reductions that are accepted by the bidding process are expected to commit and are compensated based on their bidding price as compared to the locational marginal pricing (LMP) and retail rates [125]. Unfortunately, it has been determined that consumers are likely to artificially inflate the baseline to increase their compensation [126]. Through a systematic comparison of the academic social welfare and industrial approaches to DSM, Jiang et al. [122, 123] illustrated that inaccurate baselines in industrial DSM could potentially lead to higher systems costs, wrong dispatch levels, and unachievable social welfare. Furthermore, more recent studies have shown that inflated baselines could result in more control requirements in subsequent layers of enterprise control [34, 47, 77, 78].

One emerging concept for demand-side management is called "Transactive Energy" and it is used to refer to "techniques for managing generation, consumption or flow of electric power within the electric power system through the use of economic or market-based constructs while considering grid reliability constructs" [127]. Many consider the "homeostatic utility control model" proposed by Fred Schweppe in 1980 [128] as the intellectual inspiration for transactive energy (control). Transactive energy techniques can be implemented on a localized level such as residential demand response, or on a generation to consumption level. A transactive energy project by the Pacific Northwest National Laboratories (PNNL) studied the effect of two-way communication between generation and distributed DSRs on energy balance, line congestion, and real-time prices [129] in the Olympic Peninsula in Washington State. This demonstration tested the GridWise transactive energy architecture on 100 homes in the region. This demo has since been extended to 5 states, 11 utilities, the Bonneville Power Administration (BPA), two universities, and multiple companies [130]. In this demonstration, they were able to test the performance of the control architecture on various system instabilities such as power outages, wind fluctuations, and transmission incidences such as line outages [130]. Another approach, the Transactive Energy Market Information Exchange (TeMIX), applies decentralized decision-making and control techniques at the grid periphery to allow direct interaction between consumer devices and distribution grid devices [131]. This project enables smart grid services that can quickly respond to the high penetration of variable energy resources, PEVs, and energy storage. Transactive energy platforms are enhanced by the concept of dynamic pricing and tariffs [132] which provide a trading experience for electricity markets that almost mimics the stock market. Finally, transactive energy approaches eliminate the need for demand response baselines and have the potential to avoid many of the associated negative impacts [51].

As demand-side management develops, rigorous assessment becomes an important challenge. In that regard, holistic assessment must be techno-economic as in Challenge I, and cross-multiple timescales as in Challenge II. Furthermore, in discussing demand-side management, it is important to recognize that the (economic) utility of consumed electricity is different depending on its purpose. For example, a kWh of electricity used in space heating is not equivalent to a kWh of electricity used in making silicon wafers. The latter provides much greater value to its consumers; and consequently their willingness to pay for that kWh would be quite different. To that effect, modeling the economic utility of electricity consumption is of paramount importance as it represents a large trade-off with price incentives in DSM schemes. Therefore, it will become increasingly important to revise the utility models of demand-side participants so that they more closely reflect the reality. Such an approach may quickly overwhelm the practical constraints of centralized market designs and instead may require distributed decision-making approaches. Distributed control architectures offer a middle-ground between decentralized and centralized architectures. Like decentralized architectures they have multiple controllers acting on a physical system but add coordination between controllers so as to achieve performance similar to or equal to centralized architectures [33]. Finally, it is important to recognize that while market-based approaches may result in economic efficiency, they may not guarantee physical life-cycle properties. Approaches that too closely resemble the stock market must recognize that financial markets do not necessarily exhibit stable behavior. Consequently, DSM programs find the appropriate balance of physical as well as economic signals.

5   Conclusion and Future Work

In conclusion, this work identified several long-term drivers which together cause the introduction of distributed energy resources at the grid's periphery. This, in turn, poses significant long-term challenges. Power grid assessment must be increasingly holistic considering technical and economic trade-offs as well as variations that span multiple layers. Such techniques demand multilayer approaches that represent hybrid dynamic phenomena which are difficult to design formally. Demand-side resources (DSRs) are also expected to play a significant role in promoting grid reliability. Utility modeling as well as multilayered, scalable, and distributed control algorithms will enhance the integration of DSRs. Moving forward, power systems' design and operation must adapt to the changing needs and interests of new and old stakeholders, be they in the electric power grid or in interdependent infrastructures. Finally, the newly evolved "smart grid" must ultimately demonstrate resilient self-healing operation which will likely be enabled by distributed control and/or multi-agent systems. This work has highlighted some of the recent contributions with respect to these areas and identified areas where many challenges still remain.

References

1. A. Von Meier, *Electric Power Systems: A Conceptual Introduction* (Wiley, 2006)
2. P. Schavemaker, L. Van Der Sluis, *Electrical Power System Essentials* (Wiley, 2008)
3. European Commission, *A Roadmap for Moving to a Competitive Low Carbon Economy in 2050* (European Commission, Brussel, 2011)
4. M. Haller, S. Ludig, N. Bauer, Decarbonization scenarios for the EU and MENA power system: considering spatial distribution and short term dynamics of renewable generationnnamics of renewable generation. Energy Policy **47**, 282–290 (2012)
5. T. Ackermann, E.M. Carlini, B. Ernst, F. Groome, A. Orths, J. O'Sullivan, M. de la Torre Rodriguez, V. Silva, Integrating variable renewables in Europe: current status and recent extreme events. IEEE Power Energy Mag. **13**(6), 67–77 (2015)
6. J. Rogelj, M. Den Elzen, N. Höhne, T. Fransen, H. Fekete, H. Winkler, R. Schaeffer, F. Sha, K. Riahi, M. Meinshausen, Paris agreement climate proposals need a boost to keep warming well below 2 °C. Nature **534**(7609), 631–639 (2016)
7. W. Obergassel, C. Arens, L. Hermwille, N. Kreibich, F. Mersmann, H.E. Ott, H. Wang-Helmreich, Phoenix from the ashes—an analysis of the paris agreement to the united nations framework convention on climate change. Wuppertal Instit. Climate Environ. Energy **1**, 1–54 (2016)
8. M.A. Delmas, M.J. Montes-Sancho, US state policies for renewable energy: context and effectiveness. Energy Policy **39**(5), 2273–2288 (2011)
9. C.P.U. Commission, California renewables portfolio standard (RPS) (2011), http://www.cpuc.ca.gov/PUC/energy/Renewables/index.htm. Accessed 10 Nov 2012
10. IEA, World Energy Outlook, Energy and Air Pollution. Technical report, International Energy Agency, Paris France (2016)
11. F. Birol et al., World energy outlook 2010. Int. Energy Agency **1** (2010)
12. F. Birol, E. Mwangi, U. Global, D. Miller, S. Renewables, P. Horsman, Power to the people. IAEA Bull. **46**, 9–12 (2004)
13. C.W. Gellings, The concept of demand-side management for electric utilities. Proc. IEEE **73**(10), 1468–1470 (1985)
14. P. Palensky, D. Dietrich, Demand side management: demand response, intelligent energy systems, and smart loads. IEEE Trans. Ind. Inform. **7**(3), 381–388 (2011)
15. P. Siano, Demand response and smart grids—a survey. Renew. Sustain. Energy Rev. **30**, 461–478 (2014)
16. G. Pasaoglu, M. Honselaar, C. Thiel, Potential vehicle fleet CO_2 reductions and cost implications for various vehicle technology deployment scenarios in Europe. Energy Policy **40**, 404–421 (2012)
17. T. Litman, *Comprehensive Evaluation of Transport Energy Conservation and Emission Reduction Policies Victoria Transport Policy Institute*, Victoria, BC (2012)
18. D. Anair, A. Mahmassani, *State of Charge: Electric Vehicles' Global Warming Emissions and Fuel-Cost Savings Across the United States* (Union of Concerned Scientists, 2012)
19. R. Al Junaibi, A. Viswanath, A.M. Farid, Technical feasibility assessment of electric vehicles: an Abu Dhabi example, in *2013 International Conference on Connected Vehicles and Expo (ICCVE)* (IEEE, 2013), pp. 410–417
20. M.D. Galus, R.A. Waraich, F. Noembrini, K. Steurs, G. Georges, K. Boulouchos, K.W. Axhausen, G. Andersson, Integrating power systems, transport systems and vehicle technology for electric mobility impact assessment and efficient control. IEEE Trans. Smart Grid **3**(2), 934–949 (2012)
21. A.M. Farid, A hybrid dynamic system model for multi-modal transportation electrification. IEEE Trans. Control Syst. Technol. **PP**(99) (2016) 1–12
22. A.M. Farid, Electrified transportation system performance: conventional vs. online electric vehicles, in *Electrification of Ground Transportation Systems for Environment and Energy Conservation*, ed. by N.P. Suh, D.H. Cho (Springer, Berlin, Heidelberg, 2016), pp. 1–25

23. S. Stoft, *Power Systems Economics: Designing Markets for Electricity* (2002)
24. W.W. Hogan, Electricity wholesale market design in a low-carbon future, in *Harnessing Renewable Energy in Electric Power Systems: Theory, Practice, Policy*, vol. 113 (2010)
25. W.W. Hogan, Multiple market-clearing prices, electricity market design and price manipulation. Electr. J. **25**(4), 18–32 (2012)
26. A. Garcia, L. Mili, J. Momoh, *Modeling Electricity Markets: A Brief Introduction* (2008)
27. D.S. Kirschen, G. Strbac, *Fundamentals of Power System Economics* (Wiley, 2004)
28. M. Shahidehpour, H. Yamin, Z. Li, Market overview in electric power systems, in *Market Operations in Electric Power Systems: Forecasting, Scheduling, and Risk Management* (2002), pp. 1–20
29. M.H. Albadi, E. El-Saadany, A summary of demand response in electricity markets. Electric Power Syst. Res. **78**(11), 1989–1996 (2008)
30. S. Breukers, E. Heiskanen, B. Brohmann, R. Mourik, C. Feenstra, Connecting research to practice to improve energy demand-side management (DSM). Energy **36**(4), 2176–2185 (2011)
31. G. Strbac, Demand side management: benefits and challenges. Energy Policy **36**(12), 4419–4426 (2008)
32. A.M. Farid, B. Jiang, A. Muzhikyan, K. Youcef-Toumi, The need for holistic enterprise control assessment methods for the future electricity grid. Renew. Sustain. Energy Rev. **56**(1), 669–685 (2015)
33. M. Farina, G.F. Trecate, Decentralized and distributed control, in *EECI-HYCON2 Graduate School on Control* (2012)
34. A. Muzhikyan, A.M. Farid, K. Youcef-Toumi, An enterprise control assessment method for variable energy resource induced power system imbalances Part 1: Methodology. IEEE Trans. Ind. Electron. **62**(4), 2448–2458 (2015)
35. M. Mohseni, S.M. Islam, Review of international grid codes for wind power integration: diversity, technology and a case for global standard. Renew. Sustain. Energy Rev. **16**(6), 3876–3890 (2012)
36. F. Diaz-Gonzalez, M. Hau, A. Sumper, O. Gomis-Bellmunt, F. Díaz-gonzález, Participation of wind power plants in system frequency control: review of grid code requirements and control methods. Renew. Sustain. Energy Rev. **34**, 551–564 (2014)
37. M. Wolter, Reducing the curtailment amount of renewable energy using efficiency-based allocation, in *Mathworks Energy and Utilities Virtual Conference* (2012), pp. 1–2
38. K. Heussen, S. Koch, A. Ulbig, G. Andersson, Unified system-level modeling of intermittent renewable energy sources and energy storage for power system operation. IEEE Syst. J. **6**(1), 140–151 (2012)
39. J.M. Guerrero, P.C. Loh, T.L. Lee, M. Chandorkar, Advanced control architectures for intelligent microgrids—Part II: Power quality, energy storage, and AC/DC microgrids. IEEE Trans. Ind. Electron. **60**(4), 1263–1270 (2013)
40. I. Atzeni, L.G. Ordóñez, G. Scutari, D.P. Palomar, J.R. Fonollosa, Demand-side management via distributed energy generation and storage optimization. IEEE Trans. Smart Grid **4**(2), 866–876 (2013)
41. M. Albadi, E. El-Saadany, Comparative study on impacts of wind profiles on thermal units scheduling costs. IET Renew. Power Gener. **5**(1), 26–35 (2011)
42. T. Aigner, S. Jaehnert, G.L. Doorman, T. Gjengedal, The effect of large-scale wind power on system balancing in Northern Europe. IEEE Trans. Sustain. Energy **3**(4), 751–759 (2012)
43. A. Robitaille, I. Kamwa, A.H. Oussedik, M. de Montigny, N. Menemenlis, M. Huneault, A. Forcione, R. Mailhot, J. Bourret, L. Bernier, Preliminary impacts of wind power integration in the hydro-quebec system. Wind Eng. **36**(1), 35–52 (2012)
44. K. Bruninx, E. Delarue, A statistical description of the error on wind power forecasts for probabilistic reserve sizing. IEEE Trans. Sustain. Energy **5**(3), 995–1002 (2014)
45. H. Holttinen, M. Milligan, B. Kirby, T. Acker, V. Neimane, T. Molinski, Using standard deviation as a measure of increased operational reserve requirement for wind power. Wind Eng. **32**(4), 355–377 (2008)

46. A. Muzhikyan, A.M. Farid, K. Youcef-Toumi, An a priori analytical method for determination of operating reserves requirements. Int. J. Energy Power Syst. **86**(3), 1–11 (2016)
47. A. Muzhikyan, A.M. Farid, K. Youcef-Toumi, An enterprise control assessment method for variable energy resource induced power system imbalances Part 2: Results. IEEE Trans. Ind. Electron. **62**(4), 2459–2467 (2015)
48. A. Muzhikyan, A.M. Farid, K. Youcef-Toumi, Relative merits of load following reserves and energy storage market integration towards power system imbalances. Int. J. Electr. Power Energy Syst. **74**(1), 222–229 (2016)
49. A. Muzhikyan, A.M. Farid, T. Mezher, The impact of wind power geographical smoothing on operating reserve requirements, in *IEEE American Control Conference*, Boston, MA, USA (2016), pp. 1–6
50. B. Jiang, A.M. Farid, K. Youcef-Toumi, A comparison of day-ahead wholesale market: social welfare vs industrial demand side management, in *IEEE International Conference on Industrial Technology*, Sevilla, Spain (2015), pp. 1–7
51. B. Jiang, A.M. Farid, K. Youcef-Toumi, Demand side management in a day-ahead wholesale market a comparison of industrial and social welfare approaches. Appl. Energy **156**(1), 642–654 (2015)
52. P.G. Martin, The need for enterprise control. InTech Nov/Dec 1–5 (2012)
53. ANSI-ISA, Enterprise Control System Integration Part 3: Activity Models of Manufacturing Operations Management. Technical report, The International Society of Automation (2005)
54. L.M. Sanchez, R. Nagi, A review of agile manufacturing systems. Int. J. Prod. Res. **39**(16), 3561–3600 (2001)
55. A. Gunasekaran, Agile manufacturing: enablers and an implementation framework. Int. J. Prod. Res. **36**(5), 1223–1247 (1998)
56. R. Beach, A.P. Muhlemann, D.H.R. Price, A. Paterson, J.A. Sharp, Review of manufacturing flexibility. Eur. J. Oper. Res. **122**(1), 41–57 (2000)
57. A. De Toni, S. Tonchia, Manufacturing flexibility: a literature review. Int. J. Prod. Res. **36**(6), 1587–1617 (1998)
58. H.J. Pels, J.C. Wortmann, A.J.R. Zwegers, Flexibility in manufacturing: an architectural point of view. Comput. Ind. **33**(2–3), 271–283 (1997)
59. T.J. Williams, G.A. Rathwell, H. Li, *A Handbook on Master Planning and Implementation for Enterprise Integration Programs* (Purdue University Institute for Interdisciplinary Engineering Studies, 2001)
60. K. Kosanke, F. Vernadat, M. Zelm, Cimosa: enterprise engineering and integration. Comput. Ind. **40**(2–3), 83–87 (1999)
61. ANSI-ISA, *Enterprise-Control System Integration Part 1: Models and Terminology*, Number July (Instrument Society of America, 2000)
62. F.F. Wu, K. Moslehi, A. Bose, Power system control centers: past, present, and future. Proc. IEEE **93**(11), 1890–1908 (2005)
63. Y. Yan, Y. Qian, H. Sharif, D. Tipper, A survey on smart grid communication infrastructures: motivations, requirements and challenges. IEEE Commun. Surv. Tutor. **15**(1), 5–20 (2013)
64. Anonymous, NIST Framework and Roadmap for Smart Grid Interoperability Standards Release 1.0: NIST Special Publication 1108. Technical Report NIST Special Special Publication 1108, Office of the National Coordinator for Smart Grid Interoperability, National Institute of Standards and Technology, United States Department of Commerce, Washington D.C. (2010)
65. G. Wang, A. Kowli, M. Negrete-Pincetic, E. Shafieepoorfard, S. Meyn, A control theorist's perspective on dynamic competitive equilibria in electricity markets. IFAC Proc. Vol. **44**(1), 4933–4938 (2011)
66. D.J. Shiltz, M. Cvetković, A.M. Annaswamy, An integrated dynamic market mechanism for real-time markets and frequency regulation. IEEE Trans. Sustain. Energy **7**(2), 875–885 (2016)
67. N. Li, C. Zhao, L. Chen, Connecting automatic generation control and economic dispatch from an optimization view. IEEE Trans. Control Netw. Syst. **3**(3), 254–264 (2016)

68. M. Amin, Toward self-healing energy infrastructure systems. Comput. Appl. Power (IEEE) **14**(1), 20–28 (2001)
69. M. Amin, Toward a self-healing energy infrastructure, in *Power Engineering Society General Meeting. 2006 IEEE Power Engineering Society General Meeting*, Montreal, Canada (IEEE, 2006), 7 pp. BN—1 4244 0493 2
70. M. Amin, Challenges in reliability, security, efficiency, and resilience of energy infrastructure: toward smart self-healing electric power grid, in *BT—2008 IEEE Power and Energy Society General Meeting. 2008 IEEE Power Energy Society General Meeting*, 20–24 July 2008, Pittsburgh, PA, USA (IEEE, 2008), pp. 1–5
71. S.M. Amin, Smart grid overview issues and opportunities. Advances and challenges in sensing modeling simulation optimization and control. Eur. J. Control **17**(5–6), 547–567 (2011)
72. S.D.J. McArthur, P.C. Taylor, G.W. Ault, J.E. King, D. Athanasiadis, V.D. Alimisis, M. Czaplewski, The autonomic power system—network operation and control beyond smart grids, in *2012 3rd IEEE PES Innovative Smart Grid Technologies Europe (ISGT Europe)* (Oct 2012), pp. 1–7
73. P. Wood, D. Shiltz, T.R. Nudell, A. Hussain, A.M. Annaswamy, A framework for evaluating the resilience of dynamic real-time market mechanisms. IEEE Trans. Smart Grid **7**(6), 2904–2912 (2016)
74. A.K. Bejestani, A. Annaswamy, T. Samad, A hierarchical transactive control architecture for renewables integration in smart grids: analytical modeling and stability. IEEE Trans. Smart Grid **5**(4), 2054–2065 (2014)
75. A. Muzhikyan, A.M. Farid, K. Youcef-Toumi, Variable energy resource induced power system imbalance as mitigated by real-time markets and operator actions, in *3rd MIT-MI Joint Workshop on the Reliability of Power System Operation & Control in the Presence of Increasing Penetration of Variable Energy Sources*, Abu Dhabi, UAE (2012), pp. 1–6
76. A. Muzhikyan, A.M. Farid, K.Y. Toumi, Variable energy resource induced power system imbalances, in *5th MIT-MI Joint Workshop on the Reliability of Power System Operation & Control in the Presence of Increasing Penetration of Variable Energy Sources*, Abu Dhabi, UAE (2013), pp. 1–41
77. A. Muzhikyan, A.M. Farid, K. Youcef-Toumi, Variable energy resource induced power system imbalances: a generalized assessment approach, in *IEEE Conference on Technologies for Sustainability*, Portland, Oregon (2013), pp. 1–8
78. A. Muzhikyan, A.M. Farid, K. Youcef-Toumi, Variable energy resource induced power system imbalances: mitigation by increased system flexibility, spinning reserves and regulation, in *IEEE Conference on Technologies for Sustainability*, Portland, Oregon (2013), pp. 1–7
79. E. Ela, M. O'Malley, Studying the variability and uncertainty impacts of variable generation at multiple timescales. IEEE Trans. Power Syst. **27**(3), 1324–1333 (2012)
80. A. Kiani, A. Annaswamy, A hierarchical transactive control architecture for renewables integration in smart grids, in *2012 IEEE 51st Annual Conference on Decision and Control (CDC)* (IEEE, 2012), pp. 4985–4990
81. A. Muzhikyan, A.M. Farid, K. Youcef-Toumi, An enhanced method for determination of the ramping reserves, in *IEEE American Control Conference*, Los Angeles, CA, USA (2015), pp. 1–8
82. A. Muzhikyan, A.M. Farid, K. Youcef-Toumi, An enhanced method for determination of the regulation reserves, in *IEEE American Control Conference*, Los Angeles, CA, USA (2015), pp. 1–8
83. A. Muzhikyan, A.M. Farid, K. Youcef-Toumi, An enhanced method for the determination of load following reserves, in *American Control Conference, 2014*, Portland, Oregon (2014), pp. 1–8
84. S.P. Meyn, P. Barooah, A. Bušić, Y. Chen, J. Ehren, Ancillary service to the grid using intelligent deferrable loads. IEEE Trans. Autom. Control **60**(11), 2847–2862 (2015)
85. G. Wenzel, M. Negrete-Pincetic, D.E. Olivares, J. MacDonald, D.S. Callaway, Real-time charging strategies for an electric vehicle aggregator to provide ancillary services. IEEE Trans. Smart Grid (2017)

86. H. Holttinen, M. Milligan, E. Ela, N. Menemenlis, J. Dobschinski, B. Rawn, R.J. Bessa, D. Flynn, E. Gomez-Lazaro, N.K. Detlefsen, Methodologies to determine operating reserves due to increased wind power. IEEE Trans. Sustain. Energy **3**(4), 713–723 (2012)

87. A.S. Brouwer, M. van den Broek, A. Seebregts, A. Faaij, Impacts of large-scale Intermittent Renewable Energy Sources on electricity systems, and how these can be modeled. Renew. Sustain. Energy Rev. **33**, 443–466 (2014)

88. H. Holttinen, M.O. Malley, J. Dillon, D. Flynn, Recommendations for wind integration studies—IEA task 25. Technical report, International Energy Agency, Helsinki (2012)

89. H. Holttinen, A. Orths, H. Abilgaard, F. van Hulle, J. Kiviluoma, B. Lange, M. OMalley, D. Flynn, A. Keane, J. Dillon, E.M. Carlini, J.O. Tande, A. Estanquiro, E.G. Lazaro, L. Soder, M. Milligan, C. Smith, C. Clark, IEA wind export group report on recommended practices wind integration studies. Technical report, International Energy Agency, Paris, France (2013)

90. E. Ela, M. Milligan, B. Kirby, Operating reserves and variable generation. Contract **303**(August), 275–3000 (2011)

91. Y.G. Rebours, D.S. Kirschen, M. Trotignon, S. Rossignol, A survey of frequency and voltage control ancillary services—Part I: Technical features. IEEE Trans. Power Syst. **22**(1), 350–357 (2007)

92. CIGRE, Ancillary services: an overview of international practices technical brochure 435. Technical report, CIGRE Working Group C5.06 (October 2010)

93. A.J. Wood, B.F. Wollenberg, *Power Generation, Operation, and Control*, 3rd edn. (Wiley, Hoboken, NJ, USA, 2014)

94. A.G. Expósito, A. Gomez-Exposito, A.J. Conejo, C. Canizares, *Electric Energy Systems: Analysis and Operation* (CRC Press, 2016)

95. A.E. Curtright, J. Apt, The character of power output from utility-scale photovoltaic systems. Prog. Photovolt. Res. Appl. **2008**, 241–247 (2007)

96. J. Apt, A. Curtright, The spectrum of power from utility-scale wind farms and solar photovoltaic arrays. Technical report, Carnegie Mellon Electricity Industry Center Working Paper, Pittsburgh, PA, United States (2007)

97. Federal Energy Regulatory Commission: Order No. 764. Technical report, FERC (2012)

98. W. Wei, N. Li, J. Wang, S. Mei, Estimating the probability of infeasible real-time dispatch without exact distributions of stochastic wind generations. IEEE Trans. Power Syst. **31**(6), 5022–5032 (2016)

99. X. Zhang, A. Papachristodoulou, Distributed dynamic feedback control for smart power networks with tree topology, in *American Control Conference (ACC), 2014* (IEEE, 2014), pp. 1156–1161

100. X. Zhang, A. Papachristodoulou, A real-time control framework for smart power networks with star topology, in *American Control Conference (ACC), 2013* (IEEE, 2013), pp. 5062–5067

101. A. Jokic, *Price-Based Optimal Control of Electrical Power Systems* (2007)

102. H. Zhang, T. Feng, G.H. Yang, H. Liang, Distributed cooperative optimal control for multi-agent systems on directed graphs: an inverse optimal approach. IEEE Trans. Cybern. **45**(7), 1315–1326 (2015)

103. J. Lunze, F. Lamnabhi-Lagarrigue, *Handbook of Hybrid Systems Control: Theory, Tools, Applications* (Cambridge University Press, Cambridge, UK; New York, 2009)

104. A.M. Annaswamy, M. Amin, C. DeMarco, T. Samad, J. Aho, G. Arnold, A. Buckspan, A. Cadena, D. Callaway, E. Camacho, A. Caramanis, A. Chakrabortty, A. Chakraborty, J. Chow, M. Dahleh, A.D. Dominguez-Garcia, D. Dotta, A.M. Farid, P. Flikkema, D. Gayme, S. Genc, M.G.I. Fisa, I. Hiskens, P. Houpt, G. Hug, P. Khargonekar, H. Khurana, A. Kiani, S. Low, J. McDonald, E. Mojica-Nava, A.L. Motto, L. Pao, A. Parisio, A. Pinder, M. Polis, M. Roozbehani, Z. Qu, N. Quijano, J. Stoustrup, *IEEE Vision for Smart Grid Controls: 2030 and Beyond*. Technical report, IEEE (2013)

105. A.H. Mohsenian-Rad, V.W. Wong, J. Jatskevich, R. Schober, A. Leon-Garcia, Autonomous demand-side management based on game-theoretic energy consumption scheduling for the future smart grid. IEEE Trans. Smart Grid **1**(3), 320–331 (2010)

106. Z. Baharlouei, M. Hashemi, H. Narimani, H. Mohsenian-Rad, Achieving optimality and fairness in autonomous demand response: benchmarks and billing mechanisms. IEEE Trans. Smart Grid **4**(2), 968–975 (2013)
107. M.A.A. Pedrasa, T.D. Spooner, I.F. MacGill, Scheduling of demand side resources using binary particle swarm optimization. IEEE Trans. Power Syst. **24**(3), 1173–1181 (2009)
108. B. Ramanathan, V. Vittal, A framework for evaluation of advanced direct load control with minimum disruption. IEEE Trans. Power Syst. **23**(4), 1681–1688 (2008)
109. P. Samadi, A.H. Mohsenian-Rad, R. Schober, V.W. Wong, J. Jatskevich, Optimal real-time pricing algorithm based on utility maximization for smart grid, in *2010 First IEEE International Conference on Smart Grid Communications (SmartGridComm)* (IEEE, 2010), pp. 415–420
110. H. Chen, Y. Li, R.H. Louie, B. Vucetic, Autonomous demand side management based on energy consumption scheduling and instantaneous load billing: an aggregative game approach. IEEE Trans. Smart Grid **5**(4), 1744–1754 (2014)
111. Y.Y. Hsu, C.C. Su, Dispatch of direct load control using dynamic programming. IEEE Trans. Power Syst. **6**(3), 1056–1061 (1991)
112. K. Bhattacharyya, M. Crow, A fuzzy logic based approach to direct load control. IEEE Trans. Power Syst. **11**(2), 708–714 (1996)
113. J. Hansen, J. Knudsen, A. Kiani, A. Annaswamy, J. Stoustrup, A dynamic market mechanism for markets with shiftable demand response. IFAC Proc. Vol. **47**(3), 1873–1878 (2014)
114. A.K. Bejestani, A. Annaswamy, A dynamic mechanism for wholesale energy market: stability and robustness. IEEE Trans. Smart Grid **5**(6), 2877–2888 (2014)
115. Z. Almahmoud, J. Crandall, K. Elbassioni, T.T. Nguyen, M. Roozbehani, Dynamic pricing in smart grids under thresholding policies: algorithms and heuristics (2016), arXiv:1610.07559
116. D.P. Zhou, M. Roozbehani, M.A. Dahleh, C.J. Tomlin, Stability analysis of wholesale electricity markets under dynamic consumption models and real-time pricing, in *American Control Conference (ACC), 2017* (IEEE, 2017), pp. 2048–2053
117. A. Kiani, A. Annaswamy, Wholesale energy market in a smart grid: dynamic modeling and stability, in *2011 50th IEEE Conference on Decision and Control and European Control Conference (CDC-ECC)* (IEEE, 2011), pp. 2202–2207
118. S. Jenkins, A. Annaswamy, J. Hansen, J. Knudsen, A dynamic model of the combined electricity and natural gas markets, in *Innovative Smart Grid Technologies Conference (ISGT), 2015 IEEE Power & Energy Society* (IEEE, 2015), pp. 1–5
119. M.J. Garcia, T.R. Nudell, A.M. Annaswamy, A dynamic regulation market mechanism for improved financial settlements in wholesale electricity markets, in *American Control Conference (ACC), 2017* (IEEE, 2017), pp. 1425–1430
120. Q. Huang, M. Roozbehani, M.A. Dahleh, Efficiency-risk tradeoffs in electricity markets with dynamic demand response. IEEE Trans. Smart Grid **6**(1), 279–290 (2015)
121. J. Knudsen, J. Hansen, A.M. Annaswamy, A dynamic market mechanism for the integration of renewables and demand response. IEEE Trans. Control Syst. Technol. **24**(3), 940–955 (2016)
122. B. Jiang, A. Muzhikyan, A.M. Farid, K. Youcef-Toumi, Demand side management in power grid enterprise control—a comparison of industrial and social welfare approaches. Appl. Energy **187**, 833–846 (2017)
123. B. Jiang, A. Muzhikyan, A.M. Farid, K. Youcef-Toumi, Impacts of industrial baseline errors in demand side management enabled enterprise control, in *IECON 2015—41st Annual Conference of the IEEE Industrial Electronics Society*, Yokohama, Japan (2015), pp. 1–6
124. B. Jiang, A.M. Farid, K. Youcef-Toumi, Impacts of industrial baseline errors on demand side management in day-ahead wholesale markets, in *Proceedings of the ASME Power & Energy 2015: Energy Solutions for Sustainable Future*, San Diego, CA (2015), pp. 1–7
125. R. Walawalkar, S. Fernands, N. Thakur, K.R. Chevva, Evolution and current status of demand response (DR) in electricity markets: insights from PJM and NYISO. Energy **35**(4), 1553–1560 (2010)
126. H.P. Chao, Demand response in wholesale electricity markets: the choice of customer baseline. J. Regul. Econ. **39**(1) 68–88 (2011)

127. Gridwise Architecture Council, Transactive Energy (2016), http://www.gridwiseac.org/about/transactive_energy.aspx. Accessed 18 Feb 2017
128. F.C. Schweppe, R.D. Tabors, J.L. Kirtley, H.R. Outhred, F.H. Pickel, A.J. Cox, Homeostatic utility control. IEEE Trans. Power Appar. Syst. **PAS-99**(3), 1151–1163 (1980)
129. D. Hammerstrom, R. Ambrosio, J. Brous, T. Carlon, D. Ghassin, J. DeSteese, R. Guttromson, G. Horst, O. Järregren, R. Kajfasz et al., *Pacific Northwest GridWise TM Testbed Demonstration Projects, Volume I: The Olympic Peninsula Project* (2007)
130. R. Melton, Pacific northwest smart grid demonstration project technology performance report volume 1: technology performance. Technical report, Pacific Northwest National Laboratory (PNNL), Richland, WA, US (2015)
131. E.G. Cazalet, Temix: a foundation for transactive energy in a smart grid world, in *Grid-Interop Forum* (2010)
132. D. Hammerstrom, T. Oliver, R. Melton, R. Ambrosio, Standardization of a hierarchical transactive control system. Grid Interoper. **9**, 1949–3053 (2009)

Part II
Distributed Control for DER Integration

Distributed Control of Power Grids

Jakob Stoustrup

Abstract This chapter provides a short introduction to the part of this volume dealing with distributed control of power grids. A brief description of some of the challenges facing existing power grids from a control perspective is given. The research community dealing with distributed control of power grids is highly active, and there already exists a vast literature on the topic. A coverage of this literature with any pretense of full or partial completeness would be difficult (if not impossible) and is not in any way attempted in this brief introduction.

1 Introduction

The electrical power grids of the world rely on infrastructure that emerged based on pre-digital technology more than a hundred years ago. Since the origin, the power grids have developed massively in technology in order to improve resiliency, safety, and effectiveness. As new technologies have been introduced, the grids have grown in complexity and during the past couple of decades, digital technology has been massively deployed.

Due to the recent grid evolution, however, the power grids of the world have to address challenges in terms of an urgent need for massively increased flexibility. This increased flexibility is required in order to integrate a higher penetration of renewable generation, of rooftop PV and other Distributed Energy Resources (DERs), see e.g., DNV-GL [8]. In addition to an increased level of intermittent generation, also power usage patterns are changing dramatically. The change of usage patterns on different time scales is driven, e.g., by the spread of power electronic devices and by a slow but steady increase of electric vehicles.

The combination of an increased amount of renewable generation and changed usage patterns introduce more variability and more uncertainty in the power grids, and thus, threaten to compromise grid reliability, if appropriate action is not taken.

J. Stoustrup (✉)
Automation & Control, Aalborg University, Aalborg, Denmark
e-mail: jakob@es.aau.dk

© Springer Nature Switzerland AG 2019
J. Stoustrup et al. (eds.), *Smart Grid Control*, Power Electronics
and Power Systems, https://doi.org/10.1007/978-3-319-98310-3_5

Fortunately, some of the very same changes to the power grids also constitute a potential opportunity to mitigate the challenges. By deploying advanced control solutions at various levels of the grid, there are significant possibilities for establishing flexibility to the extent required in the immediate future, but also on a longer term.

The position chapters in this part of the present manuscript encompass a number of research issues that together constitute part of a transformational grid control paradigm based on distributed control algorithms. These proposed research challenges address how to manage dynamic changes within power grids at a local or a global scale in a reliable way by leveraging additional resources in the grid. At a system level, massive deployment of distributed control technology is expected to facilitate a more efficient usage of natural resources and a significant reduction of greenhouse gas emissions. One of the underlying instruments in obtaining these goals involves matching power consumption to intermittent generation in real-time. Also, the capacity of distribution grid networks can be expected to be exploited more efficiently by employing local DERs optimally. This facilitates a potential of reduced investments in distribution grid extensions in response to increased consumption.

Also, distributed control solutions deployed massively in power distribution grid also carry the potential to substitute a proportion of the spinning reserves. This, in turn, reduces the need to curtail renewable generation and thereby also reducing, e.g., the need for fossil energy resources.

In similarity with other large-scale systems, by tradition the power grids of the world have been engineered with hierarchical control topologies, guided by separation in temporal and spatial scales. This is reflected in the well-known chain from centralized generation via transmission, sub-transmission, and finally over distribution systems to load consumption. Across the world, such chains have traditionally been managed in a strict top-down manner. During the past couple of decades, however, an increasing number of successes have been reported across various industries, where application of distributed control solutions have shown significant advantages over traditional control solutions based on hierarchical control topologies. In a similar fashion, power system operators on several continents have gained experience from experiments with various types of distributed control solutions in power grids. Results indicate that, indeed, such solutions can lead to increased efficiency in grid operation and to reduced greenhouse gas emissions, in part by offering scalable integration of DERs, especially in the medium voltage and the low voltage grids.

The path toward a fully deployed power grid solution based on advanced control theory, and in part relying on distributed control topologies, however, still requires adequate responses to a number of research challenges remaining, which is the topic of the contributions of this part of the present manuscript. The solutions proposed and in smaller scales explored experimentally range from control topologies on one extreme based on system operator management from top to bottom to the other extreme where the grids are operated as a system of weakly connected microgrids. The research challenges described in the chapters of this part of the manuscript in part encompasses the question of which control topologies better facilitate an interconnected power system with a high penetration of renewable power generation and orchestrates the operation of a large number of DERs.

2 Legacy Grids, Trends, and Enabling Technologies

From a control perspective, the legacy grids across the world have been operated according to very similar paradigms, which can be summarized by the conventional primary/secondary/tertiary control system:

Control level	Timescale	Goal	Strategy
Primary	Real-time	Stabilizing frequency and voltage	Decentralized
Secondary	Minutes	Restoring frequency	Centralized
Tertiary	Offline	Optimizing operation	Centralized/forecast

Several current trends and future expectations to grid developments are challenging whether, from a control perspective, this hierarchical structure based on temporal (and spatial) separation is still the best (or even appropriate) for future power grids, see e.g., EPRI [11].

As one major trend, physical volatility is steadily increasing in part due to the development mentioned above with increased renewable penetration and distributed generation, and in part due to growing demand in systems with an aging infrastructure. As a result, a lowered inertia and reduced robustness margins are seen on grids worldwide.

As a second major trend, a number of technological advances are seen, e.g., in terms of novel sensors and actuators (e.g., PMUs, FACTS) and access to grid-edge resources (e.g., flexible loads). Further, advanced control is being introduced in a vast number of grid-connected cyber–physical systems. This collectively facilitates a future cyber-coordination layer for smart grid solutions.

Several other emerging technologies constitute a basis for accelerating the transition toward a smarter grid. In the following, we shall emphasize a few among several enabling technologies for the potential grid transformation.

2.1 Advanced Metering Infrastructure (AMI)

Since deregulation and market-driven pricing were introduced in a majority of power grids across the world, utilities have been pursuing technologies that could assist in matching power generation to power consumption. Thus, an advanced metering infrastructure has been deployed in grids worldwide, in part consisting of *smart meters* deployed at individual customer nodes. The capabilities of smart meters vary significantly, but usually more than just automatic reading of energy deliveries are offered. Some smart meters offer real-time or near real-time power usage. Additional services include notifications of power outages and power quality measurements. A majority of smart meters are equipped with technology for two-way communication which makes them a significant potential enabler for advanced control solutions.

As part of an Advanced Metering Infrastructure (AMI), smart meters can be seen as communication hubs for systems that can measure, collect and analyze energy usage, but also receive signals to activate and manage flexible power consuming and/or generating (e.g., rooftop PV systems) at the user's end.

Two major classes of distributed control approaches for power distribution grids based on AMI can be discriminated based on the type of signal, they would send to the smart meters. In *direct control* approaches, meters would receive a *command signal*, either in terms of an ON/OFF signal for specific devices, or in terms of a reference power signal that select devices would have to follow. In *indirect control* approaches, meters would receive a *price signal*, leaving to the consumers' discretion, how the consumer would like to respond to the price signal by increasing or decreasing consumption. A variation of this is the Transactive Control and Coordination approach, see below, where a market-like structure lets consumers negotiate delivery of a certain quantity of energy at a certain priced based on two-way communication.

2.2 Internet of Things (IoT)

Across the world, massive research investments are being made in the Internet of Things (IoT), possibly under the conviction that "...*the 'Industrial Internet' [will] start the next Industrial Revolution*" (Joe Salvo, GE).

IoT is expected to transform a large number of industries, including Manufacturing, Agriculture, Mining, Transportation, Oil and gas, etc. However, IoT also holds a huge potential for transforming the Energy and Power area and in particular the electrical power grids. Leveraging the immense intelligence at the edge of the grid, however, requires a paradigm shift with a transition from centralized to decentralized decision-making.

In order to benefit from the expected future access to a huge number of grid-edge resources, the legacy grid control architecture is further challenged. The system will never obtain sufficient bandwidth for accumulating, storing, and processing the immense amount of data. The inherent latency involved in centralized processing will prohibit decisions to be made on timescales required by the grid.

IoT integration in the grid requires data processing to be performed as close to the data collection nodes as possible. It is also necessary to allow these nodes to make decisions (semi-)autonomously.

2.3 Advanced Inverter Technology

During the past couple of decades, converters based on power electronics have been vastly deployed in the large continental power grids, see, e.g., Blaabjerg et al. [3] and references therein. The literature has had a strong emphasis on potential challenges for this major change to the grid. From a controls perspective, it is possible, however, also to take the opposite point of view and pursue opportunities embarking from the significantly extended control capabilities offered by such devices.

From a controls perspective, these opportunities would involve a power grid with an extensive deployment of controllable power electronic devices and sufficient resources for real or virtual storage. Such a power grid needs to have a potential for at least the same resilience and reliability as the legacy grid, even if the penetration of intermittent renewable generation is significantly increased. In order to realize most of the potential of widely deployed controllable power electronic devices, advanced control techniques are required, and the legacy grid structure is not able to facilitate this. In order to achieve this, there is a need for developing control algorithms that aggregate and disaggregate control capabilities from power electronic devices across levels in the power system. The solutions obtained will be used to investigate to which extent the new capabilities will facilitate increased penetration of intermittent renewable generation in terms of assessing the added control authority in various frequency ranges relevant for addressing this type of intermittency.

The applications of scalable control systems involve two fundamental issues: (i) the realization of the scalable control systems with power electronics apparatus, and (ii) the assessment of control performance at different system levels.

Power electronics systems are basically a hybrid system of the discrete switching events of power semiconductor devices and the continuous dynamics of passive components. Power electronics converters with different power scales operate with different switching speeds for reduced power losses, which consequently sets the upper limit for the response time of the control system, and further challenges the realization of the scalable control systems for power converters with different power ratings.

The small time constants of power converters and wider bandwidth of their control systems complicate the dynamic coupling and interactions between the converters and power grids at different system levels, implying more electromagnetic transient oscillations. Hence, the dynamic characterization of power electronic components or subsystems equipped with scalable control systems is essential for the performance assessment, Rocabert et al. [54].

2.3.1   Virtual Inertia

By deploying a scalable control system for grid-wide coordination of converters, it will be possible to transform any converter-controlled energy storage to a unit that contributes to the overall stabilization of the power grid. In particular, such a unit can be controlled to emulate rotating mechanical energy of conventional generators, so-called virtual inertia. Actual units could be either at the generating side, e.g., an electrical storage at a wind farm, or on the load side (smart grid), e.g., a Heating, Ventilation, and Air-Conditioning system in a large commercial building.

2.3.2   Harmonic Stability

The increasing penetration of power electronics systems aggravate harmonic distortions in the power grid, due to the nonlinear switching operations of electronic devices. The harmonics tend to trigger the electrical resonance frequencies of power systems, and are further coupled with the fast control dynamics of power converters causing harmonic oscillations. This instability phenomenon has recently been reported in large-scale renewable power plants and cable-based transmission grids. To prevent harmonic instability, advanced control theory could be used first to reshape the dynamic behavior of power converters with positive damping characteristics, and then to synthesize the damping over a wide frequency range by system-wide coordination of converters.

3   Distributed Control Paradigms

Distributed control has been proposed as part of a novel control paradigm for power grids for various parts of the grids. In particular, however, distributed control has been proposed as an approach to enable load-side participation. Load-side participation as a supplement to control on the generation side is interesting for a large number of reasons, including:

- Load-side control might be faster as there is little or low inertia
- The huge number of devices on the load side has the potential of making the system more reliable by a spreading approach
- A large number of sensing and actuating nodes offers the potential of a better ability to localize disturbances
- As the need for control capacity reduces on the generator side, generation can be made more efficient
- No additional emissions or use of fossil resources are required for load-side control.

According to Lu and Hammerstrom [38], residential power loads account for approximately one-third of peak demand, but 61% of these devices are *'Grid Friendly'*, i.e., they have a potential for participating in load-side control. In the US, the operating reserve is 13% of the peak, whereas the total 'Grid Friendly' capacity is 18%.

A distributed control paradigm must be able to provide voltage control. One problem in that context is that several distributed control topologies might lead to steady-state voltage deviations. So, on one hand, voltage regulation constraints must be built into a feasible distributed grid control solution. On the other hand, a distributed control solution should be expected to provide a reasonable sharing of loads between available DERs. Unfortunately, there is a fundamental conflict between these two objectives, and a feasible power grid distributed control solution must be able to provide an acceptable compromise between the two.

An important enabler for applying distributed control as part of frequency control is the notion of *Grid Friendly Appliances (GFA)*. GFA is a specific standard for interfacing between grid frequency and individual devices. Some recent work on GFA can be found in Lian et al. [36], Williams et al. [62] , Elizondo et al. [10], Lian et al. [35], Moya et al. [46] and references therein.

A large number of distributed control approaches have been proposed in the control literature. Several of these have been suggested as candidates for being part of an advanced power systems control solution. In the following, however, among this large group of solutions, we shall only briefly introduce three categories that have been emphasized in the challenge chapters of this volume. It should be emphasized, however, that several other approaches are relevant for distributed control of power grids. Examples of such approaches are passivity-based and port-Hamiltonian methods, see e.g., Schiffer et al. [55] and references therein.

3.1   Transactive Control and Coordination

An approach to massive activation of DERs that has gained significant attention is the so-called *Transactive Control and Coordination (TC2)* approach, see, e.g., Subbarao et al. [58] and references therein (please, refer also to the part of this volume that deals with markets). Some further recent references are Subbarao et al. [59], Li et al. [31–33].

TC2 offers to manage generation, power flows and consumption with reliability constraints by market-like constructs. This is achieved by using global information and local control decisions at nodes where the power flow can be affected. Each node communicates with the network via transactive incentives and feedback signals. TC2 is a flexible design in the sense that it allows deployment at all levels of the energy hierarchy.

TC2 offers a distributed approach based on self-organized market-like constructs. Thereby TC2 has the potential to overcome the challenge formed by a huge number of controllable assets, which make centralized optimization unworkable. TC2 has a simple information protocol, which is common between all nodes at all levels of a system, comprised by quantity, price/value, and time. This makes TC2 a candidate solution for challenges related to interoperability. In terms of security and privacy, TC2 attempts to minimize sensitivities by limiting the required amount of data exchange to the triple mentioned above. Finally, TC2 potentially achieves scalability by being self-similar at all grid scales. The TC2 paradigm for control and communication is common across all nodes of the system. A proposed ratio of supply nodes to served nodes is 10^3.

3.2 Consensus-Based Distributed Control and Coordination

Consensus-based control algorithms are a subclass of cooperative control algorithms. The main idea is that each local agent generates decisions for a local subset of the total system states based on a global objective. Each agent has the ability to communicate subject to a given communication topology with a small number of neighboring agents with a certain bandwidth, limiting the amount of information exchanged.

The consensus algorithm now proceeds by all agents communicating an estimate of one or several global variables with their nearest neighbors based on their own state information and past values of their neighbors' estimates of said variable(s). It can be shown that under mild observability assumptions and a simple connectivity assumption on the communication topology that all local estimates will converge to the global value(s).

Consensus-based control algorithms have several potential applications for electrical power grids. Maybe the obvious candidate is consensus-based control applied to a grid configuration consisting of weakly connected microgrids that collectively have to provide a certain grid objective, e.g., voltage stabilization. There are, however, many other examples, e.g., coordination of units below a substation and between substations, on and between individual radials, etc.

It has been widely claimed, but not proven, that consensus algorithms are scalable. Actually, some experimental evidence suggests that scalability is not straightforward, so future work might be needed. Other issues have been related to integration of distributed generation and especially storage in the algorithms (a recent breakthrough to that end has been published in Wu et al. [63]). To speed up consensus-based control/coordination algorithms, an important contribution can be found in Olshevsky [47].

3.3 Distributed Control Based on Distributed Optimization Algorithms

A huge research effort in the optimization community has been dedicated to distributed optimization. A significant proportion of available distributed optimization algorithms, including but not limited to subgradient algorithms, can be applied as the basis for distributed control.

One large class of distributed optimization techniques are based on augmented Lagrangian decompositions. These approaches include dual decomposition, alternating direction method of multipliers, and analytical target cascading. Another relevant class of techniques are based on decentralized solutions of Karush–Kuhn–Tucker (KKT) conditions. Examples of these are consensus/innovation methods (see also above) and the optimality condition decomposition.

A recent survey of distributed control approaches based on distributed optimization has been published in Molzahn et al. [45]. Please, refer to this paper for further literature on the topic.

4 Overview of Challenge Chapters

The remainder of this part of the present volume contains three separate challenge chapters. The topics include how to control flexible loads in order to make them behave like virtual storages, how to model low-inertia inverter-dominated power systems, how to systematically distribute the *design* of local controllers, and how to employ stochastic control theory for smart grid solutions, in particular, for microgrids.

In the following, each of the three challenge chapters will be described shortly.

4.1 *Virtual Energy Storage from Flexible Loads: Distributed Control with QoS Constraints by Prabir Barooah*

This chapter discusses the concept of virtual energy storages. By manipulating demand around a nominal baseline, power consuming devices in our infrastructure can contribute to accommodating situations with excess power or with power deficiency, thus acting in ways that resemble an electrical storage device. However, to make loads exhibit appropriate charging and discharging patterns that are useful to the power grid requires the solution of a number of complex control problems. In essence, the main challenge is to achieve a good compromise between providing adequate grid services while maintaining Quality of Service for power consumers.

One of the challenges discussed in the chapter relates to the difficulty of providing good capacity estimates due to variations over time, especially those caused by exogenous factors such as weather, which is not conveniently captured in simple models. Also, the power baseline for a device acting as VES is very difficult to estimate.

Another line of challenges relate to dispatch. An optimal dispatch of VES resources should be based on cost, but cost of operation is very difficult to quantify or estimate without tedious modeling of every available system. As part of estimating cost, round-trip efficiency should be established, which is also generally difficult to capture.

Finally, the chapter includes a short discussion of challenges associated with the interplay between control and communication topologies.

4.2 Distributed Design of Local Controllers for Future Smart Grids by Tomonori Sadamoto, Takayuki Ishizaki, Takuro Kato, and Jun-ichi Imura

In this chapter, the authors describe a new notion of distributed controller design where local controllers are individually designed by using partial models of the system, e.g., single-machine-infinite-bus models, and their control actions are individually determined by local feedbacks from corresponding neighborhoods.

One of the challenges discussed in this chapter is that this approach can cause transient instability of power systems with large-scale photovoltaic generators (PVs) integration. Motivated by this, a distributed design problem with consideration of node clustering is formulated, and it is discussed how to find clusters from the perspective of controllability with respect to the corresponding local inputs.

Another challenge discussed in this chapter relates to design of interchangeable components. In order to facilitate the addition of new components, it would be desirable that newly added components have interchangeability or plug-and-play capability. A significant challenge, however, relates to identifying the class of such components as well as which portfolio constraints to be imposed on the interconnection of the components to preexisting grids.

4.3 Smart Grid Control: Opportunities and Research Challenges A Decentralized Stochastic Control Approach by Maryam Khanbaghi

In this chapter, a future grid structure with massive presence of microgrids and nanogrids is discussed. As one of several salient features, in such a grid structure, resilience can be pursued by appropriately alternating between islanded and grid-integrated mode for microgrids and nanogrids. In challenging situations, where a conventional grid architecture could risk wide-area blackouts, an architecture dominated by large numbers of micro- and nanogrids could keep critical parts of the grid operational by transitioning these to islanded operation.

One of the control challenges discussed in the paper relates to how to ensure that the performance obtained for each nano-/microgrid in islanded mode is inherited to a suitable extent, when most or all of these are operated in grid-integrated mode. The chapter proposes to address this control challenge by (1) pursuing to design a robust control strategy to maintain stability, and (2) by reevaluating system requirements in order to maintain optimality.

NB! This list includes a number of seminal contributions on distributed control of power grids that are not singled out above.

References

1. P.P. Barker, R.W. De Mello, Determining the impact of distributed generation on power systems. i. radial distribution systems, in *Proceedings IEEE Power Engineering Society*, vol. 3, pp. 1645–1656 (2000). https://doi.org/10.1109/PESS.2000.868775
2. B. Biegel, P. Andersen, J. Stoustrup, L.H. Hansen, D.V. Tackie, Information modeling for direct control of distributed energy resources, in *Proceedings of the 2013 American Control Conference* (Washington, DC, USA, 2013), pp. 3498–3504. https://doi.org/10.1109/ACC.2013.6580372
3. F. Blaabjerg, R. Teodorescu, M. Liserre, A.V. Timbus, Overview of control and synchronization of three phase distributed power generation systems. IEEE Trans. Industr. Electr. **53**(5), 1398–1409 (2006)
4. R.E. Brown, Impact of smart grid on distribution system design, in *Proceedings IEEE Power Energy Society General Meeting* (Pittsburgh, PA, 2008). https://doi.org/10.1109/PES.2008.4596843
5. D.S. Callaway, I.A. Hiskens, Achieving controllability of electric loads. Proc. IEEE **99**(1), 184–199 (2011). https://doi.org/10.1109/JPROC.2010.2081652
6. K. De Brabandere, K. Vanthournout, J. Driesen, G. Deconinck, R. Belmans, Control of microgrids, in *Proceedings IEEE Power Engineering and Society General Meeting*, pp. 1–7 (2007). https://doi.org/10.1109/PES.2007.386042
7. A.L. Dimeas, N.D. Hatziargyriou, Operation of a multiagent system for microgrid control. IEEE Trans. Power Syst. **20**(3), 1447–1455 (2005). https://doi.org/10.1109/TPWRS.2005.852060
8. DNV-GL (2014) A Review of Distributed Energy Resources, DNV-GL for the New York Independent System Operator
9. J. Driesen, F. Katiraei, Design for distributed energy resources. IEEE Power Energy Mag. **6**(3), 30–40 (2008). https://doi.org/10.1109/MPE.2008.918703
10. M. Elizondo, K. Kalsi, C. Moya, W. Zhang, Frequency responsive demand in U.S. western power system model. In: *Proceedings of, IEEE Power and Energy Society General Meeting* (Denver, CO, USA, 2015)
11. EPRI, The integrated grid: realizing the full value of central and distributed energy resources (2014)
12. J. Fan, S. Borlase, The evolution of distribution. IEEE Power Energy Mag. **7**(2), 63–68 (2009). https://doi.org/10.1109/MPE.2008.931392
13. J. Goellner, M. Prica, J. Miller, S. Pullins, J. Westerman, J. Harmon, T. Grabowski, H. Weller, B. Renz, S. Knudsen, D. Coen, Demand dispatch—intelligent demand for a more efficient grid. Tech. Rep. DOE/NETL-DE-FE0004001 (National Energy Technology Laboratory, 2011)
14. F.V. Gomes, S.J. Carneiro, J.L.R. Pereira, M.P. Vinagre, P.A.N. Garcia, L.R. Araujo, A new heuristic reconfiguration algorithm for large distribution systems. IEEE Trans. Power Syst. **20**(3), 1373–1378 (2005). https://doi.org/10.1109/TPWRS.2005.851937
15. F.V. Gomes, S. Carneiro Jr., J.L.R. Pereira, M.P. Vinagre, P.A.N. Garcia, L.R. de Araujo, A new distribution system reconfiguration approach using optimum power flow and sensitivity analysis for loss reduction. IEEE Trans. Power Syst. **21**(4), 1616–1623 (2006). https://doi.org/10.1109/TPWRS.2006.879290
16. A. Gusrialdi, Z. Qu, M.A. Simaan, Distributed scheduling and cooperative control for charging of electric vehicles at highway service stations. IEEE Trans. Intell. Transp. Syst. **18**(10), 2713–2727 (2017)
17. R. Harvey, Z. Qu, Cooperative control and networked operation of passivity-short systems, in *Recent Advances and Future Directions on Adaptation and Control*, ed. by K.G. Vamvoudakis, S. Jagannathan (Elsevier, Cambridge, MA, 2016), pp. 499–518
18. G.T. Heydt, The next generation of power distribution systems. IEEE Trans. Smart Grid **1**(3), 225–235 (2010). https://doi.org/10.1109/TSG.2010.2080328
19. R. Hidalgo, C. Abbey, G. Joós, A review of active distribution networks enabling technologies, in *Proceedings IEEE Power Energy and Soceity General Meeting*, pp. 1–9 (2010). https://doi.org/10.1109/PES.2010.5590051

20. C.A. Hill, M.C. Such, D. Chen, J. Gonzalez, W.M. Grady, Battery energy storage for enabling integration of distributed solar power generation. IEEE Trans. Smart Grid **3**(2), 850–857 (2012). https://doi.org/10.1109/TSG.2012.2190113
21. J. Huang, C. Jiang, R. Xu, A review on distributed energy resources and Microgrid. Renew. Sust. Energy Rev. **12**(9), 2472–2483 (2008). https://doi.org/10.1016/j.rser.2007.06.004
22. A. Ipakchi, F. Albuyeh, Grid of the future. IEEE Power Energy Mag. **7**(2), 52–62 (2009). https://doi.org/10.1109/MPE.2008.931384
23. R. Johansson, A. Rantzer, Distributed decision making and control (Springer, 2012)
24. H. Karimi, H. Nikkhajoei, R. Iravani, Control of an electronically-coupled distributed resource unit subsequent to an islanding event. IEEE Trans. Power Deliv. **23**(1), 493–501 (2008). https://doi.org/10.1109/TPWRD.2007.911189
25. F. Katiraei, M. Iravani, Power management strategies for a microgrid with multiple distributed generation units. IEEE Trans. Power Syst. **21**(4), 1821–1831 (2006). https://doi.org/10.1109/TPWRS.2006.879260
26. F. Katiraei, M.R. Iravani, P.W. Lehn, Micro-grid autonomous operation during and subsequent to islanding process. IEEE Trans. Power Deliv. **20**(1), 248–257 (2005). https://doi.org/10.1109/TPWRD.2004.835051
27. F. Katiraei, M. Iravani, P.W. Lehn, Small-signal dynamic model of a micro-grid including conventional and electronically interfaced distributed resources. IET Gener. Transm. Dis. **1**(3), 369–378 (2007). https://doi.org/10.1049/iet-gtd:20045207
28. R. Lasseter, A. Akhil, C. Marnay, J. Stephens, J. Dagle, R. Guttromson, A.S. Meliopoulous, R. Yinger, J. Eto, *Integration of Distributed Energy Resources. The CERTS Microgrid Concept* (Lawrence Berkeley National Laboratory, 2002)
29. S. Li, W. Zhang, J. Lian, K. Kalsi, Market-based coordination of thermostatically controlled loads-Part I: a mechanism design formulation. IEEE Trans. Power Syst. **31**(2), 1170–1178 (2016a). https://doi.org/10.1109/TPWRS.2015.2432057
30. S. Li, W. Zhang, J. Lian, K. Kalsi, Market-based coordination of thermostatically controlled loads-Part II: Unknown parameters and case studies. IEEE Trans. Power Syst. **31**(2), 1179–1187 (2016b). https://doi.org/10.1109/TPWRS.2015.2432060
31. S. Li, W. Zhang, J. Lian, K. Kalsi, Market-based coordination of thermostatically controlled loads-part i: a mechanism design formulation. IEEE Trans. Power Syst. **31**(2), 1170–1178 (2016c)
32. S. Li, W. Zhang, J. Lian, K. Kalsi, Market-based coordination of thermostatically controlled loads-part ii: unknown parameters and case studies. IEEE Trans. Power Syst. **31**(2), 1179–1187 (2016d)
33. S. Li, W. Zhang, J. Lian, K. Kalsi, Social optima in non-cooperative mean field games, in *Proceedings of 2016 IEEE Conference on Decision and Control* (Las Vegas, NV, USA, 2016)
34. Y.W. Li, C.N. Kao, An accurate power control strategy for power-electronics-interfaced distributed generation units operating in a low-voltage multibus microgrid. IEEE Trans. Power Electr. **24**(12), 2977–2988 (2009). https://doi.org/10.1109/TPEL.2009.2022828
35. J. Lian, Y. Sun, L. Marinovici, K. Kalsi, Improved controller design of grid friendly appliances for primary frequency response, in *Proceedings of, IEEE Power and Energy Society General Meeting* (Denver, CO, USA, 2015)
36. J. Lian, J. Hansen, L. Marinovici, K. Kalsi, Hierarchical decentralized controller design for demand-side primary frequency response, in *Proceedings of, IEEE Power and Energy Society General Meeting* (MA, USA, Boston, 2016)
37. Y. Lin, P. Barooah, S. Meyn, T. Middelkoop, Experimental evaluation of frequency regulation from commercial building hvac systems. IEEE Trans. Smart Grid **6**, 776–783 (2015). https://doi.org/10.1109/TSG.2014.2381596
38. N. Lu, D. Hammerstrom, Design considerations for frequency responsive grid friendly appliances, in *Proceedings of 2005/2006 IEEE/PES Transmission and Distribution Conference and Exhibition* (Institute of Electrical and Electronics Engineers, Dallas, TX, USA, 2006), pp. 647–652

39. O. Ma, N. Alkadi, P. Cappers, P. Denholm, J. Dudley, S. Goli, M. Hummon, S. Kiliccote, J. MacDonald, N. Matson, D. Olsen, C. Rose, M.D. Sohn, M. Starke, B. Kirby, M. O'Malley, Demand response for ancillary services. IEEE Trans. Smart Grid **4**(4), 1988–1995 (2013). https://doi.org/10.1109/TSG.2013.2258049

40. A. Maknouninejad, Z. Qu, F. Lewis, A. Davoudi, Optimal, nonlinear, and distributed designs of droop controls for dc microgrids. IEEE Trans. Smart Grid **5**(5), 2508–2516 (2014)

41. S. Massoud Amin, B.F. Wollenberg, Toward a smart grid: power delivery for the 21st century. IEEE Power Energy Mag. **3**(5), 34–41 (2005). https://doi.org/10.1109/MPAE.2005.1507024

42. J. Medina, N. Muller, I. Roytelman, Demand response and distribution grid operations: opportunities and challenges. IEEE Trans. Smart Grid **1**(2), 193–198 (2010). https://doi.org/10.1109/TSG.2010.2050156

43. S. Meyn, P. Barooah, A. Busic, Y. Chen, J. Ehren, Ancillary service to the grid using intelligent deferrable loads. IEEE Trans. Auto. Control **60**(11), 2847–2862 (2013)

44. A.H. Mohsenian-Rad, A. Leon-Garcia, Optimal residential load control with price prediction in real-time electricity pricing environments. IEEE Trans. Smart Grid **1**(2), 120–133 (2010). https://doi.org/10.1109/TSG.2010.2055903

45. D.K. Molzahn, F. Drfler, H. Sandberg, S.H. Low, S. Chakrabarti, R. Baldick, J. Lavaei, A survey of distributed optimization and control algorithms for electric power systems. IEEE Trans. Smart Grid **8**(6), 2941–2962 (2017)

46. C. Moya, W. Zhang, J. Lian, K. Kalsi, A hierarchical framework for demand-side frequency control, in *Proceedings of 2014 American Control Conference* (Portland, OR, USA, 2014), pp. 52–57

47. A. Olshevsky, Linear time average consensus on fixed graphs and implications for decentralized optimization and multi-agent control (2014). arXiv:14114186

48. P. Palensky, D. Dietrich, Demand side management: Demand response, intelligent energy systems, and smart loads. IEEE Trans. Industr. Inf. **7**(3), 381–388 (2011). https://doi.org/10.1109/TII.2011.2158841

49. J.A. Peças Lopes, C.L. Moreira, A.G. Madureira, Defining control strategies for microgrids islanded operation. IEEE Trans. Power Syst. **21**(2), 916–924 (2006). https://doi.org/10.1109/TPWRS.2006.873018

50. M.A.A. Pedrasa, T.D. Spooner, I.F. MacGill, Coordinated scheduling of residential distributed energy resources to optimize smart home energy services. IEEE Trans. Smart Grid **1**(2), 134–143 (2010). https://doi.org/10.1109/TSG.2010.2053053

51. P. Piagi, R.H. Lasseter, Autonomous control of microgrids, in *Proceedings IEEE Power Engineering and Society General Meeting*, p. 8 (2006). https://doi.org/10.1109/PES.2006.1708993

52. Z. Qu, M.A. Simaan, Modularized design for cooperative control and plug-and-play operation of networked heterogeneous systems. Automatica **50**(9), 2405–2414 (2014)

53. F. Rahimi, A. Ipakchi, Demand response as a market resource under the smart grid paradigm. IEEE Trans. Smart Grid **1**(1), 82–88 (2010). https://doi.org/10.1109/TSG.2010.2045906

54. J. Rocabert, A. Luna, F. Blaabjerg, P. Rodrguez, Control of power converters in ac microgrids. IEEE Trans. Power Electr. **27**(11), 4734–4749 (2012)

55. J. Schiffer, R. Ortega, A. Astolfi, J. Raisch, T. Sezi, Conditions for stability of droop-controlled inverter-based microgrids. Automatica **50**, 2457–2469 (2014)

56. A.M. Leite da Silva, L.C. Nascimento, M.A. da Rosa, D. Issicaba, J.A. Peças Lopes, Distributed energy resources impact on distribution system reliability under load transfer restrictions. IEEE Trans. Smart Grid **3**(4), 2048–2055 (2012). https://doi.org/10.1109/TSG.2012.2190997

57. J.W. Simpson-Porco, F. Dörfler, F. Bullo, Synchronization and power sharing for droop-controlled inverters in islanded microgrids. Automatica **49**(9), 2603–2611 (2013)

58. K. Subbarao, J. Fuller, K. Kalsi, A. Somani, R. Pratt, S. Widergren, D. Chassin, Transactive control and coordination of distributed assets for ancillary services. Technical report (Pacific Northwest National Laboratory (PNNL), Richland, WA, US, 2013)

59. K. Subbarao, J. Fuller, K. Kalsi, J. Lian, E. Mayhorn, Transactive control and coordination of distributed assets for ancillary services: controls, markets and simulations. Tech. Rep. PNNL-23764 (Pacific Northwest National Laboratory, 2015)

60. US Department of Energy, The potential benefits of distributed generation and rate-related issues that may impede their expansion (2007)
61. S.E. Widergren, K. Subbarao, J.C. Fuller, D.P. Chassin, et al., AEP Ohio gridSMART® demonstration project real-time pricing demonstration analysis. Tech. Rep. PNNL-23192, (Pacific Northwest National Laboratory, 2014)
62. T. Williams, K. Kalsi, M. Elizondo, L. Marinovici, R. Pratt, Control and coordination of frequency responsive residential water heaters, in *Proceedings of, IEEE Power and Energy Society General Meeting* (MA, USA, Boston, 2016)
63. D. Wu, T. Yang, A.A. Stoorvogel, J. Stoustrup, Distributed optimal coordination for distributed energy resources in power systems. IEEE Trans. Auto. Sci. Eng. **14**(2), 414–424 (2017). https://doi.org/10.1109/TASE.2016.2627006

Virtual Energy Storage from Flexible Loads: Distributed Control with QoS Constraints

Prabir Barooah

Abstract Loads are expected to help the power grid of the future in balancing the highs and lows caused by intermittent renewables such as solar and wind. With appropriate intelligence, loads will be able manipulate demand around a nominal baseline so that the increase and decrease of demand appears like charging and discharging of a battery, thereby creating a virtual energy storage (VES) device. An important question for the control systems community is: how to control these flexible loads so that the apparently conflicting goal of maintaining consumers' quality of service (QoS) and providing reliable grid support are achieved? We advocate a frequency domain thinking of handling both of these issues, along the lines of a recent paper. In this article, we discuss some of the challenges and opportunities in designing appropriate control algorithms and coordination architectures in obtaining reliable VES from flexible loads.

1 Introduction

A future power grid is likely to experience significant intermittency in generation from renewable sources such as solar and wind. This intermittency is illustrated in Fig. 1; the data comes from BPA (http://www.bpa.org), a balancing authority (BA) in the Pacific Northwest. The net demand, which is the difference between demand for power and renewable power generated, must be supplied by controllable generation resources. The sharp ramps and fast variations in the net demand are a cause of concern for conventional generators. They are not designed to track such a fast varying signal. Inability to track the net demand can seriously degrade reliability of the power grid: if demand–supply imbalance becomes too large, the grid frequency deviates far from the nominal value of 60 Hz, and cascading blackouts can occur.

Additional resources are needed to mitigate the volatility created by solar and wind. One possibility is to employ sufficient standby generation that can ramp up and down quickly, such as hydro and gas. Hydro is limited by geography, while the

P. Barooah (✉)
University of Florida, Gainesville, FL, USA
e-mail: pbarooah@ufl.edu

© Springer Nature Switzerland AG 2019
J. Stoustrup et al. (eds.), *Smart Grid Control*, Power Electronics
and Power Systems, https://doi.org/10.1007/978-3-319-98310-3_6

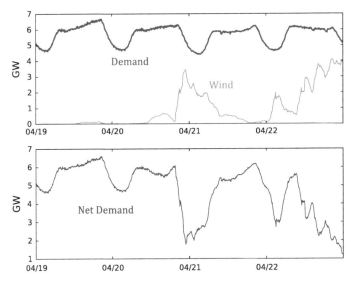

Fig. 1 (Top) Total demand and renewable generation, and (bottom) net demand in BPA (Bonneville Power Administration: http://www.bpa.gov), April 19–22, 2016

use of additional fossil plants as backup will negate the environmental benefits of renewables, apart from increasing the overall cost of energy. The business case for the power plant owners is also questionable since the plants would not sell much energy, which is already causing a few power plants to close [1]. Another possibility is to employ sufficient energy storage resources such as batteries, flywheels, pumped hydro, and compressed air systems. At present, this is a prohibitively expensive option. We discuss the cost of batteries in Sect. 4. The third possibility is to equip loads with intelligence so that their demand can be varied in such a way that mismatch between demand and generation is reduced. In fact, with the help of appropriate control algorithms, loads that have some flexibility in their power demand can be made to provide the same service as that of a battery. We call this virtual energy storage (VES) from flexible loads; see Fig. 2 for a schematic. This is to be contrasted with real energy storage (RES), which include batteries, pumped hydro, flywheels, compressed air, etc.

This paper describes some of the technological challenges and opportunities in obtaining VES from flexible loads. Any technological solution to obtaining grid support from loads must consider its effect on consumers. After all, all loads are used by consumers to provide a certain function, and they have certain expectation of the quality of service (QoS) from those loads.

There is a fast-growing literature on the control of flexible loads to provide grid support services. A dominant paradigm in this literature is control and coordination of loads through real-time prices of electricity, or some other market-based mechanism; see [2, 3] and references therein. These viewpoints have several weaknesses. One, real-time prices subject consumers to high levels of risk. Real- time prices of

Fig. 2 Virtual energy storage (VES) from flexible loads: demand is varied around a baseline with the help of a control algorithm so that the demand deviation from the baseline is akin to the charging and discharging of a battery

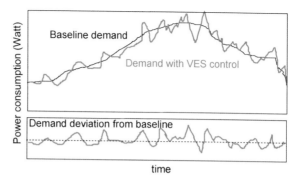

electricity are volatile even without high penetration of intermittent renewables; see [4] for examples from around the world. In fact, [4] shows that these volatilities persist even in an idealized market with participants having no market power ("price takers"), but occur purely as a result of uncertainty and ramp rate constraints. Two, they require consumers to assign a dollar value to a change in consumption with an uncertain QoS loss, e.g., "how much payment is adequate to compensate for a 1 kW decrease in power consumption?", such as in [3]. However, the answer to this question is likely to change frequently for the same consumer, depending on the context (during a party, after a workout session), and also depending on how long the loss of QoS will have to be endured. More recent work on market-based "demand response" has sought to address some of these issues by moving away from real-time prices; but using price as a coordination signal meant to help reach an equilibrium; such as [5]. However, these works also require complex information, such as specification of utility functions (utility of consumers as a function of consumption). If deployed at scale, market-based mechanisms may not lead to a reliable service that grid operators can rely on.

Evidence from existing demand response programs indicate that long- term contracts reduce the risk to consumers while providing a more reliable service to the balancing authorities. Florida Power & Light has 760,000 residential consumers enrolled in their *On Call* demand response program [6]. In return for a monthly rebate, these consumers allow FP&L to turn off their pool pumps and air conditioners a few times in a year. This program has been in place for more than a decade, and has been effective since consumers are getting a reliable return for a known loss of QoS. We, therefore, argue that a control architecture based on long-term contracts between consumers and BAs, with negotiated QoS bounds, offer a reliable consumer engagement. The control system must ensure that QoS never deviates outside of the pre-negotiated bounds. Although the rest of the paper is not dependent on long-term contracts being the only form of payment, we use that assumption to remove market considerations.

It was argued in [7] that Fourier decomposition provides a convenient framework to assign grid's needs to all supply side resources, including traditional generators, loads providing VES service, and batteries providing RES services. In this paper, we further explore the frequency domain thinking. We emphasize that current grid operation and planning already is based on a similar framework, by breaking down the requirements by timescale. Base-load power generation is scheduled based on predictions of the net demand at the slowest timescale (lowest frequency), load following and frequency regulation at intermediate and fast time scales is performed by automatic generation control that adjusts generation set points [8]. However, current taxonomy of generation-side services, such as "frequency regulation" and "load following" are inadequate in a renewable-rich power grid. In the future, "renewable following" may be as important a service as load following. Therefore, we avoid using that taxonomy in the paper.

The rest of the paper is structured as follows. Section 2 describes the VES idea in detail, and summarizes the main challenges in developing local control algorithms for a load to deliver VES to the grid with guaranteed bounds on its QoS. Section 3 discusses the challenges in developing architectures for distributed coordination of millions of loads to meet the VES service needed by the grid. Section 4 discusses cost of battery-alone storage and what it means for cost targets of VES technology.

2 Virtual Energy Storage from Flexible Loads

A load's power consumption can be varied around a baseline to provide a battery-like service. Let $p_b(t)$ be the baseline power demand of a load (or a collection of loads). Suppose its (their) demand is varied through the use of appropriate control software to be $p(t)$ so that the demand deviation from the baseline:

$$p_{\text{ves}}(t) := p(t) - p_b(t) \tag{1}$$

is zero mean: $\lim_{T \to 0} \frac{1}{T} \int_0^T p_{\text{ves}}(t)dt = 0$; cf. Fig. 2. We can then say that the load is providing VES, or, that it is acting like a virtual battery. The demand deviation $p_{\text{ves}}(t)$ is the charging power consumption of the virtual battery. Positive $p_{\text{ves}}(t)$ means the load is drawing more power from the grid than what it would have under baseline conditions; so the virtual battery is charging. Conversely, negative $p_{\text{ves}}(t)$ means it is discharging. The zero-mean nature of the demand deviation means the net energy consumption/generation of the virtual battery is 0, just like a real battery.

Two questions arise:

- For a specific load and a bound on change of its QoS, what kind of demand deviation ("virtual charge/discharge signal") $p_{\text{ves}}(t)$ is allowable that ensures the QoS bound is satisfied? And, how does this vary from load to load?
- How is the net demand signal to be apportioned among the loads so that together they can supply it, while each load maintains its QoS bound?

2.1   Constraining Loss of QoS via Constraining Bandwidth

QoS measures vary depending on load type. There are a large variety of flexible loads, such as refrigeration systems, electric vehicles, pool pumps, water heaters, data centers, municipal pumping systems, HVAC systems, etc. Each has their own QoS metrics, and a distinct degree of flexibility. For HVAC, measures of QoS include indoor temperature and ventilation rate (as a surrogate for indoor air quality). Hot water heaters—and pool pumps in some areas—are also large sources of demand. A QoS measure for a pool pump is the average number of hours the pump is on (as a surrogate for water cleanliness) [9, 10]. For hot water heaters, it is the availability of hot water that is critical. For an aluminum plant, a measure of QoS is the temperature of the smelter [11]. For all loads, whether commercial, residential or industrial, QoS metrics include the cost of energy used[1] and equipment lifetime.

The diversity of QoS metrics among distinct load types is a challenge in developing control algorithms to exploit their demand flexibility. We argue that, in fact, a unifying framework can be developed based on the spectral content of the demand variation, a viewpoint first expounded in [7]. For every load type, maintaining a specific bound on the QoS can be translated to maintaining a bound on the *bandwidth* of its demand deviation. For instance, a small and fast variation of power consumption of a commercial HVAC system can be obtained by a small and fast variation of airflow. The resulting temperature deviations will be small since the large thermal inertia of the building will act like a low-pass filter to such airflow variations. However, even a small amplitude airflow variation can lead to large deviation in indoor temperature if the variation persists for a long time, i.e., the frequency is small enough. For a given amplitude, the higher the frequency of airflow variation, the smaller the effect on QoS metrics of indoor temperature and average ventilation rate. However, above a certain frequency, QoS will reduce since equipment life will degrade. Figure 3 illustrates this idea. For loads that can only be turned on or off, such as hot water heaters, again limiting the frequency of turning on and off is needed to reduce short-cycling and ensure delivery of hot water.

In essence, the VES capacity of a load can be characterized in terms of the power spectral density (PSD) $P_{ves}(\omega)$ of the demand variation, $p_{ves}(t)$. The PSD must lie in a specific region to meet a given QoS constraint, which can be parameterized by, say, a scalar q. For every value of q, there is a curve $c_q(\omega)$ so that that QoS will be respected only if the PSD of p_{ves} lies under the curve $c_q(\omega)$. *The curve corresponding to the minimum acceptable QoS q^* determines the load's VES capacity. We call $c_{q^*}(\omega)$ the load's capacity curve.*

An illustration of the curve $c_q(\omega)$, for some q, is shown in Fig. 3. For a specific load, or load class, determination of the curve $c_q(\omega)$ can be determined either through modeling or experimental evaluation [12].

Challenges and opportunities A weakness of the frequency domain characterization of VES capacity is that variations over time, especially due to exogenous factors such

---

[1] For some large consumers, "utility bill" is a better measure since their peak demand charges may constitute a large part of the bill.

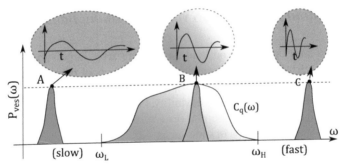

Fig. 3 Constraint on QoS is a constraint on bandwidth of demand variation. The x-axis is frequency and the y-axis is the PSD of demand variation. The PSD must lie in the region under the curve $c_q(\omega)$ to meet the QoS measure q. For a different value of q, this curve would change. The low and high limits of the frequency in which this particular load class can provide VES service are denoted as ω_L and ω_H, respectively. The three signals shown in A, B, and C, have PSDs that have the same total power (i.e., the integral of their PSDs are the same), but distinct bandwidths. The signals A and C violate the QoS metric q, because their bandwidths are too low and too high, respectively. The signal B satisfies the bandwidth requirement

as weather are not conveniently captured. For instance, during afternoon hours of very hot days, an HVAC system may have to run at peak power, and in that case a zero-mean deviation from the baseline is not possible. An alternate way of quantifying capacity that has been explored is a time-varying range (upper and lower bound) of total power consumption so that as long as power consumption stays within that bound, QoS metrics will be satisfied [13, 14]. These approaches necessarily lead to conservative estimates since a constant power deviation from a baseline that still maintains QoS constraints must be allowed in this framework. A general framework that combines the advantage of frequency-based characterization, but is capable of modeling the effect of exogenous factors on VES capacity is still lacking.

Another challenge in this approach is its dependence on baseline for its definition. The baseline is not possible to measure if a load is providing VES services, only the total power is, leading to the issues of estimating the baseline and associated estimation errors [14, 15].

2.2   Matching VES Resources to Grid's Needs

The grid needs controllable resources to meet the net demand. The net demand[2] $p_d(t)$ at time t is defined as

$$p_d(t) := p_b(t) - g_r(t) \tag{2}$$

---

[2]Usually called net-load, but we avoid that term since "load" in this paper refers to physical entities that consume power.

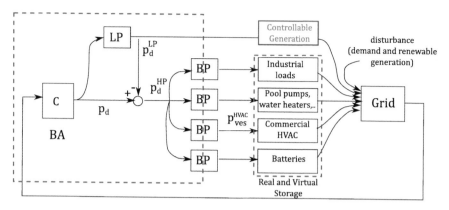

Fig. 4 A potential control architecture for the smart grid with VES based on spectral decomposition. The "Grid" block represents everything other than controllable generation and storage resources, such as loads (baseline), transmission and distribution networks, etc.

where $p_b(t)$ is the *baseline* power demand (in MW) in the grid and $g_r(t)$ is the uncontrollable renewable generation (in MW). The word baseline refers to the nominal demand from all loads, when loads are operated without employing any of the algorithms designed to extract flexibility. The net demand is the signal the grid's remaining resources will have to provide, which include traditional generators, flexible loads providing VES, and other energy storage (ES) devices such as pumped hydro, flywheels, and batteries.

How to ensure that available resources together supply the total needs of the grid, i.e., how do they together track the net demand? Our approach is based on a spectral decomposition of the net demand into distinct frequency bands, by passing it through a number of bandpass filters, as shown in Fig. 4. The "C" block at the BA computes/predicts the net demand p_d, which serves as a reference command to the aggregate controllable resources in the grid. Its low-pass component, $p_d^{LP}(t)$, is obtained by passing p_d through a low-pass filter ("LP" in Fig. 4). As long as the low-pass filter LP is designed by keeping the ramping abilities of the controllable generators in mind, the bandwidth of the signal $p_d^{LP}(t)$ will be low enough that controllable generators will be able to track it. The remaining high-pass component of the net demand is $p_d^{HP}(t) := p_d(t) - p_d^{LP}(t)$, which is *zero mean*. Because of the zero-mean property, $p_d^{HP}(t)$ can be tracked by controllable storage resources (whether real or virtual), by charging when $p_d^{HP}(t)$ is positive and discharging when $p_d^{HP}(t)$ is negative. The bandpass filters (BPs in Fig. 4) can be located either in a centralized manner at the BA, or in a distributed manner at the resources, or in some combination thereof, depending on the control architecture chosen.

To match to resources of appropriate ability, the zero-mean component of the net demand is passed through a number of bandpass filters to create reference signals for various energy storage resources: the "BP"s in Fig. 4. Each of the reference signals is band-limited to a particular frequency band that is suitable for a distinct class of resource. For instance, the highest frequency component of the net demand can be

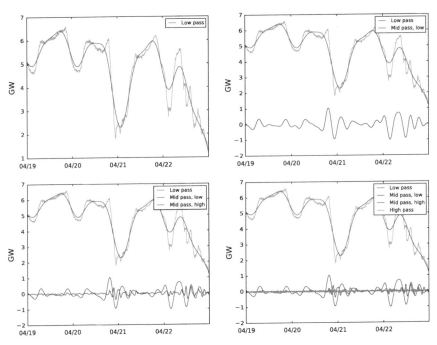

Fig. 5 Frequency decomposition of the net demand of Fig. 1: each bandpass- filtered component is a reference for a distinct class of resource that is appropriate for that frequency band

the reference signal for batteries, while the one with a slightly lower frequency can be the reference for HVAC loads providing VES. The sum of all these reference signals is the net demand. Thus, the needs of the grid are met, and yet no resource (including a conventional generator and a battery) is asked to provide a service that is not appropriate for it. Figure 5 shows an example of the frequency decomposition of the net demand based on data from BPA.

Challenges and opportunities

- *VES capacity characterization*: Based on experiments in a commercial building in the University of Florida reported in [12], we know that variable speed fans in HVAC systems can provide VES service in the frequency range of [1/(10 min) 1/(1 min)] and up to 30% of their average power without any perceptible change in indoor climate. Simulations with calibrated models show that with both chillers and fans engaged, HVAC systems can provide VES service in a slower frequency range of [1/(1 hr)]; [1/(10 min)] and up to 50% of its rated power, with an indoor temperature deviation of 2 °C [15]. Collection of pool pumps can provide VES in lower frequencies of hours [10], and so can residential air conditioners and heat pumps [16]. Industrial loads may be able to provide much lower frequency VES—than, say, HVAC—by deferring production in a timescale of days or weeks.

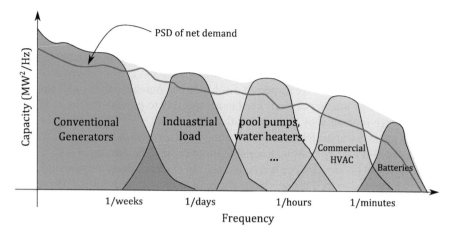

Fig. 6 Power spectral density of the net demand versus total capacity of the grid's resources

An important open question is to provide a complete characterization of the VES capacity of various classes of loads—especially industrial and residential loads—as a function of QoS within the frequency domain framework introduced here. Even for HVAC, which has been more thoroughly examined, VES capacity is likely to vary depending on the thermal load it experiences. A purely frequency domain framework may not be suited to characterize these variations [14].

Information on VES capacity as a function of QoS constraints is essential for loads to enter into contractual agreements with BAs. The appropriate payment structure is not clear yet, but at the simplest form it can be a fixed monthly payment depending on the load's QoS bound q. For more sophisticated loads such as industrial loads or large commercial HVAC, the payment can also consist of a "milage" payment depending on the actual VES service the load provided [17].

- *Ensuring resource adequacy*: The combined capacity of various resources (generation, VES, and RES resources) must be larger than the net demand. The grid's needs can again be quantified by the PSD of the net demand. Figure 6 illustrates a hypothetical scenario in which resources are adequate: the capacity curves of each category of resources—limited to various frequency bands due to QoS constraints—including conventional generators, VES resources, and RES resources, together cover the PSD at all frequencies. In this case, we can say that adequate resources exist.

- *Optimal allocation of VES and battery storage*: The cost of various types of VES resources are likely to be distinct. How much of each kind should a BA recruit to meet its requirements with sufficient margin at the minimum cost? Methodologies for answering such questions are essential to the BAs for planning purposes. Currently, a bottleneck in answering this question is the lack of estimates of VES cost. Section 4 discusses cost of battery storage that provides an upper bound for allowable cost of VES before VES becomes noncompetitive with battery-based energy storage.

3   Coordination of Loads to Obtain Required VES

To obtain VES service without violating QoS constraints, a two-tier strategy is
required: local control and coordinated control. The local controller ensures that
the load's QoS constraints are respected. Each load can provide only a small amount
of VES, so a large number of loads need to act together to provide the desired VES
service, which is ensured by the coordination algorithm.

In this paper, we consider *loads with continuously variable demand (LCVD)* such
as commercial HVAC systems with variable speed drives. The demand of such a load
can be varied to be any number within a range. In contrast, many residential loads
can only be turned on or off; their demand cannot be continuously varied. However,
if a load aggregator is used, the aggregator becomes a LCVD from the BA's point of
view even if the all loads managed by the aggregator are on/off type [16].[3]

3.1   Local Intelligence

Here, the task is for the power demand deviation (from the baseline) $p_{ves}(t)$ of a load
to track an external reference. The external reference must satisfy the bandwidth
constraint described in Sect. 2 to maintain QoS, which can be ensured by locally
bandpass filtering a grid-level reference.

Challenges and opportunities

- *Baseline uncertainty*: The challenges in designing the local intelligence to ensure
  tracking is measuring the output, the power deviation from the baseline, since the
  baseline, by definition, cannot be measured. In [12], this challenge was addressed
  by exploiting timescale separation between the VES reference to be tracked and the
  baseline. Since the baseline power consumption is dictated by the normal climate
  control system, it is of lower frequency than the high-frequency VES reference
  the system was designed to track. As a result, the baseline can be recovered by
  low-pass filtering the power consumption measurement.
  When the VES reference signal is of the same timescale as the baseline, the problem
  of separating the baseline becomes quite challenging. In our prior work [15] as well
  as in [13, 14], the baseline was prespecified by solving an optimization problem
  that ensured QoS (indoor climate) constraints were satisfied. The local controller
  was then tasked with tracking the total power: baseline plus VES reference.
- *Continuously variable demand from on/off actuators*: Chillers in commercial
  buildings are a much bigger load than fans, but they are predominantly on/off
  actuators, since their motors do not have variable speed drives. It is still possible
  to vary their power demand continuously in a range by indirect means, such as

---

[3]The problem of controlling an aggregate of on/offloads so that the power consumption of the
collective tracks a smooth signal while respecting every load's QoS constraints has a different set
of challenges that we do not go into in this paper; see [10].

airflow rate, due to the inlet guide vane controls. However, models of appropriate complexity that can be used to design and study local controllers for such equipment are lacking. Existing dynamic models of chillers are too complex for control design; e.g., [18]. A similar issue exists for packaged air conditioning units used in small commercial buildings, which may have variable speed fans but constant speed compressor motors. For chillers, especially larger ones, avoiding short-cycling is a key QoS requirement.

- *Round trip efficiency*: For thermal loads such as air conditioners, it is not clear if there is a loss of efficiency in varying their demand over a baseline instead of running them at their baseline. In other words, what is the "round trip efficiency" of the virtual battery? Work in this direction is preliminary [19].

3.2 Coordination

How does one break up the grid-level reference signal among many LCVD, each with its own QoS constraints? For the purpose of exposition, let us limit our attention to one particular frequency band, say, the component—p_{ves}^{HVAC} in Fig. 4—that will be supplied by commercial HVAC systems.

One possibility is for the grid to broadcast p_{ves}^{HVAC} and each load locally bandpass filters it to compute its own VES reference signal. This architecture is shown in Fig. 7: the goal is to ensure $y(t) = r(t)$, where $r(t)$ is the grid-supplied reference signal for demand deviation. The bandpass filter $F_i(s)$ at load i has to be designed so that load i's QoS is satisfied and the grid-level tracking goal, $y = r$, is also satisfied. Load i's QoS will be satisfied if the PSD of its local reference signal lies within its capacity curve $c_i(\omega)$. Recall that capacity curve was defined in Sect. 2.1. Note that if $P_{ves}^{HVAC}(\omega)$ is the PSD of the grid-level reference signal $p_{ves}^{HVAC}(t)$, then the PSD of the i-th load's local reference is $|F_i(j\omega)|^2 P_{ves}^{HVAC}(\omega)$. The CL_i block in Fig. 7 represents the closed-loop system consisting of a load and its local intelligence that

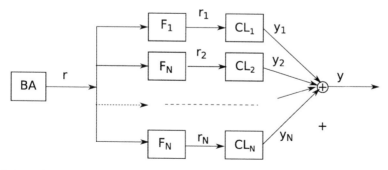

Fig. 7 Part of an open-loop coordination architecture for an aggregate of LCVDs to track a grid-level VES reference. Only the forward path between the BA and the loads are shown; the outer loop feedback between the "Grid" block of Fig. 4 and the BA is omitted

can track a reference signal for its demand deviation. The load belonging to the LCVD class is crucial; only such a load can track a reference other than a square-wave. Assuming the local intelligence at each load i is such that it tracks the local reference signal $r_i(t)$ perfectly, i.e., $y_i(t) = r_i(t)$, the equation $\sum_i r_i(t) = r(t)$ must be satisfied for the grid-level tracking goal to be satisfied. That is, if there are N loads supplying VES in the "high pass" category, then the following must hold to ensure that the loads together track the grid-level reference:

$$\sum_i^N F_i(j\omega) = 1, \quad \omega \in [\omega_L^{(HP)} \ \ \omega_H^{(HP)}]. \tag{3}$$

When the grid operator enters into an agreement with a load to obtain VES resource, it obtains the load's VES capacity curve $c_i(\omega)$, either through modeling or through a system identification test. The local bandpass filter F_i is mutually agreed upon at that time. The grid operator must engage enough loads to ensure that (3) holds.

Even though this architecture satisfies the needs of both the grid and the loads, it lacks robustness to uncertainty due to its open-loop nature. There are many sources of uncertainty: the number of loads providing service at any given time, the capacity of some of the loads, etc., are all likely to vary over time in less-than predictable manner.

An alternate, more robust, architecture using feedback is proposed in [20], in which load coordinate their actions by using a global feedback signal that can be measured locally. Figure 8 shows this architecture. In particular, each load measures the grid frequency, which can be locally measured at loads [21, 22]. Since the deviation of the grid frequency from its nominal value (60 Hz) is a measure of demand–supply mismatch, it can estimate the demand–supply mismatch from this measurement. Since total supply is conventional plus renewable generation, the demand–supply mismatch—total demand minus total supply—is precisely the net demand minus conventional generation, so it is the zero-mean component of the net demand after the low-pass component is removed. The load computes the appropriate VES reference for itself by passing the estimated demand–supply imbalance with its local bandpass filter.

The control algorithm proposed in [20] goes one step further, and assumes that the BA broadcasts a prediction (for the next hour) of the demand–supply imbalance. The BA is in a unique position to predict this signal, since it has statistical models to predict grid-level baseline demand $d_b(t)$ and renewable generation $g_r(t)$, and it can predict the power generation by conventional generators $g_c(t)$ based on the contracts in place. The VES controller at each load uses an MPC scheme to compute appropriate power deviation (VES reference) subject to a QoS constraint expressed in terms of the Fourier transform of its local reference. High gain feedback due to the actions of other loads is avoided by estimating the VES supplied by other loads from the estimating the grid-level demand–supply imbalance and its own VES signal. The grid-level demand–supply imbalance is estimated from locally measured grid frequency.

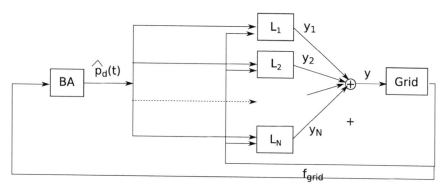

Fig. 8 A potential control architecture for coordination among VES resources by using local feedback (on grid frequency f_{grid}) and broadcast from BA on the predicted demand–supply imbalance $\hat{p}_d$

An extremely simplified schematic representation of this architecture, with N loads $L_1, L_2, \ldots, L_N$, is shown in Fig. 8. The goal is not for the aggregate response y to track some BA-supplied reference. Rather, it is to determine y_i's so that the aggregate response y minimizes demand-generation mismatch and each y_i satisfies the QoS constraint of load i.

The advantage of this architecture is that it is much more robust to uncertainty in how many loads are providing VES service at a given time and what their capacities are. In addition, distributed coordination among loads is achieved without any sort of inter-load communication. Only one-way broadcast from the BA to the loads is needed. Simulation studies reported in [20] shows the architecture is effective in providing robust tracking in presence of uncertainty.

Resource adequacy can be ensured by the BA by signing enough contracts so that the following holds:

$$|\sum_{i}^{N} c_i(j\omega)| > 1, \quad \omega \in [\omega_L^{(HP)} \ \omega_H^{(HP)}], \tag{4}$$

where $c_i(\omega)$ is the capacity curve of the ith load. The subscript q^* in $c_{q^*}(\omega)$, which was used in defining the capacity curve in Sect. 2.1 is suppressed here to avoid clutter. The advantage is that the inequality (4) is far easier to ensure than the equality (3), especially when a large number of loads are involved.

Challenges and opportunities

- *Communication architecture*: A large body of literature exist on distributed control, and the architectures discussed above are not the only possible ones. Most of the distributed coordination architectures proposed in the literature rely on inter-agent communication *within a neighborhood* for meeting network-wide goals. With the recent push toward an Internet of Things (IoT) paradigm, it is likely that smart loads will be part of the IoT. In that case, it is not clear what an appropriate notion of neighborhood is. All to all communication may be infeasible, but there is no rationale for limiting to a geographically defined neighborhood. Communicating with very far off (in a geographic sense) agents may be possible over the Internet. That may help with certain performance metrics, but may introduce larger delays. Determining these tradeoffs for distributed control in the age of IoT remains an important open question, one that is particularly relevant to the smart grid.

- *Contract/mechanism design*: A load may not provide the maximum capacity that was used at the time of signing contracts. That may not be malicious; if all of them provide maximum capacity at all times that may, in fact, cause demand–supply mismatch. If some loads bear a much larger share of the burden of required storage, it is reasonable they should be incentivized more than others. It is not clear what is an appropriate incentive to loads providing VES in such a scenario. Currently, generators in many ISOs are paid based on a two-part scheme based on capacity and mileage, but such a scheme may not be scalable to millions of loads.

- *Characterizing loads on-line*: The capacity of a load needs to be known to ensure that the loads together have enough bandwidth to track the reference. This can be done through a system identification experiment, as was done in [12] for the fan motor of an HVAC system. However, such a method may not be scalable to a large number of loads, and it may fail to identify slow variations in load's VES capacity over long time periods. Is it possible for the BA to be sure—without examining every single load—that the loads together have enough capacity to meets its need?

4   Cost

Without a cost advantage over real energy storage, virtual energy storage has little justification. Cost of VES is hard to estimate. On one hand, VES involves a change of software, with little change in hardware. Yet, the cost of large-scale deployment of VES may vary a lot depending on the kind of communication infrastructure and hardware retrofits needed. Cost of retrofitting existing consumer loads to make them VES-friendly is likely to be prohibitive, but it is equally likely that the additional cost of equipping loads with the required hardware and software at the factory will be negligible. However, precise estimates are lacking at this point.

Although the cost of VES may be hard to estimate at this point, we can establish an upper bound on the cost of VES beyond which VES loses its economic advantage. This upper bound is the minimum cost of the main competitor of VES, that of battery storage.

To estimate the cost of battery-based storage, we examine how the levelized cost of electricity (LCOE) will change if a battery is used to store the average daily generation of energy from an intermittent renewable source, say solar. The LCOE is the total cost incurred in the lifetime of the generator, divided by the total energy generated over the same period.

Consider a renewable generator with peak generation capacity 1 kW. Suppose the capacity factor of the generator is f, so that the average energy it produces in a day is $24f$ kWh. Let the lifespan of the generator be t_{lg} years. The total energy generated by the generator in its lifespan is $365t_{lg}24f$ kWh.

Suppose a battery is added to the generator so that it can store the average daily energy produced. That is, the energy capacity of the battery is $24f$ kWh. Let the lifetime of the battery be t_{lb} years, and its cost be c \$/kWh. Then, the cost of batteries over the life of the renewable generator is $24fct_{lg}/t_{lb}$ \$.

Since adding a battery does not change the energy generated, the *additional* LCOE due to the battery is the total cost of battery over the lifetime of the generator divided by the total energy generated during the same period:

$$\Delta\text{LCOE}_{\text{battery}} = \frac{24fc\frac{t_{lg}}{t_{lb}}}{365t_{lg}24f} = \frac{c}{365t_{lb}}. \quad (\$/\text{kWh}) \qquad (5)$$

Among the myriad types of batteries, Sodium Sulfur (NaS) batteries have had a lead in terms of grid storage, but the cost of Li-ion batteries—used in mobile phones and electric cars—is decreasing the fastest: at an annual rate of approximately 14% per year during 2009–2014 [23]. The cheapest Li-ion batteries in 2015 cost about \$300/kWh (batteries used in Tesla's model S electric car [23]), and they have a lifetime of approximately 5000 charge–discharge cycles [24]. If the battery undergoes one charge–discharge cycle every day, its lifespan will be $5000/365 = 13.7$ years.

Plugging $c = 300$ and $t_{lb} = 13.7$, we see that the additional LCOE due to batteries is ≈ 6 ¢/kWh. Since several important costs are ignored here, especially the cost of balance of systems and the cost of capital, the true cost will be higher than this estimate. A more thorough cost estimate can be performed using the methodology in [24]. Even this low estimate of battery cost is quite high compared to the mean retail electricity rate in the U.S., which in December 2016 was 12.2 ¢/kWh (from https://www.eia.gov/electricity/). If we take the estimate, 6 ¢/kWh, as the true cost of battery storage, the cost of VES must be less than 6 ¢/kWh for it be competitive with battery-based energy storage.

In comparing batteries with VES, one should keep in mind that battery-based energy storage is likely to be much more reliable than VES. Availability of VES may depend on time of day, weather, etc., while batteries are a firm resource. Therefore, an optimal solution will probably consist of expensive but highly reliable batteries as well as inexpensive but less reliable VES.

5   Summary

Loads can vary their power around a baseline in a zero-mean fashion to effectively act like batteries, thereby providing virtual energy storage (VES) to help the grid. A frequency domain framework for characterizing loads flexibility vis-a-vis consumer's QoS is advocated, following [7]. The framework is powerful enough to handle not just flexible loads but also conventional generators and batteries. However, it is highly simplified: issues of transmission constraints, distribution network and voltage support, contingency reserves are not considered yet, which are worthwhile avenues for further refinement. Some results on local control and distributed coordination of loads within this framework, are mentioned. Challenges and opportunities in extending this framework to design reliable VES services, including some of the open problems, are summarized.

Acknowledgements Financial support from NSF (through grant 1646229) and DOE (through a BTO-GMLC project titled *virtual batteries*) is gratefully acknowledged.

References

1. G. Parkinson, UBS: closures coal and gas fired power plants in Europe accelerating (2015). http://www.energypost.eu/ubs-closures-coal-gas-fired-power-plants-europe-accelerating/. Accessed 4 May 2015
2. J.H. Yoon, R. Baldick, A. Novoselac, Dynamic demand response controller based on real-time retail price for residential buildings. IEEE Trans. Smart Grid **5**(1), 121–129 (2014)
3. D.J. Hammerstrom, R. Ambrosio, J. Brous, T.A. Carlon, D.P. Chassin, J.G. DeSteese, R.T. Guttromson, G.R. Horst, O.M. Jrvegren, R. Kajfasz, S. Katipamula, L. Kiesling, N.T. Le, P. Michie, T.V. Oliver, R.G. Pratt, S. Thompson, M. Yao, Pacific northwest gridwise testbed demonstration projects. Part I. Olympic peninsula project. Pacific Northwest National Laboratory, Tech. Rep. PNNL-17167, Oct 2007
4. G. Wang, M. Negrete-Pincetic, A. Kowli, E. Shafieepoorfard, S. Meyn, U. Shanbhag, Real-time prices in an entropic grid, in *IEEE PES Innovative Smart Grid Technologies (ISGT)*, Jan 2012, pp. 1–8
5. J. Knudsen, J. Hansen, A.M. Annaswamy, A dynamic market mechanism for the integration of renewables and demand response. IEEE Trans. Control Syst. Technol. **24**(3), 940–955 (2016)
6. FPL on call saving program (2013). http://tinyurl.com/k3ldwe3
7. P. Barooah, A. Bušić, S. Meyn, Spectral decomposition of demand side flexibility for reliable ancillary service in a smart grid, in *48th Hawaii International Conference on Systems Science*, Jan 2015 (invited paper)
8. B. Kirby, Ancillary services: technical and commercial insights (prepared for Wärtsilä North America Inc, 2007)
9. A. Nayyar, M. Negrete-Pincetic, K. Poolla, P. Varaiya, Duration-differentiated energy services with a continuum of loads. IEEE Trans. Control Netw. Syst. **3**(2), 182–191 (2016)
10. S. Meyn, P. Barooah, A. Bušić, Y. Chen, J. Ehren, Ancillary service to the grid from intelligent deferrable loads. IEEE Trans. Auto. Control **60**, 2847–2862 (2015)
11. D. Todd, M. Caufield, B. Helms, A. Generating, I. Starke, B. Kirby, J. Kueck, Providing reliability services through demand response: a preliminary evaluation of the demand response capabilities of alcoa inc. ORNL/TM **233** (2008)
12. Y. Lin, P. Barooah, S. Meyn, T. Middelkoop, Experimental evaluation of frequency regulation from commercial building HVAC systems. IEEE Trans. Smart Grid **6**, 776–783 (2015)
13. H. Hao, D. Wu, J. Lian, T. Yang, Optimal coordination of building loads and energy storage for power grid and end user services. IEEE Trans. Smart Grid **PP**(99), 1–1 (2017)
14. E. Vrettos, F. Oldewurtel, G. Andersson, Robust energy-constrained frequency reserves from aggregations of commercial buildings (2015). arXiv:1506.05399v1 [cs.SY]
15. Y. Lin, P. Barooah, J. Mathieu, Ancillary services through demand scheduling and control of commercial buildings. IEEE Trans. Power Syst. (2017)

16. B. Biegel, P. Andersen, J.S. Mathias, B. Madsen, L. Henrik, H. Lotte, H. Rasmussen, Aggregation and control of flexible consumers—a real life demonstration. IFAC, 1–6 (2014)
17. FERC, Frequency Regulation Compensation in the Organized Wholesale Power Markets, FERC 755, Docket Nos. RM11-7-000 and AD10-11-000 , Oct 2011, pp. 1–123
18. S. Bendapudi, J. Braun, E. Groll, A dynamic model of a vapor compression liquid chiller, in *International Refrigeration and Air Conditioning Conference* (2002)
19. I. Beil, I. Hiskens, S. Backhaus, Round-trip efficiency of fast demand response in a large commercial air conditioner. Energy Build. **97**, 47–55 (2015)
20. J. Brooks, P. Barooah, Demand control for maintaining grid frequency without affecting consumers quality of service through decentralized bandwidth-constrained MPC, in *American Control Conference* (2017, accepted)
21. Y. Liu, A US-wide power systems frequency monitoring network, in *Power Engineering Society General Meeting* (2006)
22. P.J. Douglass, R. Garcia-Valle, P. Nyeng, J. Stergaard, M. Togeby, Demand as frequency controlled reserve: implementation and practical demonstration, in *2nd IEEE PES International Conference and Exhibition on Innovative Smart Grid Technologies*, Dec 2011, pp. 1–7
23. B. Nykvist, M. Nilsson, Rapidly falling costs of battery packs for electric vehicles, in *Nature Climate Change*, vol. 5 (2015), pp. 329–332
24. B. Zakeri, S. Syri, Electrical energy storage systems: a comparative life cycle cost analysis. Renew. Sustain. Energy Rev. **42**, 569–596 (2015)

Distributed Design of Smart Grids for Large-Scalability and Evolution

Tomonori Sadamoto, Takayuki Ishizaki and Jun-ichi Imura

Abstract Due to the massive complexity and organizational differences of future power grids, the notion of *distributed design* becomes more significant in a near future. The distributed design is a new notion of system design in which we individually design local subsystems and independently connect each of them to a pre-existing system. In this article, we discuss challenges and opportunities for solving problems of the distributed design of smart grids so that they are flexible to incorporate regional and organizational differences, resilient to undesirable incidents, and able to facilitate addition and modifications of grid components.

Keywords Distributed design · Controllability · Interoperability
Resiliency · Power system evolution · Plug-and-play capability

1 Introduction

Toward the realization of low-carbon society, activities for the development of smart grids have been growing in the world. Although there is no exact definition of smart grids, anticipated benefits and requirements of smart grids are shown in the report from National Institute of Standards and Technology (NIST) [1]. Following this report, in this article, we focus on the three requirements below.

T. Sadamoto (✉) · T. Ishizaki · J. Imura (✉)
Department of Systems and Control Engineering, School of Engineering,
Tokyo Institute of Technology, and Japan Science and Technology Agency,
CREST, Tokyo, Japan
e-mail: sadamoto@sc.e.titech.ac.jp

T. Ishizaki
e-mail: ishizaki@sc.e.titech.ac.jp

J. Imura
e-mail: imura@sc.e.titech.ac.jp

© Springer Nature Switzerland AG 2019
J. Stoustrup et al. (eds.), *Smart Grid Control*, Power Electronics
and Power Systems, https://doi.org/10.1007/978-3-319-98310-3_7

1. Smart grids should be flexible to incorporate regional and organizational differences.
2. Smart grids should be operated resiliently to disturbances, attacks, and natural disasters.
3. Smart grids should facilitate addition and modification of system components.

How should we design smart grids satisfying these requirements? One may consider two approaches: *centralized design* and *distributed design*.

The centralized design is a notion of system design in which we design an overall system from the scratch. Examples of the centralized design include stabilizing controller design based on the entire power system model. The centralized design needs both of the full knowledge of the entire system and powerful authority that can construct the entire system from scratch. However, due to the massive complexity and organizational differences of power grids, this approach is impossible.

The distributed design is another notion of system design in which we individually design local subsystems and independently connect each of them to a preexisting system. Examples of the distributed design include local-controllers design based on partial models of a power system, e.g., single-machine-infinite-bus models in which the machine's behavior depending on grid behavior variation is completely neglected. The notion of this distributed design in the control community has been proposed in [2]. We note here that the distributed design is a different notion of traditional distributed control in the literature [3, 4]. In the distributed control, input signals of local controllers are determined individually, but the controllers are designed jointly, with access to the entire system model. On the other hand, in the distributed design of local controllers, not only the controllers' actuation, but also their design is performed individually. Thus, owing to the distributed nature of the design process, this distributed design approach would suit for the construction of large-scale and complex network systems like power grids [5]. Nevertheless, the distributed design theory has not yet been established, and there exist open problems.

This article aims to summarize challenges and opportunities to fulfill the aforementioned three requirements from a viewpoint of the distributed design. This article is organized as follows.

In Sect. 2, we discuss a problem of the distributed design of local controllers to satisfy requirement 1 for a power system with photovoltaic (PV) integration. First, we show a numerical simulation illustrating how the PV integration has an influence on the centralized and distributed design of controllers. The PV penetration causes reduction of system inertia [6, 7], resulting in the enhancement of the system controllability. As a result, in the distributed design case, the transient instability tends to be induced as the inertia reduction because negative effects of the unmodeled dynamics can be more strongly stimulated by the controllers due to higher controllability. Thus, we need a systematic mechanism enabling us the distributed design of local controllers to accomplish a global objective without causing transient instability of power grids. Next, we discuss challenges and opportunities for solving this issue.

In Sect. 3, we consider a problem to make power grids resilient, described as requirement 2. Following [8], in this article, we define resilient systems as systems that can maintain an acceptable level of operation in face of spatially local undesirable incidents. In this section, first, the significance of resilient system design is shown through numerical simulation of wind-integrated power systems. More specifically, we show that faults at wind farms cause serious oscillation in power flow of the entire power system due to a resonance mode of wind farms. Next, we consider a problem of the distributed design of local controllers so that each of them can make the associated local subsystem resilient, and discuss challenges and opportunities to solve the problem.

In Sect. 4, we discuss a problem to satisfy requirement 3. Examples of this power system evolution include penetration of distributed energy resources (DERs), upgrade of power electronics facilities such as transmission lines, and construction of new power plants. The entire power system must adopt such evolution while maintaining transient stability and performance without any additional configuration of the preexisting power system. However, so far, control theory does not have paid much attention to the characteristics of systems' long-term evolution, but studied evolved system. In this section, we focus on evolution aspects of power grids, and briefly discuss challenges and opportunities for such power grid evolution.

Finally, concluding remarks are provided in Sect. 5.

2 Distributed Design of Local Controllers to Incorporate Regional and Organizational Differences

2.1 Motivating Example

In this subsection, we compare the centralized and distributed design of local controllers for two types of power systems having different system inertia caused by different levels of PV penetration [6, 7].

First, we consider a simple power system example composed of five synchronous generators and two loads without any PV farms, as shown in Fig. 1. The synchronous generators are modeled as the combination of electromechanical swing dynamics

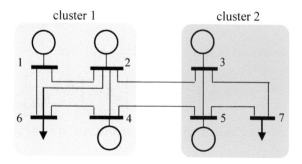

Fig. 1 Power system model composed of five generators and two loads, each of which is denoted by the circles and arrows, respectively

cluster 1   cluster 2

with a second-order governor, and the loads are modeled as dynamical loads whose dynamics are also swing dynamics. For $l \in \{1, 2\}$, we denote the lth cluster dynamics by

$$
\begin{aligned}
\Sigma_{[1]} &: \begin{cases} \dot{x}_{[1]} = A_{[1]}x_{[1]} + A_{[1,2]}x_{[2]} + B_{[1]}u_{[1]} \\ y_{[1]} = C_{[1]}x_{[1]} \end{cases} \\
\Sigma_{[2]} &: \begin{cases} \dot{x}_{[2]} = A_{[2]}x_{[2]} + A_{[2,1]}x_{[1]} + B_{[2]}u_{[2]} \\ y_{[2]} = C_{[2]}x_{[2]}. \end{cases}
\end{aligned}
\tag{1}
$$

For this power system model, we consider quantifying influence of the lth cluster dynamics on the kth cluster from a viewpoint of controllability. To this end, we define Q_{lk} as the $\mathcal{H}_2$-norm of the system whose input is the lth cluster subsystem and the output is the frequency of all generators and loads inside the kth cluster. Note that the value of Q_{lk} can be regarded as a measure of the controllability of u_l on Σ_k. The values of Q_{lk} for $l = 1$ are

$$
Q_{11} = 0.13, \quad Q_{12} = 0.007.
$$

By comparing them, we can see that the controllability of $u_{[1]}$ on $\Sigma_{[2]}$ is much lower than that on $\Sigma_{[1]}$. Similarly, we have found that the controllability of $u_{[2]}$ on $\Sigma_{[1]}$ is also much lower than that on $\Sigma_{[2]}$. These results imply that $\Sigma_{[1]}$ and $\Sigma_{[2]}$ are almost decoupled from the viewpoint of the controllability with respect to $u_{[1]}$ and $u_{[2]}$.

We compare the control performance achieved by the following two types of controllers:

- An ensemble of two controllers each of which is designed based on the lth isolated cluster dynamics ($\Sigma_{[l]}$ with $A_{[l,k]} = 0$, $k \neq l$), i.e., $K_{\mathrm{D}} := \{K_{[1]}, K_{[2]}\}$, where

$$
K_{[l]} : \begin{cases} \dot{\xi}_{[l]} = A_{[l]}\xi_{[l]} + B_{[l]}u_{[l]} + H_{[l]}(y_{[l]} - C_{[l]}\xi_{[l]}) \\ u_{[l]} = F_{[l]}\xi_{[l]} \end{cases}, \quad l \in \{1, 2\}.
\tag{2}
$$

In (2), $F_{[l]}$ and $H_{[l]}$ are found such that $A_{[l]} + B_{[l]}F_{[l]}$ and $A_{[l]} - H_{[l]}C_{[l]}$ are Hurwitz, respectively.

- A controller based on the entire system model, i.e.,

$$
K_{\mathrm{C}} : \begin{cases} \dot{\xi} = A\xi + Bu + H(y - C\xi) \\ u = F\xi, \end{cases}
\tag{3}
$$

with $u := [u_{[1]}^{\mathsf{T}}, u_{[2]}^{\mathsf{T}}]^{\mathsf{T}}$, $y := [y_{[1]}^{\mathsf{T}}, y_{[2]}^{\mathsf{T}}]^{\mathsf{T}}$ and

$$
A := \begin{bmatrix} A_{[1]} & A_{[1,2]} \\ A_{[2,1]} & A_{[2]} \end{bmatrix}, \ B := \begin{bmatrix} B_{[1]} & \\ & B_{[2]} \end{bmatrix}, \ C := \begin{bmatrix} C_{[1]} & \\ & C_{[2]} \end{bmatrix},
$$

where F and H are found such that $A + BF$ and $A - HC$ are Hurwitz, respectively.

Fig. 2 Trajectories of
frequency of all generators
and loads

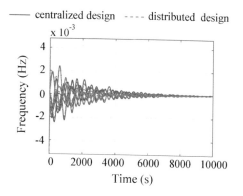

As we have described in Sect. 1, the former partial-model-based controller design
is distributed design of local controllers while the latter full-model-based controller
design is centralized design of a centralized controller. In Fig. 2, the red dotted lines
and blue solid lines show the frequency of all generators and loads in the case where
the decentralized controller ensemble K_D and the centralized controller K_C are used,
respectively. We see that K_D and K_C can achieve comparable damping performance.
To quantify this, we define a performance measure

$$J(K) := \sup_{x(0) \in \mathbb{B}} \|\omega(t)\|_{\mathcal{L}_2}, \quad K \in \{K_D, K_C\}, \tag{4}$$

where $\omega \in \mathbb{R}^7$ is the stacked version of the frequency of all generators and loads, and
$\mathbb{B}$ is the unit ball such that all state variables excluding angles are confined to zero.
The resultant values are

$$J(K_D) = 22.26, \quad J(K_C) = 22.26, \tag{5}$$

which are, in fact, comparable. This fact stems from the aforementioned decoupled
property from a viewpoint of controllability.

Next, we investigate what happens when a large amount of PVs are penetrated.
We suppose that PV farms, each of which is considered to be an aggregation of PV
generators inside each farm, share buses with preexisting generators; see Fig. 3a. In
this article, we suppose that the influence of this PV penetration is modeled as the
decrement of the value of inertia of generators in order to reflect the fact that the
large-scale penetration of PVs can cause the reduction of inertia of the overall power
system [6, 7]. In this case, the metric Q_{lk} for $l = 1$ and $k \in \{1, 2\}$ are

$$Q_{11} = 8.8, \quad Q_{12} = 0.4.$$

Next, we design K_D in (2) and K_C in (3). We plot the resultant frequency trajectories
in Fig. 3b. Furthermore, the resultant values of $J(\cdot)$ in (4) are

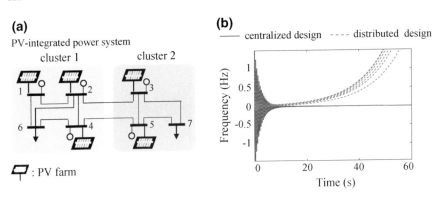

(a)
PV-integrated power system

cluster 1   cluster 2

(b)

centralized design   - - - - distributed design

: PV farm

Fig. 3 (Left) Power system with large-scale PV integration. (Right) Trajectories of frequency of all generators and loads

$$J(K_\mathrm{D}) = \infty, \quad J(K_\mathrm{C}) = 16.62.$$

We can see that the value of $J(K_\mathrm{C})$ is smaller than that in (5). This performance improvement stems from the fact that the usual notion of controllability is higher than that in the previous case owing to smaller inertia constants. However, K_D destabilizes the system. This discrepancy comes from the fact that the interference between the two clusters, neglected in the design of each decentralized controller, is more strongly induced due to the higher controllability of $u_{[1]}$ on $\Sigma_{[2]}$ and that of $u_{[2]}$ on $\Sigma_{[1]}$.

2.2 Challenges and Opportunities

Through the above example, we found that

1. the reduction of the generator inertia enhances system controllability,
2. the centralized design of a centralized controller has the potential to utilize enhanced controllability for control performance improvement, and
3. the distributed design of decentralized controllers may induce instability due to the interference among clusters when the controllability with respect to each cluster is not sufficiently small.

Fact 3 is more significant toward the implementation of smart grids because typical controller design approaches taken in power community are some special cases of this distributed design. An example of such approaches is power system stabilizer (PSS) design [9, 10] based on a single-machine-infinite-bus model, where the behavior of that model is isolated from the other grid dynamics by neglecting the bus voltage variation. However, as we have shown in numerical simulation, such distributed design may pose serious threat to the power system stability in a future smart grid. Therefore, it is crucial to develop a method of distributed design of decentralized controllers so that they can accomplish a global objective such as damping performance of all generators' frequency while guaranteeing the stability of the entire closed-loop system.

Furthermore, it would be meaningful to discuss how to make clusters, i.e., how to partition a system of interest into clusters, for achieving better control performance. To find an optimal (or suboptimal) cluster set, it would be important to consider the controllability as discussed in the motivation example. Furthermore, it would be also significant to clarify what information is needed for finding an optimal (or suboptimal) cluster set.

An open problem related to this optimal clustering is optimal DERs allocation in power grids. The location of DERs has an influence on steady-state power flow of a grid, thereby influencing the grid characteristics such as transient stability and controllability. Thus, in order to solve optimal allocation problems, it would be necessary to reveal the relationship between those dynamical characteristics and power flow.

3 Resilient System Design

3.1 Motivating Example

In this subsection, we numerically investigate how the wind farm dynamics has an impact on the resilience of wind-integrated power systems. We consider an IEEE 68-bus power system with the integration of a single wind farm, as shown in Fig. 4. The generator model is the combination of the standard flux-decay model [9] and an automatic voltage regulator (AVR) with PSS, and the loads are modeled as constant power loads. Following [11], the wind farm can be regarded as an aggregated wind generator whose output power is the total power of wind generators inside the farm. The aggregated model consists of a wind turbine, doubly fed induction generator (DFIG), and an internal controller; see [12] for the modeling details.

Figure 5 shows the trajectories of all generators' frequency when a fault happens at the wind farm. The blue solid and red dotted lines are the cases where the number of wind generators inside the farm is small and large, respectively. We can see from this figure that the entire power system becomes more oscillatory when a larger scale wind farm is penetrated. In other words, a power grid with large-scale wind penetration has less resilience against a fault at the wind farm. This observation was also shown in a slightly different context in [13]. This oscillation induction is due to the fact that the impact caused by the fault is more strongly stimulated by the resonance mode of DFIG, thereby causing the oscillation of power flow of the entire system; see [12] for more detail discussion.

One option to combat this oscillatory behavior is to tune the PI gains of the internal controller in the wind farm. However, such tuning must be done extremely carefully with full knowledge of the entire closed-loop model, because both low and

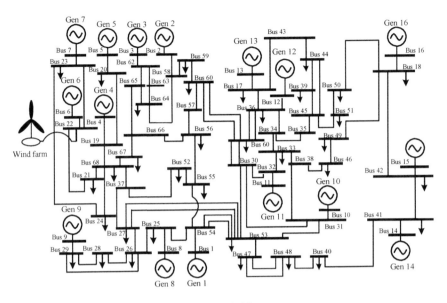

Fig. 4 IEEE 68-bus power system with a single wind farm

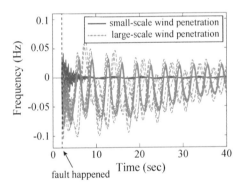

Fig. 5 Trajectories of all generators' frequency when fault happened at wind farm

high values of these gains can jeopardize closed-loop stability. These observations motivate us for building a much more systematic method of the distributed design of controllers by which the resilience of the wind farm can be enhanced in a desired way.

3.2 Challenges and Opportunities

The concept of resilient system design has been introduced in [14]. In [8], the authors have discussed a conceptual property of resilient control systems. As a similar concept to resilience, the authors in [15] have proposed a notion called intelligent balancing authority which is a portfolio of power system equipment responsible for having adequate control to ensure stability and good dynamic response of their own areas. However, it is still an open problem how we design resilient control systems.

Although there is no common definition of resilience, one can define resilient systems as systems that can maintain an acceptable level of operation in the face of spatially local undesirable incidents. As a related work for enhancing resilience in this sense, in [16] the authors have proposed a method called *retrofit control* to improve a performance of spatially local subsystems against local faults. An advantage of this method is that the retrofit controller can be designed based on the model of a local subsystem of interest without any knowledge about the entire system model. Furthermore, retrofit controllers do not have any influence on each other. Indeed, when a fault happens at a certain subsystem, only the corresponding retrofit controller improves damping as soon as it is activated while the other retrofit controllers are inactivated. Therefore, the distributed design of retrofit controllers enjoys a natural decoupling property from one subsystem to another. Future works of this retrofit control include robustness analysis of retrofit controllers against uncertainty of the local subsystem.

4 Challenges and Opportunities for Power Grid Evolution

We deal with power grids' long-term evolution such as the penetration of DERs, as shown in Fig. 6. The entire power system must adopt such evolution while maintaining transient stability and functions of the entire system, without additional configuration of the preexisting power system. So far, control theory does not have paid much attention to the characteristics of systems' long-term evolution. However, in order to establish a mechanism so that power grids can facilitate addition and modification of system components, it would be necessary to develop a theory for explicitly dealing with systems' long-term evolution. We briefly discuss challenges and opportunities for this issue.

So far, power grids have been evolved with the advance of human civilization. However, large-scale blackouts sometimes happen around the world, which shows the vulnerability of power systems. Cascading failure is regarded as one of the main mechanism of large blackouts [17]. Toward the development of power grids that can decrease the risk of cascading failures, in [18], the authors have proposed power grid

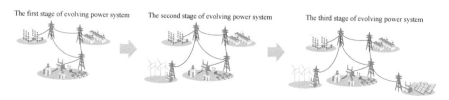

The first stage of evolving power system   The second stage of evolving power system   The third stage of evolving power system

Fig. 6 Illustrative example of power system evolution

evolution models where new power plants and substations are constructed according to a rule reflecting practical power system planning, and have evaluated the probability of cascading failures from a viewpoint of Self-Organized Criticality (SOC) theory in complex network analysis. However, since the proposed models represent the transition of static power systems without any dynamics such as generator dynamics, we cannot evaluate dynamical characteristics such as transient stability. How to describe the long-term transition of dynamical systems is a question that deserves to be further studied.

In order to facilitate the addition of new components, it would be desirable that newly added components have interchangeability or plug-and-play capability [1]. A challenge for the systematic design of interchangeable components is to reveal the class of such components as well as a portfolio to be imposed on the interconnection of the components to preexisting grids. Related works on this topic are as follows. In [19], the authors have proposed a strategy for constructing a large-scale network system while keeping an expanding system stable. It is shown that the entire closed-loop system is stable as long as strictly passive subsystems are interconnected via a passive interconnection. One approach for a broader class of systems can be found in [20]. In this approach, we consider a module consisting of the newly added component satisfying a matching condition and a compensator. It is shown that the evolving network system keeps its stability as long as each module is connected to the associated subsystem such that the local closed-loop system composed of these two systems is stable. These approaches show particular sufficient conditions of components and interconnection rules for guaranteeing the stability of evolving network systems. Further studies to reveal the class of interchangeable components and interconnection rules are necessary.

The evolution of practical power systems is not a self-organized process in a strict sense, but is a process containing feedback mechanism performed by human industrial activities. For example, when a new power plant is constructed today, the amount of power consumption around the new power plant will increase tomorrow, resulting in the need of further evolution of the power system such as upgrade of the transmission line. Such feedback mechanism needs to be introduced to evolution process in order to realize the intelligent power grid's adaptation [18].

5   Conclusion

Smart grid can be regarded as an electric system integrated across electricity generation, transmission, substations, distribution, and consumption, to achieve that the grid is not only clean and stable, but also interoperable, resilient, and changeable. Due to the massive complexity and organizational differences of future power grids, it is impossible to construct such smart grids from the scratch. Instead, an approach what we can take practically is the distributed design of grid components. In view of this, in this article, we have discussed challenges and opportunities for solving the following three problems. First, to make grids flexible to incorporate regional and organizational differences, we have considered a problem of the distributed design of local controllers accomplishing a global objective by cooperating each other. Next, we have considered distributed design of decentralized controllers so that they can individually enhance resilience of associated subsystems. Furthermore, we have discussed long-term grids' evolution caused by addition or modifications of grid components, and have considered the distributed design of components having plug-and-play capability.

Acknowledgements This work was supported by JST CREST Grant Number JPMJCR15K1, Japan.

References

1. National Institute of Standards and Technology (NIST): NIST Framework and Roadmap for smart grid interoperability standards, Release 3.0 (2014), http://nvlpubs.nist.gov/nistpubs/SpecialPublications/NIST.SP.1108r3.pdf
2. C. Langbort, J. Delvenne, Distributed design methods for linear quadratic control and their limitations. IEEE Trans. Auto. Control **55**(9), 2085–2093 (2010)
3. N. Sandell, P. Varaiya, M. Athans, M. Safonov, Survey of decentralized control methods for large scale systems. IEEE Trans. Auto. Control **23**(2), 108–128 (1978)
4. L. Bakule, Decentralized control: an overview. Ann. Rev. Control **32**(1), 87–98 (2008)
5. F.L. Lagarrigue, A. Annaswamy, S. Engell, A. Isaksson, P. Khargonekar, R.M. Murray, H. Nijmeijer, T. Samad, D. Tilbury, P. Van den Hof, Systems & control for the future of humanity, research agenda: current and future roles, impact and grand challenges. Ann. Rev. Control 1–64 (2017)
6. S. Eftekharnejad, V. Vittal, G.T. Heydt, B. Keel, J. Loehr, Impact of increased penetration of photovoltaic generation on power systems. IEEE Trans. Power Syst. **28**(2), 893–901 (2013)
7. B. Tamimi, C. Cañizares, K. Bhattacharya, System stability impact of large-scale and distributed solar photovoltaic generation: the case of Ontario, Canada. IEEE Trans. Sustain. Energy **4**(3), 680–688 (2013)
8. D. Wei, K. Ji, Resilient industrial control system (RICS): concepts, formulation, metrics, and insights, in *Proceedings of International Symposium on Resilient Control Systems* (2010), pp. 15–22
9. P. Kundur, N.J. Prabha, M.G. Lauby, Power system stability and control, vol. 7 (New York, McGraw-hill, 1994)
10. E.V. Larsen, D.A. Swann, Applying power system stabilizers Part ii: performance objectives and tuning concepts. IEEE Trans. Power Apparat. Syst. 3025–3033 (1981)
11. V. Akhmatov, H. Knudsen, An aggregate model of a grid-connected, large-scale, offshore wind farm for power stability investigationsimportance of windmill mechanical system. Int. J. Electr. Power Energy Syst. **24**(9), 709–717 (2002)
12. T. Sadamoto, A. Chakrabortty, T. Ishizaki, J. Imura, Retrofit control of wind-integrated power systems. IEEE Trans. Power Syst. (in Press) (2017). https://doi.org/10.1109/TPWRS.2017.2750411
13. S. Chandra, D. Gayme, A. Chakrabortty, Time-scale modeling of wind-integrated power systems. IEEE Trans. Power Syst. **31**(6), 4712–4721 (2016)
14. C.G. Rieger, D. Gertman, M. McQueen, Resilient control systems: next generation design research, in *Proceedings of International Conference on Human System Interactions* (2009), pp. 632–636
15. S. Baros, M. Ilić, Intelligent Balancing Authorities (iBAs) for transient stabilization of large power systems, in *PES General Meeting|Conference & Exposition* (2014), pp. 1–5

16. T. Ishizaki, T. Sadamoto, J. Imura, H. Sandberg, K.H. Johansson, Retrofit control: localization of controller design and mplementation, Automatica **95**, 336–346 (2018)
17. D.P. Nedic, I. Dobson, D.S. Kirchen, B.A. Carreras, V.E. Lynch, Criticality in a cascading failure blackout model. Int. J. Electr. Power Energy Syst. **28**(9), 627–633 (2006)
18. S. Mei, X. Zhang, M. Cao, Power grid complexity (Springer Science & Business Media, 2011)
19. K. Urata, M. Inoue, S. Adachi, Passivity-based strategy for constructing large-scale and expanding network systems, in *Proceedings of European Control Conference* (2015), pp. 3554–3559
20. T. Sadamoto, T. Ishizaki, J. Imura, Hierarchical distributed design of stabilizing controllers for an evolving network system, in *Proceedings of Conference on Decision and Control* (2015), pp. 3337–3342

Smart Grid Control: Opportunities and Research Challenges a Decentralized Stochastic Control Approach

Maryam Khanbaghi

Abstract Recent challenges in power system stability and operation are due to the fact that these complex systems have not evolved in a way to deal with new forms of power generation and load types. Although the grid of the twenty-first century known as "Smart Grid" uses technologies such as two-way communication, advanced sensors, computer-based remote control, and automation, it does not adequately consider increased use of renewable energy which is becoming a major component of the power grid. Moreover, the potential for instability caused by increased frequency deviation and energy imbalance due to high penetration of Renewable Energy Sources (RES) places major constraints on their usage. Therefore, finding a solution to improve the stability of power systems with high penetration of Renewable Energy (RE) is a major challenge that needs to be addressed.

One of the key challenges of the electric power grid of the future is the management of a decentralized power system that includes multiple Distributed Energy Resources (DER) with high usage of uncertain energy sources. In fact, proliferation of microgrids (μGrid) requires utilities to revisit the existing decentralized grid management and its structure. The grid of the future is flexible where both load and generation are stochastic. Therefore, a decentralized stochastic control strategy that is capable of treating the grid as a flexible entity may be a natural solution. In this paper, a description of opportunities and challenges that the grid of the future may encounter is provided. Further, new requirements into the proposed control scheme are considered.

M. Khanbaghi (✉)
Department of Electrical Engineering, Santa Clara University,
500 El Camino Real, Santa Clara, CA 95053-0840, USA
e-mail: mkhanbaghi@scu.edu

© Springer Nature Switzerland AG 2019
J. Stoustrup et al. (eds.), *Smart Grid Control*, Power Electronics
and Power Systems, https://doi.org/10.1007/978-3-319-98310-3_8

129

1 Introduction

The development of human society and its growing economic needs were the catalyst that drove the evolution of transmission grids stage-by-stage, with the aid of innovative technologies. The shortage of fossil fuel energy over the next few decades combined with its potential damage to the environment pushed for large-scale usage of RES. Due to increased penetration of RE, the existing power grid infrastructure faces new challenges in three major areas: policy, pricing, and technology. Although the research opportunity proposed in this paper requires changes in all three areas, the focus is on technology and more specifically control system strategies.

With recent technology advances in energy storage and nonfossil fuel power generation such as fuel cell, the time is ripe to review the existing power grid control structure with key emphasis on balancing between the complexity and stability of this system.

Since the usage of RE is a global priority, intensive research is being conducted on the integration of RES into existing power generation systems. Current research direction can be divided into two main paths:

- Power grid voltage regulation and frequency control with integration of RES, and
- Control of μGrids to satisfy their power exchange and demand.

In both paths, the main objective is stability of the overall grid at the lowest cost. Further, for both research directions, centralized and decentralized control have been considered. Here, we name only a few recent publications in decentralized control. Etemadi et al. [6] consider a decentralized robust control for an islanded multi-DER μGrid. Dagdougui and Sacile [4] proposed a decentralized control of smart μGrid, where each μGrid is modeled as an inventory system locally producing RES. A decentralized stochastic control to maintain power system stability due to increase of RES is proposed by Vrettos et al. [15]. Worthmann et al. [16] compared centralized and decentralized control for a power grid with high integration of DER and distributed battery storage. Di Fazio et al. [5] use decentralized control to improve the voltage profile along the feeder of a distribution system. Kizilkale and Malhame [8] proposed using mean field control theory to manage the coordination and decentralized control of large systems. Authors in [1] also provide an overview of recent publications on distributed and decentralized voltage control of smart distribution networks.

Proposed control strategies in these studies attempt to address challenges of integrating RE to a network originally designed for one-way power flow from generation to the customer. However, considering that 90% of all power outages and disturbances originate in the distribution system [7], it is more desirable to move toward a decentralized grid that allows more flexibility and scalability. It is becoming increasingly apparent that the future of the grid will be "plug and play" [2] with optimized interoperability where μGrids play an essential role. Therefore, the new opportunity from a control theory perspective would be to revisit the grid structure concurrently with the design of the control strategy.

In Sect. 2 of this paper, we first review a grid-integrated μGrid that can provide flexibility, reliability, and resiliency. Then, we define μGrids control system structure and requirements needed for this flexible grid. Furthermore, the concept of *nanogrid* (nGrid) is introduced. In the third section, by recognizing that the grid-integrated μGrids fit the definition of complex systems, we envision a decentralized stochastic control architecture in order to meet the stability and robustness required for this complex system.

2 A Grid-Integrated μGrid

The existing electricity grid is unidirectional by design. It converts only one-third of fuel energy into electricity, without recovering the waste heat. Approximately 8% of its output is lost along transmission lines, while 20% of generation capacity exists to only meet peak demand [2].

Furthermore, the electricity industry landscape is changing due to DER, storage technology, active demand side participation, and a desire for green/renewable energy. Simultaneously, the grid infrastructure is requiring increased investment due to aging in order to maintain resiliency and reliability. These two factors combined (an inefficient grid system and changing industry landscape) create an important and unique opportunity for the control system community by revisiting the 100-year-old structure with the control objective in mind, in order to manage resiliency, reliability, and cost. Therefore, the objective here is not to propose a control strategy for the existing grid but using control theory to provide input for the grid of the future which will be inevitably redesigned. The power industry has already embarked in this path by considering DERs as a viable alternative to centralized generation [14].

The main aspect of this work is to revisit control strategies that were not viable a few years ago due to lack of required technologies. The grid of the future, "Smart Grid", is considered to have adequate sensors, computer networks, and automation in order to make an attempt to revisit its structure. We first make the assumption that residential areas can operate independently from the grid. This means that each house or residential building will have their own μGrid. The residential section of the grid will be completely disconnected from the grid and will be controlled at the household level. We will call these single home μGrids, *nanogrids* (nGrid), which always operate in an island mode and are individually controlled. Every nGrid will have an RE source (i.e., solar), a storage and a sustainable clean energy generator (i.e., fuel cell). If we first focus on a region that the sun is abundantly available, residential RE sources are rooftop solar panels. Solar generations are random. This randomness due to PVs generation that can sell back to the grid and charging demand of Electric Vehicles (EV), combined with unpredictable individual usage is becoming more and more one of the major concerns of utilities. Hence, separation of nGrid from the main grid will help to ease the randomness of the load viewed by the central generation.

Furthermore, we make the assumption that commercial, industrial, university campuses and military locations will form μGrids that can operate in islanded mode and

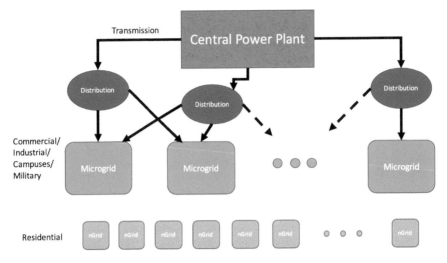

Fig. 1 A simple schematic of the grid of future

connected mode. In this configuration, we forgo the idea of bidirectional transfer of energy between the plant and µGrids. The central power plant will be connected to these µGrids through the distribution centers. Figure 1 presents a schematic description of this idea. In this configuration, central power plants may use intermittent RE sources such as solar farms or controlled RE such as hydro plants. We also assume that no storage is used at the power plant; therefore, energy produced needs to be used by the load. However, storage can be used at distribution centers and µGrids locations.

2.1 µGrids Usage

According to the U.S. Department of Energy (DOE), a µGrid is "a group of interconnected loads and DERs within clearly defined electrical boundaries that act as a single controllable entity with respect to the grid. A µGrid can connect and disconnect from the grid to enable it to operate either in a grid-connected or island mode". A µGrid is capable of supporting a predefined number of loads but it is typically not designed to operate indefinitely without being connected to the traditional utility infrastructure.

The new generation of µGrids will not only be used for supplying backup power but also a more complex configuration that contains all the essential elements of a large-scale grid such as the ability to balance electrical demand with sources, schedule the dispatch of resources and preserve grid reliability. In fact, these new µGrids will enable increased DERs utilization, limit greenhouse emission, improve local grid reliability, and reduce operating cost. Moreover, driven by declining cost of DERs, μGrids are becoming more appealing for managing the variability of RE [11].

2.2 μGrids Control

In the proposed scenarios, each μGrid needs to be controlled individually. A μGrid controller defined by DOE refers to an advanced control system, comprised of multiple components and subsystems, capable of sensing grid conditions, monitoring and controlling the operation of a μGrid in order to maintain electricity delivery to loads during all μGrid operating modes. Hence, μGrid is a concept for which the control system is the defining and enabling technology [14]. μGrid key control objectives are a seamless transition between islanded and connected mode and load management using local generation and storage for optimized performance. There is ongoing research in this area [3, 9, 10]. Bulk of the research is to design optimal control for μGrids with random generation. When generation does not meet demand, the optimal controller has for objective to connect μGrids back to the grid by minimizing cost. Here, we assume that each μGrid is optimally controlled and meets its objectives when in islanded mode. The challenge is whether the optimality can be maintained when connected back to the main grid. This concept will be explained more in the next section.

3 A Decentralized Stochastic Optimal Control System

For clarity, in this section we use the following definitions:

- Loads: μGrids connected to the grid are central power plant loads,
- Subload: load of each μGrid,
- Generation: central power plant generation,
- Subgeneration: energy sources for each μGrid,
- System: central power plant, transmission, distribution, and μGrids, and
- Subsystem: each μGrid is a subsystem.

For this study, we make the assumption that the grid of the future is a "plug and play" type of grid with the structure proposed in Fig. 1. The proposed grid is a large dynamic system composed of several subsystems. Each μGrid may be connected or disconnected randomly from the grid depending on the load, subload, subgeneration, and generation.

μGrids in this structure will operate by default in islanded mode. It may connect to the grid for two reasons:

- to meet the subload demand in case of insufficient subgeneration, or
- to compensate for overgeneration by the central power plant.

Such on and off participation of the subsystems represent changes in the system structure which may destroy stability and cause the system to collapse. To prevent collapse, systems should be built to have the desirable stability properties that are invariant under structural perturbations [12].

In stabilization of complex systems, decentralized schemes are more practical since perfect knowledge of the interconnections between different subsystems is not required, and the control strategy may guarantee robust stability among the sub-systems. That is, closed loop interconnected systems that are stabilized by local feedback laws are connectively stable [13]. Furthermore, when a system is complex, it is difficult or impossible to obtain partitions by physical reasoning. A systematic method which can be used to decompose the system by extracting information from subsystems is required [17].

Decentralized control is an appealing concept for the proposed power grid. It offers an essential reduction in communication requirements without significant loss of performance. Each subsystem is controlled on the basis of locally available infor-mation, seldom connecting to other μGrids. The control objective is to stabilize each disconnected μGrid using a stochastic optimal control and at the same time, ensure that the stability and optimality are retained when each μGrid has to connect back to the grid. Therefore, the challenge resides in two parts:

- design a robust control strategy to maintain stability and
- reevaluate system requirements in order to maintain optimality.

Furthermore, in complex dynamic systems, achieving optimality is complicated by the presence of uncertainties in the interconnections among the subsystems. For this reason, the focus will be first on stability via a robust control strategy, optimality will be addressed consequently.

In the past, much of the effort of robust decentralized control strategies was directed toward offsetting the effects of load variations in the system [17]. However, with the emergence of intermittent RE, the problem of robust control has taken on additional significance and complexity, necessitating the use of stochastic models. This is largely due to the fact that in a stochastic environment, the operating conditions become difficult to anticipate. Under such circumstances, it is important to develop robust decentralized control strategies that can protect the system against random disturbances.

One of the primary concerns in large-scale decentralized stochastic systems is the control system robustness. It has been recognized that the degree of robustness required in a high-performance design cannot be achieved by application of existing control practices. For this reason, there has been considerable effort to develop new control schemes with some form of "built in" robustness enhancement [13, 17]. The main objective is to provide a control structure, which ensures that the system performs satisfactorily under faulty conditions. Here, faulty conditions are not viewed as equipment faults, but rather as a consequence of insufficient amounts of power production due to the inherent stochasticity of the generation.

New insights into decentralized control strategies are needed. The stochastic nature of the problem and high demands on quality of service for users make this challenging. To address this problem, it is essential to create:

- adequate methods for analysis of networked systems of high complexity with stochastic inputs for decentralized control, and
- a fundamental theoretical framework and effective methodology for designing optimal robust control schemes for complex decentralized stochastic systems.

The first part requires to evaluate computational challenges associated with the scale of the problem and to model sections of loads and generations as stochastic processes. The second part entails evaluation of multilevel control structures. Further, to address optimality, appropriate optimization techniques need to be considered.

Accomplishment of these aims is expected to provide a model and a decentralized control strategy for complex power systems that take into account variable and intermittent RES.

Satisfactory decentralized design of a system with intermittent inputs will be used to design a control strategy that minimizes the cost of operation and improves stability of the grid.

4   Summary

The grid of today uses technologies such as two-way communication, advanced sensors, computer-based remote control, and automation. It also faces the challenge of increased use of renewable energy. The use of renewable energy sources can reduce greenhouse gas emissions and dependence on fossil fuels. The main problem of installations based on renewable energy is that electricity generation cannot be fully forecasted and may not follow the trend of actual energy demand.

There has been significant interest in deployment of μGrid in the last few years. Also, based on type of customers, applications, and connections, control objectives are different. Therefore, we separated the power grid load into two categories μGrid and nGrid. This paradigm shift will address the concern of load variation due to large numbers of PVs, plug-in electric vehicles, and home storage systems. Furthermore, the direction proposed is a decentralized stochastic control strategy for complex power systems that takes into account variable and intermittent RES with a redefined load.

The opportunity available now to the control community to play a key role in shaping the grid of the twenty-first century is unprecedented. We hope with this effort to bring the attention of this community to the redesign of the system where utility companies are fully aware of the difficulties involved in transitioning their infrastructure, organizations, and processes towards an uncertain future due to DER and RE.

Acknowledgements The author would like to thank Aleksander Zecevic for his support and insightful discussions.

References

1. K.E. Antoniadou-Plytaria, I.N. Kouveliotis-Lysikatos, P.S. Georgialakis, N.D. Hatziargyriou, Distributed and decentralized voltage control of smart distribution networks: models, methods, and future research. IEEE Trans. Smart Grid, **8**(6), 2999–3008 (2017)
2. S. Bahramirad, Powering the future. IEEE Power Energy Mag. **15**(4), 8–14 (2017)
3. A. Belloni, L. Piroddi, M. Prandini, A stochastic optimal control solution to the energy management of microgrid with storage and renewables, in *2016 American Control Conference*, Boston (2016)
4. H. Dagdougui, R. Sacile, Decentralized control of the power flows in a network of smart microgrids modeled as a team of cooperative agents. IEEE Trans. Control Syst. Technol. **22**(2), 510–519 (2014)
5. A.R. Di Fazio, G. Fusco, M. Russo, Decentralized control of distributed generation for voltage profile optimization in smart feeders. IEEE Trans. Smart Grid **4**(3), 1586–1596 (2013)
6. A.H. Etemadi, J.D. Davison, R. Irvani, A decentralized robust control strategy for multi-DER microgrids—Part I: Fundamental concepts. IEEE Trans. Power Deliv. **27**(4), 1843–1853 (2012)
7. H. Farhangi, The path of the Smart Grid. IEEE Power Energy Mag. **8**(5), 18–28 (2010)
8. A.C. Kizilkale, R.P. Malhame, A class of collective target tracking problems in energy systems: cooperative versus non-cooperative mean field control solutions, in *IEEE 53rd Annual Conference on Decision and Control* (2014), pp. 3493–3498
9. Y. Levron, J.M. Guerrero, Y. Beck, Optimal power flow in microgrids with energy storage. IEEE Trans. Power Syst. **28**(3), 3226–3234 (2013)
10. A. Parisio, E. Rikos, L. Glielmo, A model predictive control approach to microgrid operation optimization. IEEE Trans. Control Syst. Technol. **22**(5), 1813–1827 (2014)
11. M. Shahidehpour, L. Zhiyi, S. Bahramirad, Z. Li, W. Tian, Networked microgrids. IEEE Power Energy Mag. **15**(4), 63–71 (2017)
12. D.D. Siljak, *Large-Scale Dynamic Systems: Stability and Structure* (North-Holland, 1978)
13. D.D. Siljak, *Decentralized Control of Complex Systems* (Dover, 2012)
14. D. Ton, J. Reilly, Microgrid controller initiatives. IEEE Power Energy Mag. **15**(4), 24–31 (2017)
15. E. Vrettos, Z. Charalampos, G. Anderson, Fast and reliable primary frequency reserves from refrigerators with decentralized stochastic control. IEEE Trans. Power Syst. **32**(4), 2924–2941 (2017)
16. K. Worthmann, C.M. Kellett, P. Braun, L. Grune, S.R. Weller, Distributed and decentralized control of residential energy systems incorporating battery storage. IEEE Trans. Smart Grid **6**(4), 1914–1923 (2012)
17. A.I. Zecevic, D.D. Siljak, *Control of Complex Systems: Structural Constraints and Uncertainty* (Springer, New York, 2010)

Part III
Wide-Area Control Using Real-Time Data

Wide-Area Communication and Control: A Cyber-Physical Perspective

Aranya Chakrabortty

1 Introduction

For several decades, the traditional mindset for controlling large-scale power systems has been limited to local output feedback control, which means that controllers installed within the operating region of any utility company typically use measurements available only from inside that region for feedback, and, in fact, more commonly only from the vicinity of the controller location. Examples of such controllers include Automatic Voltage Regulators (AVR), Power System Stabilizers (PSS), Automatic Generation Control (AGC), FACTS control, HVDC, and so on. However, the US Northeast blackout of 2003, followed by the timely emergence of sophisticated GPS-synchronized digital instrumentation technologies such as Wide-Area Measurement Systems (WAMS) led utility owners to understand how the interconnected nature of the grid topology essentially couples their controller performance with that of others, and thereby forced them to look beyond this myopic approach of local feedback and instead use wide-area measurement feedback [1]. Over the past few years, several researchers have started investigating such data-driven wide-area control designs using H_∞ control [2–4], LMIs and conic programming [5], wide-area protection [6], model reduction and control inversion [7], adaptive control [8], LQR-based optimal control [9–12], etc., complemented with insightful case studies of controller implementation for various real power systems such as the US west coast grid [13], Hydro Quebec [14], Nordic system [15, 16], and power systems in China [17], Australia [18], and Mexico [19]. A tutorial on the ongoing practices for wide-area control has recently been presented in [20], while cyber-physical implementation architectures for realizing these controls have been proposed in [21, 22].

One of the biggest roadblocks for implementing wide-area control in a practical grid, however, is that the current power grid IT infrastructure is rigid and has low capacity as it is mostly based on a closed mission-specific architecture. The current push to adopt the existing TCP/IP-based open Internet and high-performance computing technologies such as the NASPInet [23] may not be enough to meet the

A. Chakrabortty (✉)
Electrical & Computer Engineering, North Carolina State University, Raleigh, NC, USA
e-mail: achakra2@ncsu.edu

© Springer Nature Switzerland AG 2019
J. Stoustrup et al. (eds.), *Smart Grid Control*, Power Electronics
and Power Systems, https://doi.org/10.1007/978-3-319-98310-3_9

requirement of collecting and processing very large volumes of real-time data produced by such thousands of PMUs. Second, the impact of the unreliable and insecure communication and computation infrastructure, especially long delays and packet loss uncertainties over wide-area networks, on the development of new WAMS applications is not well understood. For example, uncontrolled delays in a network can easily destabilize distributed estimation algorithms for wide-area oscillation monitoring using PMU data from geographically dispersed locations. Finally, and most importantly, very little studies have been conducted to leverage the emerging IT technologies, such as cloud computing, Software-Defined Networking (SDN), and Network Function Virtualization (NFV), to accelerate the development of WAMS [24]. Another major challenge is privacy of PMU data as utility companies are often shy in sharing data from a large number of observable points within their operating regions with other companies. Equally important is cybersecurity of the data as even the slightest tampering of Synchrophasors, whether through denial-of-service attacks or data manipulation attacks, can cause catastrophical instabilities in the grid. What we need is a cyber-physical architecture that explicitly brings out potential solutions to all of these concerns, answering how data from multitudes of geographically dispersed PMUs can be shared across a large grid via a secure communication medium for successful execution of critical transmission system operations, how the various binding factors in this distributed communication system can pose bottlenecks, and, how these bottlenecks can be mitigated to guarantee the stability and performance of the grid.

Motivated by these challenges, in this tutorial we review the current state-of-the-art practice for wide-area communication and control from a Cyber-Physical System (CPS) viewpoint. We study the ways in which these practices pose limitations against optimal grid performance, how modern communication technologies such as SDN, NFV, and cloud computing can be used to overcome these limitations, and how new ideas of distributed control and optimization need to be integrated with various operational protocols and wide ranges of uncertainties of these wide-area networks to ensure that the closed-loop grid operates in a stable, reliable, robust, and efficient way. We also discuss the need for advanced modeling, simulation, and control of wide-area communication to ensure cybersecurity and resilience of wide-area controllers.

The remainder of the chapter is organized as follows. In Sect. 2 we recall standard differential-algebraic models of power system networks followed by wide-area control designs in Sect. 3. Section 4 presents the potential cyber-physical architectures for implementing these wide-area controllers over a distributed communication network. Section 5 highlights research directions on how these controllers can be made aware of the various operational uncertainties in the communication network such as delays. Section 6 summarizes the potential benefits of modern communication technology such as SDN, NFV and cloud computing in realizing the envisioned CPS architecture. Section 7 highlights the importance for making wide-area controllers secure against cyberattacks. Section 8 presents the current challenges facing simulation of extreme-scale wide-area control, while Sect. 9 describes the importance of hardware-in-loop CPS testbeds that can be used for testing, verification, and validation of such emulated control loops before they are deployed in the field. Section 10 concludes the chapter.

2 Power System Models

Consider a power system network with n synchronous generators. Each generator is modeled by a *flux-decay model* assuming that the time constants of the d- and q-axis flux are fast enough to neglect their dynamics, that the rotor frequency is around the normalized constant synchronous speed, and that the amortisseur effects are negligible. The model of the ith generator can be then written as [25]:

$$\dot{\delta}_i = \omega_i - \omega_s \tag{1}$$

$$M_i \dot{\omega}_i = P_{mi} - (V_i I_{qi} \cos(\theta_i - \delta_i) + V_i I_{di} \sin(\delta_i - \theta_i)) - d_i(\omega_i - \omega_s) \tag{2}$$

$$T_{qi} \dot{E}'_{qi} = -E'_{qi} + (x_{di} - x'_{di})I_{di} + E_{fdi} \tag{3}$$

$$T_{di} \dot{E}'_{di} = -E'_{di} + (x_{qi} - x'_{qi})I_{qi} \tag{4}$$

$$T_{Ai} \dot{E}_{fdi} = -E_{fdi} + K_{Ai}(V_{ref,i} - V_i) + u_i(t). \tag{5}$$

for $i = 1, \ldots, n$. Equations (1)–(2) are referred to as the swing equations while (3)–(5) as the excitation equations. The states δ_i, ω_i, E'_{qi}, E'_{di}, and E_{fdi} respectively denote the generator phase angle (radians), rotor velocity, the quadrature-axis internal emf, the direct-axis internal emf, and the field excitation voltage of the ith generator. The voltage at the generator terminal bus is denoted in the polar representation as $\tilde{V}_i(t) = V_i(t)\angle\theta_i(t)$. $V_{ref,i}$ is the constant setpoint for V_i. The generator current in complex phasor form is written as $I_{di} + jI_{qi} = I_i\angle\phi_i$. ω_s is the synchronous frequency, which is equal to 120π rad/s for a 60-Hz power system. M_i is the generator inertia, d_i is the generator damping, and P_{mi} is the mechanical power input from the ith turbine, all of which are considered to be constant. T_{di}, T_{qi}, and T_{Ai} are the excitation time constants; K_{Ai} is the constant voltage regulator gain; x_{di}, x'_{di}, x_{qi} and x'_{qi} are the direct-axis and quadrature-axis salient reactances and transient reactances, respectively. All variables, except for the phase angles (radians), are expressed in per unit. Equations (1)–(5) can be written in a compact form as

$$\dot{x}_i(t) = g(x_i(t), z_i(t), u_i(t), \alpha_i) \tag{6}$$

where $x_i = [\delta_i \; \omega_i \; E'_{qi} \; E'_{di} \; E_{fdi}]' \in \mathbb{R}^5$ denotes the vector of state variables, $z_i = [V_i \; \theta_i \; I_{di} \; I_{qi}]' \in \mathbb{R}^4$ denotes the vector of algebraic variables, $u_i \in \mathbb{R}$ is the control input, and α_i is the vector of the constant parameters P_{mi}, ω_s, d_i, T_{qi}, T_{di}, T_{Ai}, M_i, K_{Ai}, $V_{ref,i}$, x_{di}, x_{qi}, x'_{di}, and x'_{qi}, all of which are assumed to be known. The definition of the nonlinear function $g(\cdot)$ follows from (1)–(5).

The model (6) is a completely decentralized model since it is driven by variables belonging to the ith generator only. It is, however, not a state-space model as it contains the auxiliary variables z_i. The states x_i can be estimated for this model in a completely decentralized way if one has access to $z_i(t)$ at every instant of time. This can be assured by placing PMUs within each utility area such that the generator buses inside that area become geometrically observable, measuring the voltage and currents at the PMU buses, and thereafter computing the generator bus voltage $V_i \angle \theta_i$ and current $I_i \angle \phi_i$ (or equivalently, I_{di} and I_{qi}) from those measurements. As the PMU measurements will be corrupted with noise, the estimated values of $z_i(t)$ will not be perfect. They should rather be denoted as $\tilde{z}_i(t) = z_i(t) + n_i(t)$, where $n_i(t)$ is a Gaussian noise. An unscented Kalman filter (UKF) is next designed as

$$\dot{\hat{x}}_i(t) = g(\hat{x}_i(t), \tilde{z}_i(t), u_i(t), \alpha_i), \quad \hat{x}_i(0) = \hat{x}_{i0} \tag{7}$$

producing the state estimates $\hat{x}_i(t)$ for the ith generator at any instant of time t. For details of the construction of this UKF, please see [26]. The estimator can be installed directly at the generation site to minimize the communication of signals, and made to run continuously before and after any disturbance.

The network equations that couple (x_i, z_i) of the ith generator in (6) to the rest of the network can be written as

$$0 = I_{di}(t)V_i(t)\sin(\delta_i(t) - \theta_i(t)) + I_{qi}(t)V_i(t)\cos(\delta_i(t) - \theta_i(t)) + P_{Li}(t)$$
$$- \sum_{j=1}^{m} V_i(t)V_j(t)(G_{ij}\cos(\theta_{ij}(t)) + B_{ij}\sin(\theta_{ij}(t))) \tag{8}$$
$$0 = I_{di}(t)V_i(t)\cos(\delta_i(t) - \theta_i(t)) - I_{qi}(t)V_i(t)\sin(\delta_i(t) - \theta_i(t)) + Q_{Li}(t)$$
$$- \sum_{j=1}^{m} V_i(t)V_j(t)(G_{ij}\cos(\theta_{ij}(t)) - B_{ij}\sin(\theta_{ij}(t))), \tag{9}$$

where m is the total number of buses in the network. Here, $\theta_{ij} = \theta_i - \theta_j$, P_{Li} and Q_{Li} are the active and reactive power load demand at bus i, and G_{ij} and B_{ij} are the conductance and susceptance of the line joining buses i and j. As shown in [25], the variables $z_i(t)$ in (6) can be eliminated using (8)–(9) by a process called Kron reduction. The resulting dynamic model is linearized about a given operating point, and the small-signal model for the power system is written as

$$\dot{x}(t) = A_c x(t) + B_c u(t) \tag{10}$$
$$y(t) := \Delta\omega(t) = C x(t). \tag{11}$$

In this model, $x(t) \in \mathbb{R}^{5n}$ and $u(t) \in \mathbb{R}^n$ now represent the small-signal deviation of the actual states and excitation inputs of the n generators from their pre-disturbance equilibrium. We define $y(t) \in \mathbb{R}^n$ as a *performance* variable, based on which the closed-loop performance of the overall system as well as the saturation limits on

$u(t)$ can be judged. Ideally, this can be chosen as the vector of all electromechanical states. For simplicity, y is often chosen as the vector of the small-signal generator frequency $\Delta\omega(t)$ as frequency is the most effective electromechanical state for evaluating damping. The input $u(t)$ is commonly used for designing feedback controllers such as Power System Stabilizers (PSS), which takes local feedback from the generator speed, and passes it through a lead-lag controller for producing damping effects on the oscillations in phase angle and frequency. PSSs, however, are most effective in adding damping to the fast oscillation modes in the system, and perform poorly in adding damping to the slow or inter-area oscillation modes [20]. Our goal is to design a supplementary controller $u(t)$ for the model (10)–(11) on top of the local PSS by using state-feedback from either all or selected sets of other generators. If the state vector of every generator is fed back to every other generator, then the underlying communication network must have a complete graph, i.e., an all-to-all connection topology. If the state vectors of only a selected few generators are fed back to the controllers of some other generators, then the communication network may have a sparse structure. In either case, since long-distance data transfer is involved, we refer to this controller as a wide-area controller.

3   Controller Design

Several papers such as [9, 10] have posed the wide-area control problem as a constrained optimal control problem of the form:

$$
\begin{aligned}
\min_{K} J &= \frac{1}{2} \int_0^{\infty} (x^T(t)Qx(t) + u^T(t)Ru(t))dt \\
\text{s.t.,} \quad \dot{x}(t) &= Ax(t) + Bu(t) \\
u(t) &= K\hat{x}(t), \quad K \in \mathcal{S}, \ Q > 0, \ R \geq 0,
\end{aligned} \tag{12}
$$

where the state estimate $\hat{x}(t)$ follows from (7). The choice of the objective function J depends on the goal for wide-area control. For power oscillation damping, this function is often simply just chosen as the total energy contained in the states and inputs as in (12). For wide-area voltage control, it can be chosen as the setpoint regulation error for the voltages at desired buses [27], while for wide-area protection, it can be chosen as the total amount of time taken to trigger relays so that fault currents do not exceed their maximum values [6]. The set $\mathcal{S}$ is a structure set that determines the topology of the underlying communication network. In the current state of the art, most synchronous generators operate under a completely decentralized feedback from its own speed measurement only. Thus in today's power grid, the structure set $\mathcal{S}$ is reflected in K as

$$K = \begin{bmatrix} K_1 & 0 & 0 & \cdots & 0 \\ 0 & K_2 & 0 & \cdots & 0 \\ & & \vdots & & \\ 0 & 0 & 0 & \cdots & K_n \end{bmatrix}, \quad K_i = \begin{bmatrix} 0 & 1 & 0 & 0 & 0 \end{bmatrix} \tag{13}$$

where K_i is the PSS controller for the ith generator, $i = 1, \ldots, n$, whose state vector follows from (6). However, as pointed out before, decentralized feedback can damp only the high-frequency oscillations in the power flows. Their impact on inter-area oscillations that typically fall in the range of 0.1–2 Hz is usually small [20]. Inter-area oscillations arise due to various coherent clusters in the power network oscillating against each other following a disturbance. If left undamped they can result in catastrophic failures in the grid. In fact, both the 1996 blackout on the west coast and the 2003 blackout on the eastern grid of the United States were largely caused because of the lack of communication between generators resulting from the decentralized structure of K in (13). Triggered by the outcomes of these events, over the past few years, utility companies have gradually started moving away from the structure in (13) to a slightly more global structure as

$$K = \begin{bmatrix} K^1 & 0 & 0 & \cdots & 0 \\ 0 & K^2 & 0 & \cdots & 0 \\ & & \vdots & & \\ 0 & 0 & 0 & \cdots & K^r \end{bmatrix} \tag{14}$$

where $K^i \in \mathbb{R}^{n_i \times 5n_i}$, n_i being the number of generators within the operating region of the ith utility company, and r being the total number of such companies. All the elements of the block matrix K_i may be nonzero, meaning that the generators inside an area communicate their state information with each other to compute their control signals. The resulting controller K is block diagonal, and, therefore, a better *wide-area* controller than (13) since here at least the local generators are allowed to interact. An ideal wide-area controller, however, would be one whose off-block-diagonal entries are nonzero as well, meaning that generators across the operating areas of different companies are allowed to exchange state information. In the worst-case, K can be a standard LQG controller, and, therefore, a dense matrix with every element nonzero, which means that the communication topology is a complete graph. In reality, however, all-to-all communication can be quite expensive due to the cost of renting communication links in the cloud [28], if not unnecessary. Papers such as [10, 29] have proposed various graph sparsification algorithms based on l_1-optimization to develop controllers that require far less number of communication links without losing any significant closed-loop performance. Papers such as [9], on the other hand, have proposed the use of modal participation factors between generator states and the inter-area oscillation modes to promote network sparsity in a more structured way. Papers such as [30–32] have proposed various projection and decomposition-based control designs by which a significant portion of the communication network admits a broadcast-type architecture instead of peer-to-peer connectivity, thereby saving on the number of links. The main challenge of implementing these types of wide-area controllers, however, is the fact that utilities are still shy in terms of sharing their PMU measurements and state information with other utilities due to which they prefer to stick to the block-decentralized structure of (14). Besides these network-centric approaches, other control designs based on more traditional approaches such as adaptive control [7], robust control [3], and hybrid control [8] have also been proposed for wide-area oscillation damping in presence of various model and operational uncertainties.

Several questions arise naturally from the design problem stated in (12). For example,

1. In practice, given the size and complexity of any typical grid, the exact values of the matrices A and B are highly unlike to be known perfectly. Moreover, the entries of these matrices can change from one event to another. Therefore, just one dedicated model created from a one-time system identification may not be suitable. Thus, one pertinent question is—how can one extend the model-based design in (14) to a more measurement-based approach where online PMU measurements from different parts of the grid can be used to estimate the small-signal model of the grid, preferably in a recursive way, based on which the control signals can be updated accordingly? This estimation should also be preferably carried out in a distributed way over the sparse communication topology generated from the controller. While traditional notions of adaptive control can be highly useful here, the speed of estimation may suffer if the entire model needs to be estimated. Newer ideas from reinforcement learning, adaptive dynamic programming, and Q-learning can be useful alternatives in this case as they tend to optimize the objective function directly by bypassing the need for estimating the model.

2. Another question is—whether it is necessary to base the design of the wide-area controller $u(t)$ on the entire state-space model (14), or does it suffice to design it using a reduced-order model only? For example, recent papers such as [30] have shown that one can make use of singular perturbation based model reduction for designing $u(t)$ in a highly scalable way for consensus networks using a hierarchical control architecture. This is especially true if the grid exhibits spatial clustering of generators due to coherency. Therefore, following [30] one idea can be to estimate a reduced-order network model, where every network node represents an equivalent generator representing an entire cluster, design an aggregated control input for each equivalent generator, and then broadcast the respective inputs to every generator inside that cluster for actuation. Open questions for these types of designs would include proofs for stability, sensitivity of closed-loop performance to estimation and model reduction errors, derivation of numerical bounds for closed-loop performance as a function of the granularity of model reduction, and so on.

3. The third question is—how can one robustify the wide-area controller (14) against the typical uncertainties in both the physical layer and the communication layer? Uncertainties in the physical grid model, for example, can easily

arise from the lack of knowledge of the most updated model parameters in the set α_i in (6). With increase in renewable penetration and the associated intermittencies in generation profiles, operational uncertainties are gradually increasing in today's grid models [33]. Similarly, significant uncertainties will exist in the cyber-layer models as well including uncertainties due to delays, congestion, queuing, routing, data loss, synchronization loss, quantization, etc. A natural concern, therefore, is—how can the design in (14) be made aware of these uncertainties so as to optimize the closed-loop performance of variables for the entire CPS grid?

Many other CPS-centric design and implementation questions related to scalability, centralized versus distributed implementation, speed of computation, big data analytics in the loop, and codependence of (14) on other state estimation and control loops (for example, those using SCADA) can also arise. All of these questions deserve dedicated attention from researchers with backgrounds in control theory, power systems, signal processing, machine learning, computer science, communication engineering, economics, and information theory. In the following sections, we highlight several of these CPS research challenges for wide-area control.

4 Cyber-Physical Implementation of Wide-Area Controllers

Once designed, wide-area controllers as (12) need to be implemented in a distributed way by transmitting outputs measured by PMUs over hundreds of miles across the grid to designated controllers at the generation sites. Depending on the number of PMUs, the rate of data transfer can easily become as high as hundreds of Terabytes per second. The timescale associated with taking these control actions can be in fractions of seconds. Therefore, controlling latencies and data quality, and maintaining high reliability of communication are extremely important for these applications. The media used for wide-area communications typically use longer range, high-power radios, or Ethernet IP-based solutions. Common options include microwave and 900 MHz radio solutions, as well as $T1$ lines, digital subscriber lines (DSL), broadband connections, fiber networks, and Ethernet radio. The North American Synchrophasor Initiative (NASPI), which is a collaborative effort between the U.S. Department of Energy (DOE), the North American Electric Reliability Corporation (NERC), and various electric utilities, vendors, consultants, federal and private researchers and academics is currently developing an industrial grade, secure, standardized, distributed, and scalable data communications infrastructure called the NASPI-net to support Synchrophasor applications in North America. NASPI-net is based on an IP Multicast subscription-based model. In this model, the network routing elements are responsible for handling the subscription requests from potential PMU data receivers as well as the actual optimal path computation, optimization, and recomputation and rerouting when network failures happen. The schematic diagram

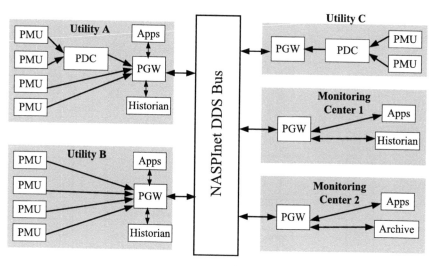

Fig. 1 Architecture of NASPInet [23]

of the NASPI-net is shown in Fig. 1. An excellent survey of NASPI-net can be found in [34].

The implementation of the state estimator (7) and the controller (12) in the NASPI-net may be done in the following way. PMU measurements $y_i(t)$ from each utility company i, as shown in Fig. 1, are first gathered in a local Phasor Data Concentrator (PDC) at the local substation using a local-area network, assuming that the PDC is located at a reasonably close geographical distance from all the PMUs in that service area. The generator bus variables $z_i(t)$ are computed from $y_i(t)$ using Kirchoff's laws at this PDC, and the decentralized state estimator (7) may be run to generate $\hat{x}_i(t)$. This state estimate then enters the NASPI-net data bus through a Phasor Gateway (PGW). The PGW has high levels of security encryption so as to prevent the flow of any malicious data between the local PDC and the data bus. The state estimates are then communicated via standard Internet protocols to the designated controllers of other generators so that they can be used for computing the control signal $u_j(t)$ for the jth generator, which finally gets actuated using the excitation control system located at this generator.

A slightly different cyber-physical architecture for implementing these types of controllers has recently been proposed in [21]. A similar distributed communication architecture for open-loop oscillation estimation using PMU data is also presented in [35]. This architecture, shown in Fig. 2, is very similar to the NASPI-net except that here the estimation of the states, and the computation of the control signals are not done at the PDC or at the generator, but entirely inside a cloud computing network. For example, PMU measurements from inside a service area are still gathered at its local PDC, but this PDC does not generate the state estimate. It rather ensures that all the measurements are properly synchronized with respect to each other, that their

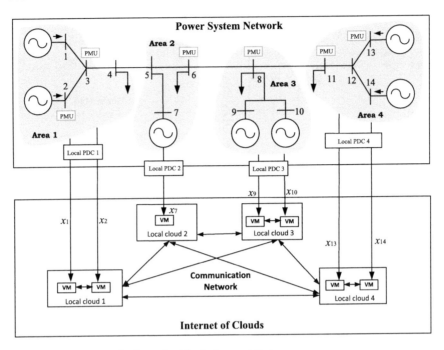

Fig. 2 Wide-area control using a cloud-in-the loop architecture

measurement noise is within acceptable limits, and that the measurements do not
consist of any *bad* data due to GPS errors, or errors in the phase-locked loop inside
the PMUs. The PDC then relays all the measurements to a local service-based private
cloud owned by the utility company, wherein they are gathered in a virtual computer
or Virtual Machine (VM), created on the fly using the available computation resources
in the cloud. The measurements are digitally represented as a periodic stream of data
points. The geographical location of the VMs can be close to that of the generators
in that area so that the latency from PDC-to-cloud communication is small. The
state estimator (7) can then be employed in this VM. The local clouds themselves
are connected to each other through an Internet of clouds, as shown in the figure.
VMs in every local cloud, depending on the sparsity structure demanded by the
controller $u(t) = K\hat{x}(t)$, are connected to other remote VMs through an advanced,
secure, third-party wide-area communication network such as SDN, an example of
which can be Internet2. The VMs can then exchange their estimated states $\hat{x}_i(t)$, and
compute their control signals $u_i(t)$ through this network in a completely distributed
way.

In reality, depending on the number of PMUs inside any service territory, the
local cloud of a utility company may have multiple VMs, each of which receives a
designated chunk of the local PMU measurements from the local PDC, as shown in
the figure. These VMs communicate with their neighboring VMs inside the cloud as
well as to those across other clouds for exchanging PMU data, and for computing

control signals via predetermined sparse feedback control laws such as (14). The control signals are, thereafter, transmitted back from the local cloud to the actuators of the corresponding generators inside their respective service regions. The resulting system is referred to as a *cloud-in-the-loop* wide-area control system. End-to-end delay specifications for wide-area control using this type of wide-area communication have been presented in [22]. Two main advantages of this architecture compared to NASPI-net are that all the computations are done inside a third-party cloud, thereby preserving the privacy of PMU data from direct exchange between the utilities, and that the communication between the clouds no longer has to be based on standard Internet protocols but can rather use more advanced networking technologies such as SDN with additional layers of network controllability.

For either architecture, the shared communication network will have delays arising from routing and queuing, besides the usual transmission delays due to the geographical distance between the VMs. Three classes of delays may be defined, namely— small delays τ_s that arise due to message queuing inside any virtual machine, thereby delaying the availability of the measured state variable assigned to that machine for computing the corresponding control signal; medium delays τ_m that arise due to communication between two virtual machines that are part of the same local cloud, and large delays τ_l that arise due to communication between any two virtual machine that are located in two different local clouds. The stochastic end-to-end delay experienced by messages in an Internet-based wide-area communication link can be modeled using three components:

1. The minimum deterministic delay, say denoted by m,
2. The Internet traffic delay with Probability Density Function (PDF), say denoted by ϕ_1, and
3. The router processing delay with PDF, say denoted by ϕ_2.

The PDF of the total delay at any time t was written in terms of these three components as

$$\phi(t) = p\,\phi_2(t) + (1-p) \int_0^t \phi_2(u)\phi_1(t-u)du, \quad t \geq 0, \tag{15}$$

Here, p is the probability of the open-period of the path with no Internet traffic. The router processing delay can be approximated by a Gaussian density function

$$\phi_2(t) = \frac{1}{\sigma\sqrt{2\pi}} e^{-\frac{(t-\mu)^2}{2\sigma^2}}, \tag{16}$$

where $\mu > m$. The Internet traffic delay is modeled by an alternating renewal process with exponential closure period when the Internet traffic is on. The PDF of this delay is given by

$$\phi_1(t) = \lambda e^{-\lambda t}, \tag{17}$$

where λ^{-1} models the mean length of the closure period. The Cumulative Distribution Function (CDF) of this delay model can then be derived as

$$P(t) = \int_{-\infty}^{t} \phi(s)ds = \frac{1}{2}\left[\text{erf}\left(\frac{\mu}{\sqrt{2}\sigma}\right) + \text{erf}\left(\frac{t-\mu}{\sqrt{2}\sigma}\right)\right]$$
$$+ \frac{(p-1)}{2}e^{(\frac{1}{2}\lambda^2\sigma^2+\mu\lambda)}e^{-\lambda t}\left[\text{erf}\left(\frac{\lambda\sigma^2+\mu}{\sqrt{2}\sigma}\right) + \text{erf}\left(\frac{t-\lambda\sigma^2-\mu}{\sqrt{2}\sigma}\right)\right]. \quad (18)$$

Random numbers arising from this CDF can be used for simulating delays of the form τ_s, τ_m and τ_l, as defined above. The challenge, however, is to translate these single-path models to multipath, multi-hop, shared network models where background traffic due to other applications may pose serious limitations in latencies. Recent references such as [36, 37] have provided interesting theoretical tools such as Markov jump process, Poisson process, multi-fractal models, and Gaussian fractional sum-difference models for modeling delay, packet loss, queuing, routing, load balancing, and traffic patterns in such multichannel communication networks.

5 Cyber-Physical Codesigns

The next question is—how can the controller in (12) be codesigned with the information about τ_s, τ_m and τ_l? The conventional approach is to hold the controller update until all the messages arrive at the end of the cycle. However, this approach may result in poor closed-loop performance, especially for damping of the inter-area oscillations. In recent literature, several researchers have looked into delay mitigation in wide-area control loops [38–41], including the seminal work of Vittal and coauthors in [3] where $\mathcal{H}_\infty$ controllers were designed for redundancy and delay insensitivity. All of these designs are, however, based on worst-case delays, which make the controller unnecessarily restrictive, and may degrade closed-loop performance. In [21], this problem was addressed by proposing a delay-aware wide-area controller, where the feedback gain matrix K was made an explicit function of delays using ideas from arbitrated network control theory [42].

For example, considering the three delays τ_s, τ_m and τ_l defined in the previous section, one may update the control input at any VM as new information arrives instead of waiting till the end of the cycle. If tweaking the protocols is difficult, then another alternative strategy will be to estimate the upper bounds for the delays using real-time calculus [43, 44]. The approach is referred to as *arbitration*, which is an emerging topic of interest in network control systems. Based on the execution of the three protocols, one may define two modes for the delays—namely, *nominal* and *overrun*. If the messages meet their intended deadlines, they are denoted as nominal. If they do not arrive by that deadline, they are referred to as overruns. Defining two parameters τ_{th1} and τ_{th2} such that $\tau_{th1} \leq \tau_{th2}$, one may define nominal, skip, and abort cases for computing the wide-area control signal as:

- If the message has a delay less than τ_{th1}, we consider the message as the nominal message of the system and no overrun strategy will be activated.

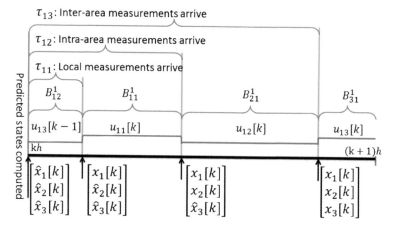

Fig. 3 Discrete-time delays $\tau_1 < \tau_2 < \tau_3$. The vector indicated next to the arrow represents the state used to compute input u_{ij} (reproduced from [21] ©2014 IEEE and used with permission)

- If the message suffers a delay greater than τ_{th1} and less than τ_{th2}, the message will be computed; however, the computations of the following message will be skipped.
- If the message suffers a delay greater than τ_{th2}, the computations of the message will be aborted, and the message is dropped. This strategy is motivated by assuming that the messages will be significantly delayed and are no longer useful.

Accordingly, a feasible way to formulate the execution rules can be: (1) if $\tau_{th1} \leq \tau_{th2} \leq \tau_{wcet}$, where τ_{wcet} is the worst-case delay in the network, both abort and skip can happen, (2) Abort Only: if $\tau_{th1} = \tau_{th2} < \tau_{wcet}$, the message will be dropped if they miss their first deadline, and (3) Skip Only: if $\tau_{th1} \leq \tau_{wcet}$ and $\tau_{th2} \geq \tau_{wcet}$. One idea will be to set $\tau_{th2} = \tau_{wcet}$ to develop a constructive strategy to determine τ_{th1}. This step can be a significant improvement over conventional network control designs in terms of both closed-loop performance and resource utilization.

To justify this approach, we cite an example from the control design presented recently in [21]. The sampling interval h of the PMUs between two consecutive control updates was broken down into three smaller intervals at which the inputs were updated as new measurements arrive, as shown in Fig. 3. If any state is unavailable, it is replaced by its predicted value. For the first generator, for example, the input $u_1(k)$ is further divided to $\left[u_{11}(k)\ u_{12}(k)\ u_{13}(k)\right]$, where $u_{ij}(k)$ denotes the input of the ith generator adjusted using the measurements of jth generator. Repeating the same logic for all generators, it was shown in [21] that the discrete-time model of the system can be written as

$$x(k+1) = Ax(k) + \sum_{i=1}^{n}\sum_{j=1}^{m(i)} B_{j1}^{i}u_{ij}(k) + \sum_{i=1}^{n} B_{i2}^{i}u_{ik(i)}(k-1), \qquad (19)$$

where $m(i)$ shows the number of times that the inputs are updated in each generator, and $k(i)$ is the index of the largest delay, or equivalently as

$$x(k+1) = Ax(k) + B_2u(k-1) + B_1u(k). \tag{20}$$

In other words, the excitation controller needs feedback from the current state samples as well as the past input samples to stabilize the closed-loop swing dynamics with communication delays. An open problem is to derive the equivalent expressions of $u(k)$ for various typical protocols used for wide-area communication, develop tractable and scalable ways for tuning the control gains to guarantee closed-loop stability and performance while promoting sparsity in the network structure as indicated by the set $\mathcal{S}$ in (14), and, most importantly, validating these communication and control laws using realistic cyber-physical testbeds. These points are explained in the following sections.

6 SDN, NFV, and Cloud Computing

As mentioned earlier, typically the Internet cannot provide the required latency and packet loss performance for grid operation under high data rates. Moreover, the network performance is highly random, and therefore, difficult to model accurately. That is why the cloud-based WAMS architecture proposed in Fig. 2 is currently garnering a lot of attention from power system engineers. However, limited studies have been conducted so far to leverage all possible benefits of cloud computing, Software-Defined Networking (SDN), and Network Function Virtualization (NFV), to accelerate this development [24]. With the recent revolution in networking technology, these new communication mechanisms can open up several degrees of freedom in programmability and virtualization for computation and communication platforms. However, customized SDN control and protocols, and sufficient experimental validation using realistic testbeds are still missing in almost all wide-area control applications.

Latency and data loss rate are important factors in the performance of all wide-area control and protection applications. Software such as Real-Time Dynamics Monitoring System (RTDMS), Phasor Grid Data Analyzer (PGDA), and GridSim are used for online oscillation monitoring using Synchrophasors. A list of related open-source software can be found in [45]. These simulation engines need to be integrated into executable actions so that results from the monitoring algorithms can be exported to a custom SQL database that can be set to trigger alerts or alarms whenever damping levels of oscillatory modes fall below prespecified thresholds. These alarm signals need to be communicated to the operator through a reliable communication network so that the operator can take manual actions to bring the damping back to acceptable levels [46]. In recent years, simulation platforms such as ExoGENI-WAMS [47] have been developed to emulate such communication platforms. The computation and communication planes are entirely shifted away from the physical infrastructure, similar to the architecture proposed in Fig. 2. Another example of a CPS simulator is GridSim [48]. The data delivery component in this simulation platform, also referred to as *GridStat*, is a publish-subscribe middleware, which allows for encrypted multicast delivery of data. GridStat is designed to meet the requirements of emerging control and protection applications that require data delivery latencies on the order of 10–20 ms over hundreds of miles with extremely high availability.

Similar to GridStat, the VMs in any cloud-in-the-loop CPS simulator may consist of two communication planes, namely a data plane and a management plane. The data plane is a collection of Forwarding Engines (FEs) designed to quickly route received messages on to the next VMs. The FEs are entirely dedicated to delivering messages from publishers to subscribers. Routing configuration information is delivered to the FEs from the management plane. The forwarding latency through an FE implemented in software is generally on the order of $100\,\mu s$, and with network processor hardware, it is less than $10\,\mu s$. The management plane, on the other hand, is a set of controllers, called QoS brokers, that manage the FEs of the data plane for every VM. The Quality-of-Service (QoS) brokers can be organized in a hierarchy to reflect the natural hierarchy in the physical infrastructure of the grid model. When a subscriber wishes to receive data from a publisher, it communicates with a QoS broker that designs a route for the data, and delivers the routing information to the relevant FEs and VMs, creating the subscription.

However, simulation platforms such as GridStat are just starting points for research. Much more advanced cloud computing and SDN protocols as well as their emulation software need to be developed for the future grid. One outstanding challenge is to develop a reliable communication software that will enable timely delivery, high reliability, and secure networking for these emulators. Timeliness of message requires guaranteed upper bounds on end-to-end latencies of packets. Legacy networking devices do not provide such guarantees, neither for commodity Internet connections nor for contemporary proprietary IP-based networks that power providers may operate on. Moreover, direct communication lacks rerouting capabilities under real-time constraints, and resorts to historic data when communication links fail. A promising solution for both timeliness and link failures can be the idea of *Distributed Hash Tables* (DHT), which was recently introduced for wide-area control applications in [49].

7 Cybersecurity

Another critical CPS challenge for both wide-area monitoring and control is the issue of cybersecurity. While security is an universally growing concern for applications at every layer of the grid, ranging from distribution grids to power markets, the challenge for WAMS is especially critical since the stakes here are much higher. The integration of cyber components with the physical grid introduces new entry points for malicious attackers. These points are remotely accessible at relatively low risks to attackers compared to physical intrusions or attacks on substations. They can be used to mount coordinated attacks to cause severe damages to the grid, resulting in catastrophic blackouts with billions of dollars worth of economic loss. One eye-opening example of such an attack was the Stuxnet [50]. Attacks can be originated on the cyber-layer as well to trigger cascading events leading to damages on physical facilities, leading to major outages. While mathematical models have been developed to model electrical faults and device failures, there are far less reliable ways of modeling and simulating realistic scenarios for different types of cyberattacks happening in a power grid. Several universities and national laboratories have only recently started developing simulation testbeds to emulate these vulnerability scenarios. Research demonstration events such as Cyber Security Awareness Month, which is a student-run cybersecurity event in the US, have been introduced by the Department of Homeland Security [51]. Many organizations are working on the development of smart grid security requirements including the NERC Critical Infrastructure Protection (NERC-CIP) plan, the National Infrastructure Protection Plan (NIPP), and National Institute of Standards and Technology (NIST). The goal is for power system operators to work with these standards organizations to develop simulation software that can model, detect, localize, and mitigate cyber vulnerabilities in the grid as quickly as possible. A more detailed overview of these methods will be provided in the forthcoming chapter by Wang.

Given the size, complexity, and enormous number of devices present in a typical grid, developing one unique solution for securing the grid from cyberattacks is probably impossible. Even the attack space can easily become huge, ranging from Denial-of-Service (DoS) attack on communication links, to disabling of physical assets, to data corruption, to GPS spoofing, to eavesdropping. Every physical application, whether it be state estimation, oscillation monitoring, Automatic Generation Control (AGC), or wide-area damping control, would need its own way of dealing with each of these attacks. One common solution to make all of these applications more resilient is to switch from centralized to distributed implementation, as alluded to in the previous sections. Although distributed communication opens up more points for an attacker to enter through, it also provides resilience through redundancy. For example, the distributed architecture shown in Fig. 4 was recently used in [52] for the purpose of wide-area oscillation monitoring. The idea was to carry out distributed estimation of the eigenvalues of the small-signal model of a power system using multiple VMs following the cloud-in-the-loop architecture of Fig. 2. If any of the VMs in a local cloud is disabled by a cyberattack, then one option is to quickly assign another estimators to take up the role of the disabled estimator. An alternative option is to run distributed localization algorithms such as those proposed in [52] to identify the faulty VMs, and eliminate them from the cloud.

So far most attack mitigation and localization methods in the literature are geared towards open-loop applications such as state estimation or oscillation estimation. Much more work is needed to extend these methods to closed-loop wide-area control, such as the design in (14). For example, an LQR wide-area controller was designed in [53] using a sparse communication graph $\mathcal{G}$. The example was also cited in [54]. The nonlinear model of the IEEE 39-bus power system model was simulated with this sparse state-feedback controller using Power System Toolbox (PST) in Matlab.

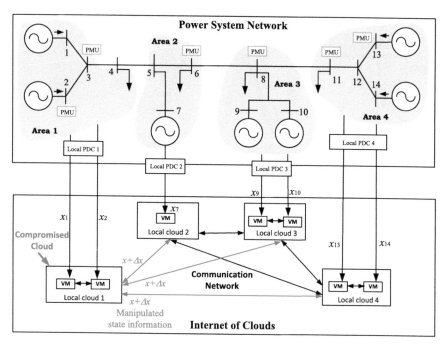

Fig. 4 Cyberattacks on wide-area monitoring and control (reproduced from [52] ©2018 IEEE and used with permission)

A fault was induced at $t = 0$, and the small-signal speed deviations of the synchronous generators were recorded, as shown in Fig. 5. The closed-loop system behavior is observed to be stable. At $t = 10$ s, a DoS attack induced on the communication link connecting generators 1 and 8, which means that these two generators are no longer capable of exchanging state information for their control actions. Instability is noticed immediately, with the frequency swings diverging with increasing amplitude at a frequency of roughly 0.05 Hz. This can be seen in Fig. 5 onwards from $t = 10$ s. At $t = 60$ s, the communication links connecting generators (2, 6) and (3, 6) are added, and the corresponding control gains are recomputed. The system is seen to regain stability, indicating that the attack has been successfully mitigated. The frequency deviations are all seen to converge to zero over time. This example shows the importance of developing formal recovery rules of wide-area controllers in face of attacks. It also highlights the need for developing effective simulation packages for investigating the impact of attack scenarios. Simulations, in fact, can reveal the most important pairs of generators that must communicate to maintain stable operating condition before and after an attack, and also the lesser important pairs that, either due to large geographical separation or weak dynamic coupling, do not necessarily add any significant contribution to stability. Software packages for illustrating other types of attack scenarios such as data manipulation attacks, jamming, eavesdropping, and GPS spoofing also need to be developed [55, 56].

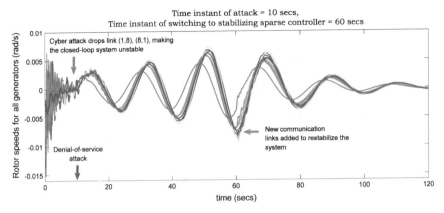

Fig. 5 Time response of generator speeds before, during, and after DoS attack (reproduced from [54] ©2017 IEEE and used with permission)

8 Simulation of Wide-Area Controllers

Besides design and implementation, an equally important challenge for WAMS engineers is to develop simulation platforms where extreme-scale power grid models, and their associated wide-area monitoring and control algorithms can be simulated for various contingencies against various system-level uncertainties. As described in the introduction, the grid itself is changing because of the proliferation of new technologies. New sources of renewable generation that are intermittent and have no rotating inertia are being added rapidly. New types of loads such as electric vehicles and smart buildings are also proliferating. Power electronic converters and controllers are being introduced to connect these new generation sources, load and storage devices to the grid. The grid is being overlaid with more PMUs, communications, computers, information processors, and controllers. The challenge is that to fully showcase the capability of any wide-area controller, all these new technologies must be modeled and simulated together. Growth in computing power shows no signs of slowing down, so we do not foresee any limitations on model size or algorithm speed. However, the ability to utilize such powerful simulations depends on how easy their handling can be made to the engineers. Moreover, power systems today are gradually becoming an integral part of other interconnected infrastructures such as gas networks, transportation networks, communication networks, water networks, economics, and food chain networks. Thus, interoperability of simulation programs will become a key to minimizing the manual effort needed to set up and run co-simulations of these systems with PMU-assisted monitoring and control loops in grid models. We highlight two potential ways by which these co-simulations can be handled.

8.1 Parallel Computing

An obvious approach to speed up simulations is to utilize parallel or distributed computing. Although specially written programs for particular parallel architectures can provide high speed-ups, the rapidly changing hardware and software makes it impossible to keep modifying the simulation programs to keep up. The trend today is to use multiprocessor computers with compilers that can distribute the computation optimally to multiple processors. For wide-area monitoring and control simulations, some applications are much more amenable to parallelization than others. For example, any simulation that requires running many contingencies can run these hundreds of contingency cases in separate processors. The dynamics of individual generators can be run in parallel but the network that connects the generators has to be solved simultaneously. It turns out, however, that the algebraic equations representing the network cannot be parallelized very efficiently and become the main bottleneck for speeding up power system simulations. Parallel computing can also be used for gain-scheduling of robust wide-area controllers.

8.2 Hybrid Simulations

In order to implement fast wide-area control strategies, it is highly desirable to have a faster than real-time simulation of that model in hand. One promising solution is hybrid mixed-signal hardware emulation. In these emulations, hardware-based digital simulation and analog simulation can be used in an integrative way to achieve a massively parallel and scale-insensitive emulation architecture. There have been several attempts at building hardware accelerators in the past [57, 58]. For example, the grid models can be emulated in hardware by a coupled set of oscillators, resistors, capacitors, and active inductors. A higher frequency can be used so as to suit the scale of on-chip elements and permit faster than real-time operation. These elements then need to be built on chip, and connected with a customizable switch matrix, allowing a large portion of the grid to be modeled in real-time. Researchers have proposed the use of Verilog-AMS model, which is a hardware modeling language that includes features for Analog and Mixed Signal (AMS) elements [59]. It can incorporate equations to model analog subcomponents. The AMS model can be designed to emulate open-loop models of very large-scale power systems with tens of thousands of buses with built-in equations for AC power flow, electromagnetic and electromechanical dynamics of synchronous generators and induction generators, AC load models, power oscillation damping control, voltage control, droop control, AGC, and PSS. Several challenges that still stand on the way for developing at-scale faster than real-time simulations are:

1. How to synthesize the transmission network without unnecessary and unrealistic assumptions, and approximations using state-of-the-art microelectronics design technology?

2. How to develop a scalable mixed-signal emulation architecture capable of large power systems with tens of thousands nodes?
3. How to design configurable units of emulation on a chip so that any large power system with realistic transmission connections can be realized via software-based configuration?

Research activities are underway on building VLSI chips that can scale to large grid models for faster than real-time emulators, especially directed towards transient wide-area AC emulations. In [58], for example, predictive simulations were shown to be very useful for real-time path rating.

9 Simulation Testbeds

Gaining access to realistic grid models and PMU data owned by utility companies can be difficult due to privacy and nondisclosure issues. More importantly, in many circumstances, even if real data are obtained they may not be sufficient for studying the detailed operation of the entire system because of their limited coverage. To resolve this problem, several smart grid simulation testbeds have recently been developed to facilitate hardware-in-loop simulation of different grid applications without the need for gaining access to real data. Selected examples of such testbeds in the United States include CPS testbeds at Washington State University using GridStat [48], cloud-assisted wide-area control testbed at North Carolina State University using ExoGENI [47], cybersecurity testbeds at Iowa State [60], TCIPG in University of Illinois [61], DETER-lab testbed at University of Southern California [62], CPS testbeds at Idaho National Lab, Cornell University, and Pacific Northwest National Labs, and a big data hub at Texas A&M [63]. A comprehensive list of many other smart grid testbeds and their CPS capabilities was recently presented in the survey paper [64]. Two key questions that most of these testbeds are trying to answer are—(1) is it possible to design sufficiently general CPS standards and protocols to support a mass plug-and-play deployment of a wide-area grid without sacrificing reliability, data privacy, and cybersecurity, and (2) if so, then what standards and protocols are required to transform today's grid into an end-to-end enabler of electric energy services.

9.1 Hardware Components

Generally speaking, the physical components of these testbeds are comprised of Real-Time Digital Simulators (RTDS) and Opal-RT. These are power system simulator tools that allow real-time simulation of both transmission and distribution models with a time-step of $50 \mu s$. The RTDS comes with its own proprietary software known as RSCAD, which allows the user to develop detailed dynamic models of various components in prototype power systems. The RTDS also comes with digital cards that allow external hardware to interface with the simulation. For example, the Gigabit Transceiver Analog Output (GTAO) card allows the user to view low-level signals proportional to voltages and currents at different buses of the system in real time. The GTAO card generates voltage and current waveforms, and communicates them to sensors such as relays, circuit breakers, and PMUs. The PMUs measure these signals, and send the resulting digitized phasor data calculations to the PDCs. The PDC time-stamps and collects the data from all the PMUs, and sends them to the server for display and archival, when requested. The hardware and the software layers of these testbeds are integrated with each other to create a substation-like environment within the confines of research laboratories. The two layers symbiotically capture power system dynamic simulations as if these measurements were made by real sensors installed at the high-voltage buses of a real transmission substation.

9.2 Cyber and Software Components

The cyber-layer, on the other hand, is generally emulated by either a local-area network or a local cloud service. The ExoGENI-WAMS testbed at NC State [47], for example, is connected to a state-funded, metro-scale, multilayered advanced dynamic optical network testbed called Breakable Experimental Network (BEN) that connects distributed cloud resources in local universities [65]. It allows one to set up dynamic multilayer connections of up to 10 Gbps between different sites. One may simulate different types of disturbance events in power system models in RTDS, collect the emulated grid responses via PMUs and other sensors, communicate these data streams via BEN, and run virtual computing nodes at various sites in the ExoGENI cloud overlaid on top of BEN to execute distributed estimation and control algorithms. Some open questions for these types of setups in the future are, for example—where to deploy the computing facilities to better facilitate data collection and processing, and how to design better communication topologies.

9.3 A Network of Remote Testbeds

One pertinent question is whether these different simulation testbeds at different locations should be conjoined with each other to create a much bigger nationwide network of CPS testbeds. And if yes, then what are the most common challenges for such remote testbed federation? Developing protocols for usability by different users, and potential safety hazards are two important challenges, for example. Researchers are also contemplating making their testbeds open to public for accelerating research in the power and CPS community, but a robust economic and ethics model for sharing access to private resources still needs to be developed. Should there be a common centralized simulation testbed for accessing power system model and data, one must also resolve standardization issues, communication issues, maintenance costs, and strategies for sustainability.

9.4 *Interoperability: Databases and User Interfaces*

One of the major challenges for the users of the hundreds of existing simulations is that none of them are compatible with each other. Using two different simulations require keeping up two different databases of input data, and being familiar with two different sets of graphical outputs. This is not only true for different types of applications but also of the same application marketed by different vendors. Thus, in the current state of the art, it is impossible to integrate these different simulation programs. The easiest way to encourage interoperability is to standardize the databases. The data that go into these databases are proprietary to the utilities, and if the utilities can agree on using a standard database the simulation vendors will have to adopt it. In the USA, the National Institute for Standards and Technology (NIST) has been tasked to develop such standards. An earlier standard called the Common Information Model (CIM) is now an IEC standard, and is slowly being adopted at different rates in different countries. A similar effort should be made to standardize user interfaces for wide-area monitoring and control. If the database and user interface for a simulation are standardized, the ability to integrate different simulators as mentioned above would become much simpler.

10 Summary and Conclusions

This tutorial serves as a technical invitation to engineers for entering the challenging and attractive research field of wide-area control of power systems. We presented an overview of the main research ideas related to the cyber-physical aspects of this topic, and established a strong dependence of control on various properties of the underlying communication, computing, and cybersecurity. Evidently, there are several challenges that need to be surmounted in order to implement the proposed designs, requiring a strong knowledge of stochastic modeling, estimation theory, distributed control, system identification, model reduction, robust control, optimization, and related topics in signal processing and computer science. It is, therefore, our hope that this topic will be viewed by budding and established control theorists as a challenging and attractive opportunity. We hope that the compelling societal importance of power and energy systems will also serve as an additional motivation to enter this endeavor.

While this tutorial overviewed the general research landscape of WAMS, in the following three chapters, we present a more detailed overview of some specific research challenges associated with this technology. A preview of these chapters is as follows.

The chapter by *Chakrabortty* enlists six research challenges that need to be resolved before wide-area control can transition from a concept to a reality. These challenges include scalability of modeling and control, system identification and online learning of models from streaming or stored PMU data followed by model validation, wide-area communication and its associated uncertainties and architectural bottlenecks, cost allocation strategies for renting links in wide-area communication networks where game-theoretic algorithms can play a significant role, cybersecurity of WAMS, and finally, WAMS simulation testbeds where real-time interaction between hardware PMUs and software emulation of communication networks can be tested with high fidelity.

The chapter by *Wang* highlights several research problems on the signal processing aspects of WAMS. It presents an overview of the data-centric challenges that the power industry is currently facing while it transitions from SCADA-based monitoring and control to wide-area monitoring and control using gigantic volumes of PMU data. Data quality is an inevitable issue for the control room incorporation of these data. Because of the communication networks that were not conventionally designed to carry high-speed PMU data and the early deployment of older PMUs, data losses and data quality degradations happen quite often in practice. Different applications have diverse requirements on the data quality. This chapter provides a detailed list of challenges associated with data quality, and also cites recommendations on how trust scores can be assigned to data for taking control actions.

The chapter by *Sun* addresses the topic of wide-area protection and mitigation of cascading failures using PMU data. In the current state-of-the-art, cascading failures in the grid are prevented by separating the grid into disconnected islands. The chapter proposes the use of controlled islanding using real-time PMU data by which system efficiency can still be maintained at an acceptable level. It specifically highlights the three main challenges on when to island, where to island, and how to island. For the latter, it proposes the use of advanced numerical algorithms for predicting transient instability using PMU data, and lists a number of open challenges on how these predictive commands can be made use of in taking appropriate control and decision-making actions.

References

1. A.G. Phadke, J.S. Thorp, *Synchronized Phasor Measurements and Their Applications* (Springer, 2008)
2. N.R. Chaudhuri, D. Chakraborty, B. Chaudhuri, Damping control in power systems under constrained communication bandwidth: a predictor corrector strategy. IEEE Trans. Control Syst. Technol. **20**(1), 223–231 (2012)
3. S. Zhang, V. Vittal, Design of wide-area power system damping controllers resilient to communication failures. IEEE Trans. Power Syst. **28**(4) (2013)
4. N.R. Chaudhuri, B. Chaudhuri, S. Ray, R. Majumder, Wide-area phasor power oscillation damping controller: a new approach to handling time-varying signal latency. IET Gener. Trans. Distrib. **4**(5), 620–630 (2010)
5. R.A. Jabr, B.C. Pal, N. Martins, A sequential conic programming approach for the coordinated and robust design of power system stabilizers. IEEE Trans. Power Syst. **25**(3), 1627–1637 (2010)
6. M. Zima, M. Larsson, P. Korba, C. Rehtanz, G. Andersson, Design aspects for wide-area monitoring and control systems. Proc. IEEE **93**(5), 980–996 (2005)

7. A. Chakrabortty, Wide-area damping control of power systems using dynamic clustering and TCSC-based redesigns. IEEE Trans. Smart Grid **2**(3) (2012)
8. J.H. Chow, S.G. Ghiocel, An adaptive wide-area power system damping controller using synchrophasor data, in *Control and Optimization Methods for Electric Smart Grids* (2012), pp. 327–342
9. A. Jain, A. Chakrabortty, E. Biyik, An online structurally constrained LQR design for damping oscillations in power system networks, in *American Control Conference*, Seattle, WA (2017)
10. F. Dorfler, M. Jovanovic, M. Chertkov, F. Bullo, Sparsity-promoting optimal wide-area control of power networks. IEEE Trans. Power Syst. **29**(5), 2281–2291 (2014)
11. X. Wu, F. Dorfler, M. Jovanovic, Input-output analysis and decentralized optimal control of inter-area oscillations in power systems. IEEE Trans. Power Syst. **31**(3), 2434–2444 (2016)
12. M.E. Raoufat, K. Tomsovic, S.M. Djouadi, Virtual actuators for wide-area damping control of power systems. IEEE Trans. Power Syst. **31**(6) (2016)
13. C.W. Taylor, D.C. Erickson, K.E. Martin, R.W. Wilson, V. Venkatasubramanian, Wide-area stability and voltage control system: R&D and online demonstration. Proc. IEEE **93**(5), 892–906 (2005)
14. I. Kamwa, R. Grondin, Y. Hebert, Wide-area measurement based stabilizing control of large power systems–A decentralized/hierarchical approach. IEEE Trans. Power Syst. **16**(1), 136–153 (2001)
15. N.R. Chaudhuri, A. Domahidi, B. Chaudhuri, R. Majumder, P. Korba, S. Ray, K. Uhlen, Power oscillation damping control using wide-area signals: a case study on Nordic equivalent system, in *IEEE PES T&D Conference and Exposition* (2010)
16. R. Majumder, B.C. Pal, C. Dafour, P. Korba, Design and real-time implementation of robust FACTS controller for damping inter-area oscillation. IEEE Trans. Power Syst. **21**(2), 809–816 (2006)
17. C. Lu, X. Wu, J. Wu, P. Li, Y. Han, L. Li, Implementations and experiences of wide-area HVDC damping control in China Southern power grid, in *Proceedings of the IEEE PES General Meeting*, San Diego, CA (2012)
18. Y. Pipelzadeh, B. Chaudhuri, T.C. Green, Wide-area power oscillation damping control through HVDC: a case study on Australian equivalent system, in *Proceedings of the IEEE PES General Meeting* (2010)
19. E.M. Martinez, L. Vanfretti, F.R. Sevilla, Automatic triggering of the interconnection between Mexico and Central America using discrete control schemes, in *IEEE PES ISGT Europe* (2013)
20. A. Chakrabortty, P. Khargonekar, Introduction to wide-area monitoring and control, in *American Control Conference*, Washington DC (2013)
21. D. Soudbaksh, A. Chakrabortty, A. Annaswamy, Delay-aware codesigns for wide-area control of power grids, in *55th IEEE Conference on Decision and Control*, Los Angeles, Dec 2014
22. N. Anh, L. Vanfretti, J. Driesen, D.V. Hertem, A Quantitative method to determine ICT delay requirements for wide-area power system damping controllers. IEEE Trans. Power Syst. **30**(4), 2023–2030 (2015)
23. North American Synchrophasor Initiative (NASPI), http://www.naspi.org
24. M. Jarschel, S. Oechsner, D. Schlosser, R. Pries, S. Goll, P. Tran-Gia, Modeling and performance evaluation of an openflow architecture, in *Proceedings of the 23rd International Teletraffic Congress* (2011)
25. P.M. Anderson, A.A. Fouad, *Power System Control and Stability*, 2nd edn. (Wiley Interscience, 2002)
26. A.K. Singh, B.C. Pal, Decentralized dynamic state estimation in power systems using unscented transformation. IEEE Trans. Power Syst. **29**(2) (2014)
27. A. Chakrabortty, J.H. Chow, A. Salazar, Interarea model estimation for radial power systems transfer paths with intermediate voltage control using synchronized phasor measurements. IEEE Trans. Power Syst. **24**(3), 1318–1326 (2009)
28. S. Malik, F. Huet, Virtual cloud: rent out the rented resources, in *International Conference for Internet Technology and Secured Transactions* (2011)
29. M. Wytock, Z. Kolter, A fast algorithm for sparse controller design (2013)

30. A.M. Boker, T.R. Nudell, A. Chakrabortty, On aggregate control of clustered consensus networks, in *American Control Conference*, Chicago, IL, USA (2015), pp. 5527–5532
31. T. Sadamoto, T. Ishizaki, J.I. Imura, Hierarchical distributed control for networked linear systems, in *53rd IEEE Conference on Decision and Control* (2014), pp. 2447–2452
32. S. Hara, J.I. Imura, K. Tsumura, T. Ishizaki, T. Sadamoto, Glocal (global/local) control synthesis for hierarchical networked systems, in *IEEE Conference on Control Applications (CCA)* (2015), pp. 107–112
33. R. Harvey, Y. Xu, Z. Qu, T. Namerikawa, Dissipativity-based design of wide-area generation control for large-scale power systems with high penetration of renewables, in *IEEE Conference on Control Technology and Applications*, Kohala Coast, Hawaii, USA, Aug 2017
34. P.T. Myrda, J. Taft, P. Donner, Recommended approach to a NASPInet architecture, in *45th Hawaii International Conference on System Science (HICSS)* (2012)
35. S. Nabavi, J. Zhang, A. Chakrabortty, Distributed optimization algorithms for wide-area oscillation monitoring in power systems using an inter-regional PMU-PDC architectures. IEEE Trans. Smart Grid **6**(5), 2551–2559 (2015)
36. J.G. Deshpande, K.C. Budka, M. Thottan, *Communication Networks for Smart Grids* (Springer, 2014)
37. D. Anderson, W.S. Cleveland, B. Xi, Multifractal and Gaussian fractional sum-difference models for internet traffic. Perform. Eval. **107** (2017)
38. H. Wu, H. Ni, G.T. Heydt, The impact of time delay on robust control design in power systems. IEEE Power Eng. Soc. Winter Meet. 1511–1516 (2002)
39. J. He, C. Lu, X. Wu et al., Design and experiment of wide area HVDC supplementary damping controller considering time delay in CSG. IET Gener. Trans. Distrib. **3**(1), 17–25 (2009)
40. B. Chaudhuri, R. Majumder, B. Pal, Wide-area measurement-based stabilizing control of power system considering signal transmission delay. IEEE Trans. Power Syst. **19**(4) (2004)
41. J.W. Stahlhut, J. Browne, G.T. Heydt, V. Vittal, Latency viewed as a stochastic process. IEEE Trans. Power Syst. **23**(1) (2008)
42. A.M. Annaswamy, D. Soudbakhsh, R. Schneider, D. Goswami, S. Chakraborty, Arbitrated network control systems: a co-design of control and platform for cyber-physical systems, in *Control of Cyber-Physical Systems*. Lecture Notes in Control and Information Sciences, vol. 449, ed. by D.C. Tarraf (Springer International Publishing, 2013), pp. 339–356
43. S. Chakraborty, S. Künzli, L. Thiele, A general framework for analysing system properties in platform-based embedded system designs, in *DATE* (2003)
44. D. Soudbakhsh, L. Phan, O. Sokolsky, I. Lee, A. Annaswamy, Co-design of control and platform with dropped signals, in *The 4th ACM/IEEE International Conference on Cyber-Physical Systems (ICCPS'13)*, April 2013
45. SmarTS Lab, Royal Institute of Technology (KTH), Sweden, https://github.com/SmarTS-Lab
46. M. Chenine, L. Vanfretti, S. Bengtsson, L. Nordstrom, Implementation of an experimental wide-area monitoring platform for development of synchronized phasor measurement applications, in *IEEE Power Energy Society General Meeting* (2011)
47. A. Chakrabortty, Y. Xin, Hardware-in-the-loop simulations and verifications of smart power systems over an Exo-GENI Testbed, in *Proceedings of the 2nd GENI Research and Educational Experiment Workshop, GREE-2013*, Utah, March 2013
48. GridStat, www.gridstat.net
49. T. Qian, F. Mueller, Y. Xin, A real-time distributed hash table, in *IEEE 20th International Conference on Embedded and Real-Time Computing Systems and Applications (RTCSA)* (2014)
50. Stuxnet, https://en.wikipedia.org/wiki/Stuxnet
51. National Cyber Security Awareness Month, https://www.it.iastate.edu/blog/view/36
52. M. Liao, A. Chakrabortty, Optimization algorithms for catching data manipulators in power system estimation loops. IEEE Trans. Control Syst. Technol. (2018)
53. A. Jain, A. Chakrabortty, E. Biyik, Structurally constrained ℓ_1-sparse control of power systems: online design and resiliency analysis, in *Proceedings of American Control Conference*, Milwaukee, WI, Jun 2018

54. A. Chakrabortty, A. Bose, Smart grid simulations and their supporting implementation methods. Proc. IEEE **105**(11), 2220–2243 (2017)
55. F. Pasqualetti, F. Dorfler, F. Bullo, Cyber-physical attacks in power networks: models, fundamental limitations and monitor design, in *50th IEEE Conference on Decision and Control and European Control Conference*, Orlando, FL (2011)
56. G. Dan, H. Sandberg, M. Ekstedt, G. Bjrkman, Challenges in power system information security. IEEE Secur. Priv. **10**(4), 62–70 (2012)
57. I. Nagel, Analog microelectronic emulation for dynamic power system computation. Ph.D. Thesis (EPFL, 2013)
58. S. Jin, Z. Huang, R. Diao, D. Wu, Y. Chen, Parallel implementation of power system dynamic simulation, in *Proceedings of the IEEE PES General Meeting* (2013)
59. E.J. Wyers, M. Steer, C. Kelley, P. Franzon, A bounded and discretized Nelder-Mead algorithm suitable for RFIC calibration. IEEE Trans. Circuits Syst. I: Regul. Pap. **60**(7), 1787–1799 (2013)
60. Power Infrastructure Cybersecurity Laboratory, http://powercyber.ece.iastate.edu/penetintro.html
61. TCIPG: Trustworthy Cyber Infrastructure for the Power Grid, https://tcipg.org
62. The DETER Project, https://deter-project.org
63. Global Network for Synchrophasor Solutions, http://gnssconsortium.org
64. M.H. Cintuglu, O.A. Mohammed, K. Akkaya, A.S. Uluagac, A survey on smart grid cyber-physical system testbeds. IEEE Commun. Surv. Tutor. **19**(1) (First Quarter, 2017)
65. GENI, www.geni.net

Research Challenges for Design and Implementation of Wide-Area Control

Aranya Chakrabortty

The 2003 Northeast blackout uncovered the vulnerability of the US power system, and manifested the urgent need for real-time state monitoring and control of the grid leading to the development of the Wide-Area Measurement Systems (WAMS) technology. Under the auspices of the US Department of Energy and the North American Synchrophasor Initiative, development and deployment of high-resolution, GPS-synchronized PMUs have been greatly accelerated, together with the development of many new WAMS architectures and applications. However, as the number of PMUs scales up to over thousands within the next few years, Independent System Operators (ISO) and utility companies are struggling to understand how the resulting gigantic volumes of real-time data can be efficiently harvested, processed, and utilized to solve wide-area monitoring and control problems for any realistic power system. Given the complexity and scale of next-generation power grids, an important lesson that researchers have learnt is that the design of and the capability to run at-scale wide-area controllers is indispensable and yet extremely difficult. Currently, there are six main research challenges that need to be resolved before wide-area control can transition from a concept to a reality. These challenges can be listed as follows.

1. **Scalability**: The first and foremost challenge for designing tractable wide-area controllers is scalability. Any typical power system network in reality would consist of several hundreds to thousands of buses, generators, and loads that are spatially distributed over wide geographical spans. Developing tractable methods for modeling, simulation, and control of such large complex networks, and implementing those designs through affordable communication continue to be a challenge for power system engineers. Foundational work on taxonomy theory for modeling and analysis of extreme-scale models of power systems has been done in theory [1], but its translation to simulation models is still missing. Software such as Modelica and Hydra, for example, need to be exploited for modeling

A. Chakrabortty (✉)
Electrical and Computer Engineering Department, North Carolina State University,
Raleigh, NC, USA
e-mail: achakra2@ncsu.edu

© Springer Nature Switzerland AG 2019
J. Stoustrup et al. (eds.), *Smart Grid Control*, Power Electronics
and Power Systems, https://doi.org/10.1007/978-3-319-98310-3_10

scalability through modularity, composition, static correctness, implicit representations, and structural dynamics. All of these selected abstractions then need to be brought under the umbrella of a common modeling language and the front end of a compiler, followed by a library and language-level abstractions that support the needs of experimentation. Similarly, from a design standpoint, conventional state-feedback and output-feedback controllers such as Linear Quadratic Regulators (LQR) and Linear Quadratic Gaussian (LQG) control involve the computation of large matrix decompositions that can result in detrimental numerical inaccuracies without any guarantee of robustness. They also demand every node in the network to share its state information with every other node, resulting in an impractically large number of communication links. Traditionally, control theorists have addressed the problem of controlling large-dimensional systems by imposing structure on controllers. The most promising approach, for example, started with the idea of decentralized control [2], followed by techniques such as singular perturbation theory [3], balanced truncation [4], and gap reduction methods among others. These methods aim to simplify the design of controllers for large systems by exploiting weak coupling between their state variables, and by ignoring states that are "less important" than others [5, 6]. The trade-off, however, is that the resulting controllers are often agnostic of the natural coupling between the states, especially the coupling between the closed-loop states, since many of these couplings were forcibly eliminated to facilitate the design itself. Therefore, extending these methods to facilitate controller designs for networks, especially to power networks where states may be defined over highly structured topologies such as spatial clustering of generators and loads [7], is quite difficult. A significant literature exists on controllability and observability properties of power networks, but the literature for developing tangible and yet simple low-dimensional controllers that satisfy global stability and dynamic performance requirements is still, unfortunately, very sparse. Ideas on aggregate control [8], glocal control [9], and hierarchical control [10] have recently been proposed to address this challenge. The goal of these designs, however, is to guarantee global closed-loop stability by modular tuning of local controller gains. Their degrees of freedom for guaranteeing a desired closed-loop performance can be limited. Some recent papers such as [11] have used structural projection-based ideas for model reduction of large networks, but not for control designs. Attention has also been drawn to designing controllers for large systems by finding low-rank solutions of algebraic Riccati equations [12]. However, like most Krylov subspace-based reduction methods, these controllers are unstructured, and hence demand as many communication links as the full-order LQR itself. Distributed controllers using model matching [13], sparsity-promoting LQR [6], and structured LQR [14, 15] promise to reduce the communication density, but their designs inherit the same dimensionality as the full-order design. What designers are lacking is a tractable approach for constructing controllers that can facilitate both design and implementation, preferably at the scale of several thousand buses. A promising rationale behind such an approach can be, for example, to exploit the clustered structure of the controllability Gramian of the closed-loop system with LQR state-feedback control, as shown recently in [16]. Clustering of generators and other assets opens up further opportunities in consummating control theory with machine learning and computer science, and, therefore, can be a very lucrative choice for designing large-scale wide-area controllers. Yet another important challenge is to ensure robustness of these controllers so that the network can still function gracefully in case a generator equipped with an important wide-area PSS fails for any reason, or in case the power system model changes drastically between two consecutive disturbance events thereby invalidating a fixed control design, and so on.

2. **System Identification and Model Validation**: Given the large size and extraordinary complexity of any realistic power system, deriving and simulating the dynamic model for an entire network becomes extremely challenging. Constructing approximate, aggregated, reduced-order models using simplifying assumptions, therefore, becomes almost imperative in practice. Papers such as [17] have defined aggregation methods for simulation time, for generation units, and for load demand units. The performance of the aggregated models is checked against detailed models including binary effects such as minimum down-time, minimum generation, or demand side contracts. This is especially important for control designs such as model predictive control, where very large optimization problems need to be solved online. The optimization in these cases usually has to be simplified by using approximated power plant models, aggregating several assets to single units, and limiting controller foresight. The main question is—if modeling granularity is necessary, then what degree of aggregation is acceptable in the asset domain as well as in the time domain?

Not just for reducing the complexity of simulations, model aggregation may also be necessary if one is purposely interested in simulating only a certain part of the grid, or a certain phenomenon that happens only over a certain timescale. For example, one may be interested in simulating only the low-frequency inter-area oscillations of a group of synchronous generators instead of the entire spectrum of their frequency response. Identifying coherent subgroups among this group of generators, aggregating the subgroups into an equivalent hypothetical generator, and analyzing the oscillation patterns of these equivalents become necessary in that situation. For example, we often hear power system operators mentioning how "Northern Washington" oscillates against "Southern California" in response to various disturbance events. The main question here is whether we can analytically construct dynamic electromechanical models for these conceptual, aggregated generators representing Washington and California, which in reality are some hypothetical combinations of thousands of actual generators. One example for this motivating problem is the Pacific AC Intertie system in the US west coast, a five-machine dynamic equivalent mass–spring–damper model for which has been widely used in the literature [18]. The main question, again, is—how can we construct an explicit dynamic model for this conceptual figure, and that too preferably in real-time, using voltage, current or power flow signal measurements in order to establish a prototype for the nonlinear inter-area dynamics of the entire interconnection?

Recently, several papers such as [19, 20] have addressed this problem, and derived a series of results on model reduction based on Synchrophasor measurements that combines aggregation theory with system identification. Several open questions still exist, however. For example, once a baseline model is constructed, one must study how it can be updated at regular intervals using newer PMU data. Ideas from adaptive learning and decomposition theory [21] can be useful for that. How this updated model can be used to predict the slow frequencies and corresponding damping factors also needs to be formalized and validated via realistic simulations. Questions also exist on how the reduced-order model can predict the sensitivity of the power flow oscillations inside any area with respect to faults in any other area. If answered correctly, utilities can exploit this information from simulations of the aggregated model, and evaluate their dynamic coupling with neighboring companies, leading to more efficient resource planning. Significant amount of work still needs to be done in formalizing how different failure scenarios in the actual full-order grid model can be translated to the aggregated model, what kind of advanced signal processing and filtering need to be applied to PMU data for accurate identification of the aggregated model parameters, and how controllers designed based on the aggregated model can be mapped back to the original system for implementation.

An equally significant challenge is the validation of identified models, whether they be full-scale models or approximated models. This is particularly true for the dynamic models of the generators and their associated controls. After the large blackout in the US west coast in 1996, it became clear that the generator models used in the studies were not accurate enough. Since then the Western Electricity Coordinating Council (WECC) standard requires updating of the model parameters through standardized testing. Power systems in other parts of the world where stability is an issue have followed a similar approach. However, off-line testing is expensive and as PMUs have proliferated over the past decade, using PMU data during disturbances to update model parameters online has become more common. Several challenges still stand on the way. For example, new technologies such as renewable generation sources and storage that require power-electronic grid interfaces have completely new types of dynamic models. Together with HVDC and FACTS devices, developing accurate modeling methods and devising procedures to validate these models and their parameters are currently lagging behind.

3. **Wide-area communication**: Another major roadblock for implementing wide-area control in a practical grid is that the current power grid IT infrastructure is rigid and low capacity as it is mostly based on a closed-mission specific architecture. The current push to adopt the existing TCP/IP-based open Internet and high-performance computing technologies such as the NASPInet [22] would not be enough to meet the requirement of collecting and processing very large volumes of real-time data produced by such thousands of PMUs. Moreover, the impact of the unreliable and insecure communication and computation infrastructure,

especially long delays and packet loss uncertainties over wide-area networks, on the development of new WAMS applications is not well understood. The need for having accurate delay models and network synchronization rules is absolutely critical for wide-area control since the timescale of the physical control loop is in the order of tens of seconds to a few minutes, while the spatial scale can range over thousands of miles, for example, the entire west coast of the US [23]. The existing PMU standards, IEEE C37.118 and IEC 61850, only specify the sensory data format and communication requirements. They do not indicate any *dynamic* performance standard of the closed-loop system. One needs to develop a cyber-physical framework where one can explicitly show how the closed-loop dynamic responses of phase angles, frequencies, voltages, and current phasors at any part of a grid model are correlated to *real* (not simulated) network delays, that arise from transport, routing, and most importantly, from *scheduling*, as other applications are running in the shared network. Several researchers have looked into delay mitigation in wide-area control loops, with controllers designed for redundancy and delay insensitivity [24–26]. All of these designs are, however, based on worst-case delays, which makes the controller unnecessarily restrictive, and may degrade closed-loop performance. Instead, what one really needs is to understand what may be the most common queuing protocols for transmitting PMU data over a shared wide-area communication network, and how will prior knowledge of these protocols help one in estimating variable queuing delays, and subsequently use that knowledge for designing *delay-aware* wide-area control designs instead of the traditional approach of *delay tolerance*. Ideas from real-time calculus and arbitrated network control systems, both of which have recently been shown to be highly promising tools for this purpose in embedded system designs can be used for this analysis [27]. The goal here is twofold—first, to characterize closed-loop response of a large power grid in terms of distinct *performance metrics*, and second, to derive analytical expressions for the error bounds between ideal designs and delay-aware designs as explicit functions of the queuing protocols. Besides delays, other challenges in communication such as packet drops, bad data detection, synchronization issues, and problems arising from quantization of PMU data also need to be addressed.

4. **Cost allocation**: Another challenge is the economics behind wide-area control. Installing a wide-area communication and control infrastructure would require significant monetary investment from the ISO and utilities. Currently, there are no incentives or markets for wide-area control. Hence, it is not clear how these companies can jointly decide the use and deployment of communication links for achieving global control objectives in the most economical way. For example, if different generators within the balancing regions of different companies are benefiting differently in terms of oscillation damping, transient or voltage stability margins, etc., then how much cost benefit does that company gain by transcending from selfish or completely local control to a system-level wide-area control? Ideas of cooperative game theory and distributed real-time control methods need to be combined to develop efficient and robust cost-sharing mechanisms before the controllers can be implemented in reality [28]. The sensitivity of cost allocation

to controllers with and without network delays also need to be tested. In fact, the final goal can even be to create a power system market for wide-area controls, where pricing and incentives are decided not only by steady-steady power flows but also their dynamics and transient oscillations, all of which cause electrical wear and tear in the electrical excitation system inside the generators.

5. **Cybersecurity**: The next challenge is resilience, privacy, and cybersecurity. The main question here is: how can we ensure privacy of PMU data, and security and resilience of wide-area computing and communication architectures against nefarious attacks and failures at both cyber and physical layers? Existing networking solutions need to be used to evaluate distributed server-based and peer-based architectures, and their potential use in security of wide-area control [29]. Attention must be paid to all three layers of resilience, i.e., detection, localization, and mitigation of attacks [30]. With several thousands of networked PMUs being scheduled to be installed in the United States by 2020, exchange of Synchrophasor data between balancing authorities for any type of wide-area control will involve several thousands of Terabytes of data flow in real-time per event, thereby opening up a wide spectrum of opportunities for adversaries to induce data manipulation attacks, denial-of-service attacks, GPS spoofing, attacks on transmission assets, and so on. The challenge is even more aggravated by the gradual transition of WAMS from centralized to distributed in order to facilitate the speed of data processing. Several recent papers have studied how false-data injection attacks may be deceptively injected into a power grid using its state estimation loops. Others have proposed estimation-based mitigation strategies to secure the grid against some of these attacks. The fundamental approach behind many of these designs is based on the so-called idea of Byzantine consensus, a fairly popular topic in distributed computing, where the goal is to drive an optimization or optimal control problem to a near-optimal solution despite the presence of a malicious agent. In practice, however, this approach is not acceptable to most WAMS operators as they are far more interested in finding out the identity of a malicious agent if it exists in the system, disconnect it from the estimation or control loop, and continue operation using the remaining nonmalicious agents rather than settling for a solution that keeps the attacker unidentified in the loop. This basic question of how to catch malicious agents in distributed wide-area monitoring applications is still an open challenge in the WAMS literature. Ideas on differential privacy are also currently being researched to ensure privacy of PMU measurements so that sensitive information about system parameters, line flows, and load consumptions that may be embedded in these measurements cannot be deciphered accurately by malicious users [31].

6. **Simulation testbeds**: The final challenge is to create a reliable simulation testbed that can be used for verification and validation of various cyber and physical level experiments for wide-area control of very large-dimensional power system models. In current state-of-art, using PMU data for research purposes is contingent on accessing the real data from specific utility companies that own the PMUs at the locations of interest. Gaining access to such data may not always be an easy task due to privacy and nondisclosure issues. More importantly, in many

circumstances, even if real PMU data are obtained they may not be sufficient for studying the detailed operation of the entire system because of their limited coverage. WAMS researchers are in serious need for a Hardware-in-Loop (HIL) simulation framework where high-fidelity detailed models of large power systems can be simulated. These simulations, for example, can be done using Real-time Digital Simulators (RTDS) and Opal-RT, and the dynamic responses can be captured via real hardware PMUs from different vendors, that are synchronized via a common GPS reference. These types of physical-layer testbeds also need to be federated to metro-scale, multilayered dynamic optical network testbeds, an example being the Breakable Experimental Network (BEN) [32], owned by the GENI project of the US National Science Foundation. The resulting testbed infrastructure would not only be relevant for WAMS, but can also complement other emerging and well-established networking testbeds around the country for different cyber-physical applications, transportation, Internet of Things, smart manufacturing, and cybersecurity. Efforts must also be made to make these testbeds as much available to the power system research community as possible so that researchers from other institutions, both nationally and internationally, can use them for carrying out their own experiments using remote connection. Appropriate measures of privacy, security, and safety must be also imposed on such remotely accessible testbeds to ensure smooth and safe usage by multiple parallel users.

References

1. V. Venkatasubramanian, H. Schattler, J. Zaborsky, Dynamics of large constrained nonlinear systems: a taxonomy theory. Proc. IEEE **83**(11), 1530–1561 (1995)
2. D.D. Siljak, *Decentralized Control of Complex Systems* (Courier Corporation, 2011)
3. P. Kokotovic, H.K. Khalil, J. Oreilly, *Singular Perturbation Methods in Control: Analysis and Design* (SIAM, 1999)
4. K. Zhou, J.C. Doyle, *Essentials of Robust Control*, vol. 180 (Prentice Hall, NJ, 1998)
5. A. Jain, A. Chakrabortty, E. Biyik, An online structurally constrained LQR design for damping oscillations in power system networks, in *American Control Conference* (Seattle, WA, 2017)
6. F. Dorfler, M.R. Jovanovic, M. Chertkov, F. Bullo, Sparse and optimal wide-area damping control in power networks, in *American Control Conference* (Washington, DC, 2013), pp. 4295–4300
7. A. Chakrabortty, Wide-area damping control of power systems using dynamic clustering and TCSC-based redesigns. IEEE Trans. Smart Grid, **2**(3) (2012)
8. A.M. Boker, T.R. Nudell, A. Chakrabortty, On aggregate control of clustered consensus networks, in *American Control Conference* (Chicago, IL, USA, 2015), pp. 5527–5532
9. S. Hara, J.I. Imura, K. Tsumura, T. Ishizaki, T. Sadamoto, Glocal (global/local) control synthesis for hierarchical networked systems, in *IEEE Conference on Control Applications* (CCA), pp. 107–112 (2015)
10. T. Sadamoto, T. Ishizaki, J.I. Imura, Hierarchical distributed control for networked linear systems, in *53rd IEEE Conference on Decision and Control*, pp. 2447–2452 (2014)
11. N. Monshizadeh, H.L. Trentelman, M.K. Camlibel, Projection-based model reduction of multi-agent systems using graph partitions. IEEE Trans. Control Netw. Syst. **1**(2), 145–154 (2014)

12. P. Benner, Z. Bujanovi, On the Solution of large-scale algebraic riccati equations by using low-dimensional invariant subspaces. Linear Algebra Appl. **488**, 430–459 (2016)
13. M. Rotkowitz, S. Lall, A characterization of convex problems in decentralized control. IEEE Trans. Autom. Control **51**(2), 274–286 (2006)
14. X. Wu, M.R. Jovanovic, Augmented lagrangian approach to design of structured optimal state feedback gains. IEEE Trans. Autom. Control **56**, 2923–2929 (2011)
15. Y.S. Wang, N. Matni, J.C. Doyle, Localized LQR optimal control, in *53rd IEEE Conference on Decision and Control* (2014), pp. 1661–1668
16. N. Xue, A. Chakrabortty, H_2-clustering of closed-loop consensus networks under a class of LQR design, in *American Control Conference* (2016)
17. S. Deml, A. Ulbig, T. Borsche, G. Andersson, *The Role of Aggregation in Power System Simulation* (IEEE PowerTech, Eindhoven, 2015)
18. A. Chakrabortty, A. Salazar, Building a dynamic electro-mechanical model for the pacific AC intertie using distributed synchrophasor measurements, in *European Transactions on Electric Power*, ed. by C.C. Liu, M. Crow, J.H. Chow. Special Issue on PMU Applications, vol. 21, no. 4 (2011). pp. 1657–1672
19. A. Chakrabortty, J.H. Chow, A. Salazar, A measurement-based framework for dynamic equivalencing of power systems using wide-area phasor measurements. IEEE Trans. Smart Grid **1**(2), 68–81 (2011)
20. S. Nabavi, A. Chakrabortty, Topology identification for dynamic equivalent models of large power system networks, in *American Control Conference* (DC, 2013)
21. G.B. Dantzig, P. Wolfe, Decomposition principle for linear programs. Oper. Res. **8**(1), 101–111 (1960)
22. North American Synchrophasor Initiative (NASPI), www.naspi.org
23. J. Zhang, S. Nabavi, A. Chakrabortty, Y. Xin, ADMM optimization strategies for wide-area oscillation monitoring in power systems under asynchronous communication delays. IEEE Trans. Smart Grid **7**(4), 2123–2133 (2016)
24. H. Wu, H. Ni, G.T. Heydt, The impact of time delay on robust control design in power systems. IEEE Power Eng. Soc. Winter Meet. 1511–1516 (2002)
25. B. Chaudhuri, R. Majumder, B. Pal, Wide-area measurement-based stabilizing control of power system considering signal transmission delay. IEEE Trans. Power Syst. **19**(4) (2004)
26. S. Zhang, V. Vittal, Design of wide-area power system damping controllers resilient to communication failures. IEEE Trans. Power Syst. **28**(4) (2013)
27. D. Soudbaksh, A. Chakrabortty, A. Annaswamy, Delay-aware cyber-physical architecture for wide-area control of power systems. IFAC Control Eng. Pract. **60**, 171–182 (2017)
28. F. Lian, A. Chakrabortty, A. Duel-Hallen, Game-theoretic multi-agent control and network cost allocation under communication constraints. IEEE J. Sel. Areas Commun. **35**(2), 330–340 (2017)
29. J. Zhang, P. Jaipuria, A. Hussain, A. Chakrabortty, Attack-resilient estimation of power system oscillation modes using distributed and parallel optimization: theoretical and experimental methods, in *Conference on Decision and Game Theory for Security* (GameSec) (Los Angeles, CA, 2014)
30. M. Liao, A. Chakrabortty, A round-robin ADMM algorithm for identifying data-manipulators in power system estimation, in *American Control Conference* (2016)
31. J. Cortes, G.E. Dullerud, S. Han, J. Le Ny, S. Mitra, G.J. Pappas, Differential privacy in control and network systems, in *IEEE Conference on Decision and Control* (2016)
32. A. Chakrabortty, Y. Xin, Hardware-in-the-loop simulations and verifications of smart power systems over an Exo-GENI testbed, in *Proceedings of 2nd GENI Research and Educational Experiment Workshop*, GREE-2013, Utah, Mar 2013

Signal Processing in Smart Grids: From Data to Reliable Information

Meng Wang

1 Introduction

The recent proliferation of data is revolutionizing the practice of power system monitoring and control. With the Smart Grid initiative, more than two thousand multichannel phasor measurement units (PMUs) [37] have now been installed in North America [35]. PMUs can directly measure GPS-synchronized bus voltage phasors, line current phasors, and the frequency, at a rate of 30 or 60 samples per second per channel. Compared to the conventional Supervisory Control and Data Acquisition (SCADA) systems that only provide measurements every 2–5 s, which are not accurately synchronized in time, PMUs can drastically improve the system visibility and enhance the situational awareness.

The data abundance imposes significant challenges on the power industry. Currently, the transmission grid operators decide control actions based on the output of state estimation, which is carried out at multi-second intervals in correspondence to the data acquisition rate of the SCADA system. Moreover, control actions are mostly computed offline and are not optimized for diverse real-time situations. With the recent data wealth, an important and urging question is how to convert the massive amounts of data to reliable information quickly so as to facilitate the following real-time control decisions.

PMUs are envisioned to improve wide-area situational awareness and prevent blackouts [1, 7]. Ever since the initial installation, many research efforts have been devoted to exploiting the PMU data in various applications, and the continued investigation is still ongoing. The applications include but not limited to state estimation [50], oscillation detection and electromechanical mode identification [17, 29], disturbance detection and location [28, 32], and dynamic security assessment [8, 21].

M. Wang (✉)
Rensselaer Polytechnic Institute, Troy, NY 12180, USA
e-mail: wangm7@rpi.edu

© Springer Nature Switzerland AG 2019
J. Stoustrup et al. (eds.), *Smart Grid Control*, Power Electronics
and Power Systems, https://doi.org/10.1007/978-3-319-98310-3_11

Data quality is an inevitable issue for the control-room incorporation of PMU data. Because of the communication networks that were not designed to carry high-speed PMU data and the early deployment of older PMUs, data losses, and data quality degradations happen quite often in practice, especially in the Eastern interconnection [41]. Current PMU-based applications usually assume that the measurements are available and reliable. To incorporate PMU data into real-time operations, a data-conditioning component is needed to reconstruct missing data [12] and correct bad measurements. Alternatively, data analysis methods that are robust to data quality degradation are worth investigation. Moreover, different applications have diverse requirements on the data quality. The trust scores of the obtained and the recovered measurements should be computed and incorporated into the design of the control actions.

Developed when the measurements were scarce, conventional methods usually require the modeling of the power system. The proliferation of PMU data enables the development of data-driven methods for feature extraction without power system modeling. Data-driven methods are much investigated, especially for applications in which accurate and explicit models are difficult to obtain. Despite all the nice properties, the output of data-driven methods might lack a clear physical interpretation. In contrast, physical models of the power systems are well studied, and conventional methods are usually accompanied with clear physical intuitions. Moreover, data-driven methods usually require parameter tuning, and the computational time of these methods could be of concern for real-time applications.

Cyber data security cannot be ignored. Cyber operations have been integrated into smart grids to enhance control performance; however, such integration also increases the possibility of cyber attacks. Although attacking the control laws of the operator is relatively difficult, an intruder could alter the measurements to mislead the operator, resulting in wrong control actions. The detection of these cyber attacks requires efforts in both the communication level through the development of advanced encryption methods and the signal processing level through the development of methods that can detect these cyber data attacks based on the abnormality in the measurements.

2 Data Quality Improvement

Data losses happen due to network congestion or PMU malfunctions. The missing data rate is reduced in recent years, but data losses still happen. When the measurements were scarce, the missing data points were interpolated using observations in the same measurement channel. Another way was to run the state estimator on the partially obtained measurements and compute the missing data points based on the estimated system state.

Now with the large amounts of data collected by many PMUs, the missing points can be directly and accurately estimated from the data without modeling the power system. The idea is to exploit the correlations in the spatial–temporal blocks of

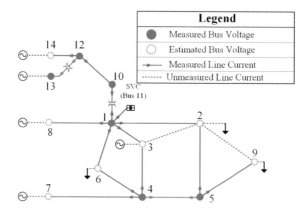

Fig. 1 Six PMUs in the Central NY Power System (reproduced from [12] ©2016 IEEE and used with permission)

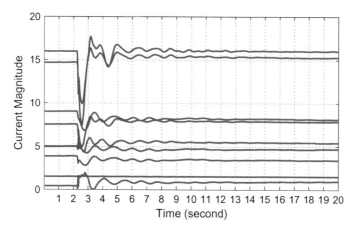

Fig. 2 Current magnitudes of PMU data (9 current phasors out of 37 phasors) (reproduced from [12] ©2016 IEEE and used with permission)

PMU data. In fact, the data correlation could be easily characterized by the low-rank property of the PMU data matrix.

Let $M \in \mathbb{C}^{m \times n}$ contain the phasor measurements (represented by complex numbers in rectangular form) from m PMU channels in n time instants. Then, M can be approximated by a low-rank matrix with a negligible error. For example, the recorded data of six multi-channel PMUs in the Central New York (NY) Power System (Fig. 1) were analyzed in [12]. M contains 37 voltage and current phasors in 20 s at a rate of 30 samples per channel per second. Figure 2 shows the current magnitudes of PMU measurements. Figure 3 shows the singular values of M. The largest 10 singular values are 894.5942, 36.8319, 20.7160, 8.3400, 3.0771, 2.4758, 1.9705, 1.3543, 0.5930, and 0.2470. We can approximate M by a rank-eight matrix with a very small error.

Fig. 3 Singular values of a 600 × 37 PMU data matrix (reproduced from [12] ©2016 IEEE and used with permission)

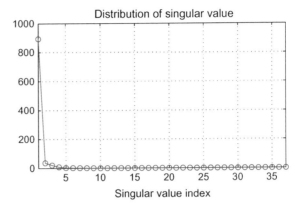

Table 1 Recovery performance of OLAP on NYISO data that contains disturbances. Computational time is the total time to recover missing points in 5-min data on a normal desktop. Relative recovery error is the ratio of recovery error to the actual value (both measured in ℓ_2-norm) [45]

	Voltage magnitude	Voltage angle	Frequency	Current magnitude	Current angle
Computational time (s)	1.305	1.327	1.239	8.121	9.113
Relative recovery error (%)	0.02	0.005	0.0015	0.24	0.05

The low-rank property enables computationally efficient methods with theoretical guarantees for various data analysis tasks. For instance, recovering missing points in a low-rank matrix M can be formulated as a convex optimization problem

$$\min_{X \in m \times n} \|X\|_* \text{ s.t. } X_{ij} = M_{ij}, \text{ for all } (ij) \in \Omega, \tag{1}$$

where Ω denotes the locations of the observed entries, and the matrix nuclear norm $\|\cdot\|_*$ is the sum of singular values. The original matrix M is proved to be the optimal solution to (1) under mild assumptions [5] and thus recovered in polynomial time.

We connected PMU data analysis with low-rank matrix theory and obtained promising results for missing PMU data recovery [9, 11, 12, 44]. We proved that the missing PMU data points can be correctly recovered under very mild assumptions [12]. The numerical evaluations of our developed online missing data recovery method, called OLAP, on recorded PMU datasets from NYISO are shown in Table 1 [45]. This 5-min dataset contains 53 voltage phasors, 53 frequencies, and 263 current phasors, with 8% missing data. Figure 4 shows the data recovery of consecutive data losses in one channel by OLAP. A capacitor-switching event during the data losses is recovered by utilizing the measurements in other channels, and this recovery is impossible by single-channel interpolations.

Fig. 4 Data recovery in one channel [45]

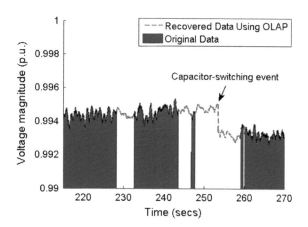

Besides missing data recovery, the low-rank property could also be exploited to detect and correct bad data. Bad data detection and identification has been an important issue for power system state estimation. It is usually integrated with the state estimation, which requires power system model. The measurements that are not consistent with the currently estimated system state are considered as bad data and removed. With the abundance of PMU data, bad data can be detected and identified directly from the data, see e.g., [10, 30, 48].

If we impose the assumption that the number of bad data is much less than the total number of measurements, the bad data detection problem can be formulated as a matrix decomposition problem. The obtained data matrix M is viewed as the sum of two matrices L^* and C^*, where the low-rank matrix L^* denotes the actual data without errors, and the sparse matrix C^* denotes the additive errors in the bad measurements. A matrix is sparse if it only has a small number of nonzero entries, while most entries are zero. The goal of matrix decomposition is to obtain L^* and C^* from M. Under mild assumptions [4], it can be achieved by solving a convex program

$$\min_{L,m\times n} \|L\|_* + \lambda \sum_{ij} |C_{ij}| \text{ s.t. } L + C = M, \tag{2}$$

where λ is a predetermined weighting factor.

The above problem formulation has been exploited to detect bad data, including injected false data by cyber attackers [10, 30]. Note that this formulation does not require any information about the system topology and the line impedances. Thus, bad data detection can be separated from state estimation. Moreover, it is shown in [10] that the topology information can be incorporated easily with minor changes to (2), resulting in a provable enhancement of the detection performance.

The above methods for data quality improvement only use PMU data. One interesting question is how to incorporate PMU data with other formats of data. For instance, conventional SCADA data provide information about power injections and power flows every 1–5 s. Is it possible to use the SCADA data to enhance the accuracy of the data recovery and error correction of PMU data? How shall we handle the different sampling rates of these data?

Another important question is how to differentiate data anomalies and system events. When a system event just starts, the affected measurements would be different from the nearby measurements. They might be treated as bad data if we directly apply (2). How shall we determine whether these measurements are bad data or resulting from system events? Is it possible to achieve this separation mostly based on data without much modeling of the system?

3   Model-Based and Data-Driven Analysis

Power system monitoring is conventionally model-based to compensate for the lack of measurements, as in dynamic state estimation [14, 43]. These methods degrade significantly when the model is inaccurate, which is a long-standing issue due to the complexity of power systems. Recent data abundance fosters the development of data-driven methods that do not require power system models.

Data-driven methods can extract information directly from data without modeling the power systems; however, completely ignoring the underlying dynamical system also has some limitations. First, data-driven methods might not perform as well as model-based methods when the model is correctly specified. Second, the computational complexity of machine-learning-based methods usually increases significantly when the data size increases. Lastly, the analyses are often lack of physical understandings of the power systems. An interesting research direction is how to incorporate the domain knowledge and engineering intuitions about the power system into the data-driven analyses.

Take disturbance identification as an example. Both model-based identification methods (see e.g., [42, 51]) and data-driven methods [3, 6, 15, 47] have been developed to identify different types of events in the system. Data-driven methods extract features (including direct features like a frequency [6] or its derivative [3], as well as indirect features like wavelet coefficients [19]) from measurements and classify those with similar features as resulting from the same event type.

We also developed a data-driven method to identify and locate events without modeling the power system [26, 27]. The key idea is to characterize an event by a low-dimensional row subspace spanned by the dominant singular vectors of the data matrix that contains spatial–temporal blocks of measurements from multiple PMUs. This subspace characterization is robust to initial system conditions and captures the system dynamics. Then an event is identified by comparing the obtained data with a pre-computed event dictionary with each dictionary atom corresponding to a row subspace of an event. The location of an event is determined based on the magnitudes of changes. Figure 5 shows the overview of this approach.

One distinctive feature of this approach is that a dictionary atom has a clear physical interpretation. It is the subspace spanned by a few dominant modes in the observation window. The subspace can be computed through Singular Value Decomposition

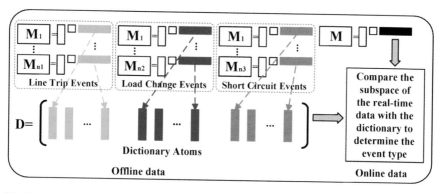

Fig. 5 Dictionary construction from historical datasets and real-time data identification through subspace comparison (reproduced from [27] ©2018 IEEE and used with permission)

Table 2 Identification results of 380 events (reproduced from [26] ©2017 IEEE and used with permission)

Type of event	IAR (%)	ELAR (%)	ALAR (%)
Line trip	100	85	94 (among 3 buses)
Short circuit	100	77	90 (among top 3 buses)
Load change	100	46	90 (among top 5 buses)

(SVD) [18]. Moreover, the dictionary size is much smaller than those of the dictionaries of time series [47] or other computed features [20]. That is because all the events will be compactly represented by a few row singular vectors to reduce the dimensionality. The reduction of the dictionary size reduces the computational complexity of both the offline training and the online event identification. The method identifies events shortly after the event starts (e.g., within 1–5 s) and can be implemented in real time, while existing methods are mostly designed for past event analysis (e.g., 30 s of data are needed in [47]).

The method is evaluated on the IEEE 68-bus test system (details see [26]). We simulate 380 events, including 160 line trip events, 100 load change events, and 120 short circuit events at different locations and with different pre-event conditions. Only one second of data is used for event identification. The constructed dictionary includes 33 events. Each event is represented by a subspace spanned by 30×6 matrix, where 30 is the number of time steps in one second and 6 is the number of dominating singular vectors.

Table 2 records the identification and location results under three criteria:

Identification Accuracy Rate (IAR): The ratio of the number of accurately identified events to the total number of events;

Approximate Location Accuracy Rate (ALAR): The ratio of the number of events with actual locations[1] among the top k buses with the most significant changes to the total number of events.

Exact Location Accuracy Rate (ELAR): A special case of ALAR when $k = 1$, i.e., the event location is exact.

The above disturbance identification method is one initial effort in incorporating physical understandings into the data-driven analyses for power system monitoring. One future direction is to extend these efforts to other aspects of power system monitoring such as state estimation and disturbance location.

4   Resilience to Cyber Data Attacks

Cyber operations have been integrated into power systems to enhance control performance; however, such integration also increases the possibility of cyber attacks. In early 2016, hackers caused a power outage for the first time in Ukraine during holiday season [36]. The development of a trustworthy power system requires developing new technologies in various aspects, such as a secured communication infrastructure and protected sensing and control devices. Here we focus on data security from a signal processing perspective.

Cyber data attacks can change the measurements obtained by the operator such that the operator would obtain a wrong estimate of the system state, resulting in harmful control actions and potential failures. These data attacks may also lead to significant financial impacts in the electricity market [49]. A malicious intruder with sufficient system configuration information can manipulate multiple measurements simultaneously, and the resulting injected false data can be are viewed as "the worst interacting bad data injected by an adversary" [25, 31].

State estimation in the presence of false data injection attacks has attracted much research attention recently. These attacks are carefully selected, and the interacting erroneous measurements cannot be detected by conventional bad data detections that only use measurements at one time instant. Many efforts have been devoted to studying the requirements to launch a cyber data attack [39] and preventing these attacks by protecting critical measurement units [2, 24]. A few recent works proposed detection methods for cyber data attacks [10, 30, 40]. Since the attacks cannot be detected only using measurements at one time instant, these methods exploit the temporal correlations in the data and detect the attacks as anomalies in time series.

Eavesdropping attacks are another form of security concerns [34]. An adversary might obtain sensitive information about the grid by monitoring the network traffic. The gathered information could be used for future crimes. Data privacy [23, 33] is an emerging issue in smart grids. PMU data are owned by regional transmission owners and considered to be private and sensitive. Privacy-guaranteed PMU data

---

[1]The location of line trip events are considered as successful if one of the two related buses are correctly identified.

communication has not yet been seriously investigated. Besides enhancing the data privacy by improving communication technologies for smart meter data [16, 22], the tools at the signal level to increase data privacy need to be developed. One can enhance data privacy by adding random noise [46] or applying quantization to the measurements [38], usually at a cost of data distortion. Some initial efforts have been devoted to developing data recovery methods from noisy and quantized measurements with a small data distortion for large amounts of PMU data [13].

Since cyber attacks can happen in various aspects of power system monitoring and control [34], it is very important to be precautious and develop the corresponding protection schemes in advance. The vulnerabilities of individual components of the system against cyber attacks should be constantly estimated, and attack prevention and detection methods should be incorporated into power system monitoring.

5 Conclusion

In summary, the data wealth brings multidisciplinary research opportunities of power engineering, signal processing, and machine learning. Data-oriented approaches, ideally incorporated with physical understandings of the power systems, can extract information from the data and enable real-time control operations. Data quality enhancement is a necessary pre-conditioning step to recover missing points and correct bad measurements. Data security issues should be taken into account in the design of these methods.

References

1. F. Aminifar, M. Fotuhi-Firuzabad, A. Safdarian, A. Davoudi, M. Shahidehpour, Synchrophasor measurement technology in power systems: Panorama and state-of-the-art. IEEE Access **2**, 1607–1628 (2014)
2. R.B. Bobba, K.M. Rogers, Q. Wang, H. Khurana, K. Nahrstedt, T.J. Overbye, Detecting false data injection attacks on DC state estimation, in *Proceedings of the First Workshop on Secure Control Systems (SCS)* (2010)
3. A. Bykhovsky, J.H. Chow, Power system disturbance identification from recorded dynamic data at the northfield substation. Int. J. Electr. Power Energy Syst. **25**(10), 787–795 (2003)
4. E.J. Candès, X. Li, Y. Ma, J. Wright, Robust principal component analysis? J. ACM (JACM) **58**(3), 11 (2011)
5. E.J. Candès, B. Recht, Exact matrix completion via convex optimization. Found. Comput. Math. **9**(6), 717–772 (2009)
6. O.P. Dahal, S.M. Brahma, H. Cao, Comprehensive clustering of disturbance events recorded by phasor measurement units. IEEE Trans. Power Del. **29**(3), 1390–1397 (2014)
7. J. De La Ree, V. Centeno, J.S. Thorp, A.G. Phadke, Synchronized phasor measurement applications in power systems. IEEE Trans. Smart Grid **1**(1), 20–27 (2010)
8. R. Diao, K. Sun, V. Vittal, R.J. O'Keefe, M.R. Richardson, N. Bhatt, D. Stradford, S.K. Sarawgi, Decision tree-based online voltage security assessment using PMU measurements. IEEE Trans. Power Syst. **24**(2), 832–839 (2009)

9. P. Gao, M. Wang, J.H. Chow, M. Berger, L.M. Seversky, Matrix completion with columns in union and sums of subspaces, in *2015 Proceedings of the IEEE Global Conference on Signal and Information Processing (GlobalSIP)* (2015), pp. 785–789
10. P. Gao, M. Wang, J.H. Chow, S.G. Ghiocel, B. Fardanesh, G. Stefopoulos, M.P. Razanousky, Identification of successive "unobservable" cyber data attacks in power systems. IEEE Trans. Signal Process. **64**(21), 5557–5570 (2016)
11. P. Gao, M. Wang, S.G. Ghiocel, J.H. Chow, Modeless reconstruction of missing synchrophasor measurements, in *Proceedings of the IEEE PES General Meeting (Selected in Best Papers Sessions)* (2014), pp. 1–5
12. P. Gao, M. Wang, S.G. Ghiocel, J.H. Chow, B. Fardanesh, G. Stefopoulos, Missing data recovery by exploiting low-dimensionality in power system synchrophasor measurements. IEEE Trans. Power Syst. **31**(2), 1006–1013 (2016)
13. P. Gao, R. Wang, M. Wang, J.H. Chow, Low-rank matrix recovery from quantized and erroneous measurements: accuracy-preserved data privatization in power grids, in *Proceedings of the Asilomar Conference on Signals, Systems, and Computers* (2016)
14. E. Ghahremani, I. Kamwa, Dynamic state estimation in power system by applying the extended kalman filter with unknown inputs to phasor measurements. IEEE Trans. Power Syst. **26**(4), 2556–2566 (2011)
15. A.K. Ghosh, D.L. Lubkeman, The classification of power system disturbance waveforms using a neural network approach. IEEE Trans. Power Del. **10**(1), 109–115 (1995)
16. J. Gomez-Vilardebo, D. Gunduz, Smart meter privacy for multiple users in the presence of an alternative energy source. IEEE Trans. Inf. Forensics Secur. **10**(1), 132–141 (2015)
17. A. Hauer, D. Trudnowski, J.G. DeSteese, A perspective on wams analysis tools for tracking of oscillatory dynamics, in *Proceedings of the IEEE Power Engineering Society General Meeting* (2007), pp. 1–10
18. R.A. Horn, C.R. Johnson, *Matrix Analysis* (Cambridge university press, 2012)
19. H. Jiang, J.J. Zhang, D.W. Gao, Fault localization in smart grid using wavelet analysis and unsupervised learning, in *Proceedings of the Asilomar Conference on Signals, Systems and Computers (ASILOMAR)*, 2012, pp. 386–390
20. H. Jiang, J.J. Zhang, W. Gao, Z. Wu, Fault detection, identification, and location in smart grid based on data-driven computational methods. IEEE Trans. Smart Grid **5**(6), 2947–2956 (2014)
21. A. Kaci, I. Kamwa, L.-A. Dessaint, S. Guillon, Synchrophasor data baselining and mining for online monitoring of dynamic security limits. IEEE Trans. Power Syst. **29**(6), 2681–2695 (2014)
22. H. Khurana, R. Bobba, T. Yardley, P. Agarwal, E. Heine, Design principles for power grid cyber-infrastructure authentication protocols, in *Proceedings of the Hawaii International Conference on System Sciences (HICSS)* (2010), pp. 1–10
23. H. Khurana, M. Hadley, N. Lu, D.A. Frincke, Smart-grid security issues. IEEE Secur. Priv. **8**(1), 81–85 (2010)
24. T. Kim, H. Poor, Strategic protection against data injection attacks on power grids. IEEE Trans. Smart Grid **2**(2), 326–333 (2011)
25. O. Kosut, L. Jia, R. Thomas, L. Tong, Malicious data attacks on smart grid state estimation: attack strategies and countermeasures, in *Proceedings of the IEEE International Conference on Smart Grid Communications (SmartGridComm)* (2010), pp. 220–225
26. W. Li, M. Wang, J.H. Chow, Fast event identification through subspace characterization of PMU data in power systems, in *Proceedings of the IEEE Power and Energy Society (PES) General Meeting* (2017), pp. 1-5
27. W. Li, M. Wang, J.H. Chow, Real-time event identification through low-dimensional subspace characterization of high-dimensional synchrophasor data. IEEE Trans. Power Syst. (Early access) (2018)
28. Y.-H. Lin, C.-W. Liu, C.-S. Chen, A new PMU-based fault detection/location technique for transmission lines with consideration of arcing fault discrimination-Part I: Theory and algorithms. IEEE Trans. Power Del. **19**(4), 1587–1593 (2004)

29. G. Liu, J. Quintero, V.M. Venkatasubramanian, Oscillation monitoring system based on wide area synchrophasors in power systems, in *Proceedings of the iREP Symposium-Bulk Power System Dynamics and Control-VII. Revitalizing Operational Reliability* (IEEE, 2007), pp. 1–13

30. L. Liu, M. Esmalifalak, Q. Ding, V.A. Emesih, Z. Han, Detecting false data injection attacks on power grid by sparse optimization. IEEE Trans. Smart Grid **5**(2), 612–621 (2014)

31. Y. Liu, P. Ning, M.K. Reiter, False data injection attacks against state estimation in electric power grids, in *Proceedings of the 16th ACM Conference on Computer and Communications Security*, 2009, pp. 21–32

32. J. Ma, Y. Makarov, C. Miller, T. Nguyen, Use multi-dimensional ellipsoid to monitor dynamic behavior of power systems based on PMU measurement, in *Proceedings of the IEEE Power and Energy Society General Meeting* (2008), pp. 1–8

33. P. McDaniel, S. McLaughlin, Security and privacy challenges in the smart grid. IEEE Secur. Priv. **7**(3), 75–77 (2009)

34. Y. Mo, T.H.-J. Kim, K. Brancik, D. Dickinson, H. Lee, A. Perrig, B. Sinopoli, Cyber-physical security of a smart grid infrastructure. Proc. IEEE **100**(1), 195–209 (2012)

35. North America Synchrophasor Initiative(NASPI), https://www.naspi.org/Badger/content/File/FileService.aspx?fileid=49CC0BEB3E3C36F3BF9C7930E0FFFD1B

36. A. Peterson, Hackers caused a blackout for the first time, researchers say, https://www.washingtonpost.com/news/the-switch/wp/2016/01/05/hackers-caused-a-blackout-for-the-first-time-researchers-say/

37. A. Phadke, J. Thorp, *Synchronized Phasor Measurements and Their Applications* (Springer, 2008)

38. A. Reinhardt, F. Englert, D. Christin, Enhancing user privacy by preprocessing distributed smart meter data, in *Proceedings of the Sustainable Internet and ICT for Sustainability (SustainIT)* (2013), pp. 1–7

39. H. Sandberg, A. Teixeira, K.H. Johansson, On security indices for state estimators in power networks, in *Proceedings of the the First Workshop on Secure Control Systems (SCS)* (2010)

40. H. Sedghi, E. Jonckheere, Statistical structure learning of smart grid for detection of false data injection, in *Proceedings of the IEEE Power and Energy Society General Meeting (PES)* (2013), pp. 1–5

41. A. Silverstein, J.E. Dagle, Successes and challenges for synchrophasor technology: an update from the north american synchrophasor initiative, in *Proceedings of the Hawaii International Conference on System Science (HICSS)* (2012), pp. 2091–2095

42. J.E. Tate, T.J. Overbye, Line outage detection using phasor angle measurements. IEEE Trans. Power Syst. **23**(4), 1644–1652 (2008)

43. G. Valverde, V. Terzija, Unscented Kalman filter for power system dynamic state estimation. IET Gener. Trans. Distrib. **5**(1), 29–37 (2011)

44. M. Wang, J.H. Chow, P. Gao, X. T. Jiang, Y. Xia, S.G. Ghiocel, B. Fardanesh, G. Stefopolous, Y. Kokai, N. Saito, A low-rank matrix approach for the analysis of large amounts of power system synchrophasor data, in *Proceedings of the IEEE Hawaii International Conference on System Sciences (HICSS)* (2015), pp. 2637–2644

45. M. Wang, G. de Mijolla, P. Gao, J. H. Chow, B. Fardanesh, G. Stefopoulos, S. Babaei, A. Ettlinger, D. Iles, D. Tran, Missing data recovery by exploiting low-dimensionality in synchrophasor measurements, in *North American Synchrophasor Initiative meetings*, March 2016

46. S. Wang, L. Cui, J. Que, D.-H. Choi, X. Jiang, S. Cheng, L. Xie, A randomized response model for privacy preserving smart metering. IEEE Trans. Smart Grid **3**(3), 1317–1324 (2012)

47. W. Wang, L. He, P. Markham, H. Qi, Y. Liu, Q.C. Cao, L.M. Tolbert, Multiple event detection and recognition through sparse unmixing for high-resolution situational awareness in power grid. IEEE Trans. Smart Grid **5**(4), 1654–1664 (2014)

48. M. Wu, L. Xie, Online detection of low-quality synchrophasor measurements: a data-driven approach. IEEE Trans. Power Syst. **32**(4), 2817–2827 (2017)

49. L. Xie, Y. Mo, B. Sinopoli, Integrity data attacks in power market operations. IEEE Trans. Smart Grid **2**(4), 659–666 (2011)

50. M. Zhou, V.A. Centeno, J.S. Thorp, A.G. Phadke, An alternative for including phasor mea-
    surements in state estimators. IEEE Trans. Power Syst. **21**(4), 1930–1937 (2006)
51. H. Zhu, G.B. Giannakis, Sparse overcomplete representations for efficient identification of
    power line outages. IEEE Trans. Power Syst. **27**(4), 2215–2224 (2012)

WAMS-Based Controlled System Separation to Mitigate Cascading Failures in Smart Grid

Kai Sun

1 Background

Catastrophic power blackouts can cause tremendous losses and influence up to tens of millions of people. Since the 1965 Northeast Blackout, many efforts have been made by power industry, but cascading power outages continued to happen. Some recent blackout events are such as the east and west coast blackouts in North America in 2003 and 2011, respectively, the 2006 European blackout and the 2012 Indian blackout events [1–5]. Blackouts are usually caused by cascading failures initiated by, e.g., natural disasters and mis-operations, which are long chains of dependent equipment failures or outages successively weakening the transmission network [6]. If not prevented or mitigated, cascading failures can break the stability and integrity of the system and result in large-area power outages. When cascading failures occur, it is hard for grid operators to manually take a real-time remedial action in a matter of tens of seconds, so automatic system-wide protection and control schemes are vitally important to stop propagation of failures towards wide areas. At present, most existing protection systems lack adaptability and system-wide coordination. Their mechanism is prone to trip equipment under a predefined fault or abnormal condition, which, however, further weakens the transmission network and may speed up propagation of failures. Therefore, effective mitigation of cascading failures requires smart grid be armed with an adaptive, system-wide protection, and control scheme.

This article will focus on the visions and research problems on a final resort against power blackouts called controlled system separation (CSS) or controlled system islanding in contrast with unintentional system islanding. The latter breaks the integrity of the network unintentionally at a late stage of cascading failures and expedites failures towards a blackout. CSS should be performed once unintentional system islanding becomes unavoidable. A power transmission system is usually

K. Sun (✉)
Department of Electrical Engineering and Computer Engineering, University of Tennessee, Knoxville, TN 37996, USA
e-mail: kaisun@utk.edu

© Springer Nature Switzerland AG 2019
J. Stoustrup et al. (eds.), *Smart Grid Control*, Power Electronics and Power Systems, https://doi.org/10.1007/978-3-319-98310-3_12

185

operated as multiple control areas. The tie lines connecting them are critical to inter-area electricity exchange as well as the integrity and stability of the system. A tie-line failure can cause oscillations of generators, overloading other lines, and increased vulnerability of the network. Due to lack of coordination at the system level, local protective actions may trip more system components, cause more oscillations, over-loaded lines and failures, and even break the network into electrical islands. That is how unintentional system islanding happens. Those islands formed not in a designed manner may have unbalanced load and generation, overloaded equipment, or unsta-ble generators, and hence are hard to survive from blackouts even if shedding lots load or generation. Compared to such unintentional islanding, CSS allows the control center to proactively separate the grid into islands. Namely, each island is strategi-cally formed with matched and stable generation so as to continue power supplies to its loads. Thus, CSS effectively prevents a blackout and the system can be restored by resynchronizing all islands, which is easier than a blackstart process from the blackout.

At present, many phasor measurement units (PMUs) have been installed in trans-mission systems as one important element of smart grid technology. PMUs provide the grid control center with time-synchronized voltage and current phasor data at a high sampling rate (e.g., 30–60 Hz for AC power systems operated at 60 Hz) through secured communication infrastructures. Based on PMUs, a wide-area measurement system (WAMS) can be established to help grid operators monitor and control the power grid and a real-time CSS scheme: the WAMS helps to monitor how control areas oscillate against each other to identify a potential separation boundary and help detect or predict out-of-step (OOS) conditions among those areas in order to decide the timing to perform CSS; once islands are formed, the WAMS can continue to monitor their frequency and voltage excursions and trigger necessary remedial actions for stabilization control; the WAMS also helps to expedite resynchronization of all islands toward power system restoration.

So far, deployment of a WAMS-based adaptive, system-wide CSS scheme for a real-life large-scale power grid has not been reported. There are still many research problems on CSS such as strategy optimization. In fact, most applications of PMUs are still in the research, development, and testing stages. Some relatively matured applications are mainly in power system visualization, oscillation and event detection, postmortem analysis, and offline model validation [7]. There are few protection systems or grid control room functions relying on PMUs, so WAMS-based grid protection and control in general is still an emerging research area. The rest of the article will first discuss some key research problems on CSS followed by a literature review on existing methods. Then, the criteria about separation strategies for CSS are discussed and accordingly, two formulations of the problem on separation strategies will be suggested. Finally, a unified framework for implementation of a WAMS-based CSS scheme in a future smart grid will be proposed with enabling techniques highlighted and existing technical gaps pinpointed.

2 Three Questions to Answer

Any practical CSS scheme has to answer three key questions: "Where to separate?" "When to separate?" and "How to separate?"

1. The first question "where" is about the locations or points of separation, i.e., which lines to be disconnected. Those points together form a cut set of the network and should properly be located to create sustainable islands.
2. Question "when" concerns the timing to disconnect the lines at separation points. Retarded disconnection at properly selected separation points may still ruin the success of CSS because the system state may change unpredictably under cascading failures. The sooner the islands are disconnected at desired points, the higher chance we have to save the grid.
3. The last question "how" is about actuation of CSS, asking what hardware devices to use, how to coordinate them at multiple locations, and what additional actions to take for stabilizing the generators, frequencies, and voltages of formed islands.

In existing literature, the three questions are rarely addressed at the same time. Most efforts focus on answering one question while assuming the others are either answered already or solvable afterward. "Where" is the most studied question. Papers [8–10] apply graph theory to model and simplify a power network and formulate a problem answering "where" and find separation strategies by methods based on the ordered binary decision diagram (OBDD). Further simulation and implementation studies on OBDD-based methods are presented in [11, 12]. Papers [13–18] present a slow-coherency-based controlled islanding approach that first groups generators by the slowest modes and then determines separation points by minimum net flow or spanning tree algorithms. Graph theory based spectral clustering and multilevel partitioning techniques are utilized in [19–25] to find separation points. References [26–31] find separation points by heuristic searching algorithms, e.g., the Particle Swarm Optimization, Genetic Algorithm, Ant Search, and Tabu Search. Mixed integer linear programming is utilized in [32–35] to optimize separation points. There are also data-driven methods [36, 37].

Not as many studies have addressed questions "when" and "how". For instance, paper [38] integrates an OBDD-based method with the extended equal area criterion (EEAC) method to ensure creating stable islands. Paper [39] takes an event-based approach that selects the strategy matching best the actual event from an offline built strategy table. Paper [25] calculates the maximal Lyapunov exponent of power swings measured by PMUs to predict an OOS condition and thus decide the timing of separation. Papers [40, 41] propose feeding real-time PMU data to offline trained decision trees to decide when to form a predesigned island. To stabilize each island formed after separation (i.e., addressing question "how"), load shedding strategies are proposed by [13, 42], in which the rate of change of frequency (ROCOF) is used as an important index on the amount of load to shed.

3 Formulation of the Problem

Although we discussed three questions separately, we have to clarify that, when system islanding is unavoidable, finding the best locations and timing to perform CSS should ideally be solved as a single CSS problem having all related questions such as where, when, and how addressed together in a systematic manner. However, two technical gaps make such a systematic resolution difficult at present:

- Lacks fast and reliable methods for real-time prediction of transient instability for general large-scale power systems under complex contingencies like cascading failures
- Lacks efficient algorithms and computational powers to overcome the computational complexities inherent with the problem.

Thus, a widely adopted approach as a compromise is to decompose that single CSS problem into subproblems solved separately, as done in past works, which inevitably sacrifices the optimality and performance of the resulting solution for CSS. With fast development of high-performance parallel computing technology and its increasing applications in smart grid, we reasonably envision that the two technical gaps may be filled or reduced and thus the single CSS problem may be tackled by parallel computers in a timely manner when CSS is needed. In the following, we introduce formulations on question "where" as an example to show computational complexities inherent with the CSS problem.

A separation strategy specifies a set of separation points where sustainable islands may form with minimum additional control to preserve frequencies and voltages. In most of the existing works, satisfaction to the following constraints is required.

- **C-I (generation coherency constraint)**: all generators after separated into each island are stable or, in other words, can keep their synchronism.
- **C-II (real power balance constraint)**: generation-load imbalance in each island is minimized or less than a tolerance to avoid frequency and voltage drifting out of acceptable ranges.
- **C-III (transmission capacity constraint)**: lines and transformers cannot be overloaded in formed islands.

C-I concerns in what manner generators are clustering or going out of step before CSS as important information for choosing separation points. Cascading failures often disturb generators to oscillate or even lose stability. Because an unstable generator will be tripped and is unable to support loads, maintaining stability of all generators in each island is the most critical criterion, whose satisfaction ensures the maximum capability to support loads. It needs to be pointed out that C-I is neither a sufficient nor necessary condition for the satisfaction of that critical criterion. The reasoning behind C-I is that if all generators of an island belong to one coherent group, they may have more chances to keep stability spontaneously or with minimum stabilization control. In fact, a coherent group of generators before CSS cannot ensure always keeping synchronism after being isolated into an island from the rest of the system; also, a generator that originally does not cohere with the others in one island is unnecessarily unstable after CSS as long as its excessive kinetic energy is absorbed timely. Therefore, requiring C-I in the problem formulation is mainly to increase the chance of successful CSS without rigorous examination of the stability of each island. That is a tradeoff due to the lack of real-time methods for predicting transient instability. If there is any theoretical breakthrough in real-time transient stability analysis, the constraint C-I and the related problem formulation should be redefined.

C-II requires allocating matched loads to the generation of each island. In practice, maintaining the frequency and the balance of real power is more important for each island since voltage and reactive power can be controlled using dispersed local reactive power compensators after the formation of the island. The satisfaction to C-II is mainly to define and enforce a threshold of real power imbalance in each island.

C-III is to avoid further overloading equipment after CSS. If a transmission line in any island is overloaded, i.e., violating the C-III, it will be tripped by protection. Thus failures may continue in the island. Since the portion of network isolated in each island becomes more vulnerable than it was in the original system, any additional equipment trip could be fatal to the survival of the island.

Of course, some other constraints can be added. For instance, a constraint may forbid disconnecting lines carrying large power flows unless they are necessary for separating the system. The main purpose of adding this constraint is to avoid large power flow redistribution caused by CSS, which may have a significant impact on the stabilities of islands.

Finding separation strategies addressing question "where" can be formulated as two problems below. The first problem is actually a relaxation of the second since once the optimal solution is found, it will usually be acceptable unless the tolerance is too small. From the computational complexity point of view, both problems are NP-hard, i.e., falling into a class of most complex problems in computation.

- **Satisfiability checking problem**: finding one or multiple acceptable solutions satisfying all the above constraints. Specifically for C-II, the tolerance on power imbalance in each island should be predefined.
- **Optimization problem**: finding the optimal solution minimizing the generation-load imbalance (C-II) in each island while satisfying the other constraints.

To mathematically formulate the problem, a power system can be modeled as an undirected, node-weighted graph or a directed, edge-weighted graph. In both graphs, buses of the power system are modeled by nodes or vertices, and branches are modeled by edges. Two models respectively assign weights to nodes and edges to model power injections at buses and power flows on branches. As pointed out in [8, 9], the problem of finding separation strategy satisfying C-I and C-II can be translated into a graph partitioning problem on either graph. Then, the solutions of that problem can be further checked with C-III and other constraints. The problem finding K acceptable or optimal islands is equivalent to finding a cut set of edges creating K subgraphs: for a node-weighted graph, the total node weight of each subgraph should be minimized (i.e., an optimization problem) or less than a tolerance (a satisfiability checking problem); for the edge-weighted graph, that problem is equivalent to the total edge weight of that cut set being minimized or less than the tolerance. The problem of finding separation strategy satisfying C-I and C-II is proved NP-complete in [9]. Thus, finding strategies also satisfying other constraints is NP-hard. There are also papers, e.g., [19], formulating a problem based on the edge-weighted graph to minimize the sum of absolute edge weights of the cut set. Note that such a problem is in fact solvable in polynomial time and the solution does not guarantee the minimum generation-load imbalance in each island or restricting the imbalance within a predefined small tolerance.

Because of the computational complexity of the problem, when the CSS on a real-life power grid is studied, simplification on the power network is often needed by power system knowledge and engineering judgments. If the integrity of a control area such as a metropolis is desired in CSS, it can be merged to an equivalent bus. Also, if offline simulation studies indicate that a group of generators has strong coherency under various contingencies, they together with neighboring load buses can be merged to one equivalent generator. Graph theory also helps simplified the graph model of a power system [8].

4   Unified Framework for WAMS-Based Controlled System Separation

This section suggests a unified framework for designing and developing a WAMS-based practical CSS scheme addressing all three questions. Some assumptions are made here:

- PMUs are placed near main generators to send real-time data streams to the central EMS located in the control center. The data are used to monitor oscillations and grouping among generators as important information on locations and timing of CSS.
- Actuators of CSS are separation relays, which can be designed based on OOS relays having both blocking and tripping functions. These separation relays are placed at the separation points according to the answer of question "where", and are remotely controlled by a CSS program integrated with the central EMS (Energy Management System) located in the grid control center. The communication channels from all separation relays to the control center are highly secured and reliable. Once CSS is decided, the program trips a proper set of separation relays at the same time while blocking the others in order for desired islands to form.

As shown in Table 1, under this framework, all tasks related to CSS are divided strategically into three time stages addressing three questions as follows:

- "Where": the **Offline Analysis** stage determines a set of potential separation points to place separation relays; thus, potential separation boundaries are made up by

Table 1 Three stages of the proposed WAMS-based unified framework for CSS

Stages	Main tasks with the questions addressed indicated
Offline Analysis (in the planning stage)	Optimize the separation points of each potential island to place separation relays (*where* and *how*)
	Design and daily update a post-separation control strategy table for each potential island (*how*)
Online Monitoring (every second)	Monitor inter-area oscillations between generators by WAMS for any potential OOS condition (*where* and *when*)
	Identify probable separation boundaries made up by proper separation points according to how generators cluster (*where*)
Real-time Control (milliseconds)	Trip all separation relays on the separation boundary where the OOS condition is detected or credibly predicted (*when* and *how*)
	Perform stabilization control in islands (*how*)

some of separation points to be monitored in the **Online Monitoring** stage; the final separation boundary is determined in the **Real-time Control** stage according to the final OOS condition detected or predicted.

- "When": **Online Monitoring** stage monitors inter-area electromechanical oscillations for early warning of potential system separation; the **Real-time Control** stage determines the timing to trip the right set of separation relays once an OOS condition appears
- "How": the **Offline Analysis** stage has separation relays placed at optimized locations and constructs a strategy table for stabilization control of each potential island; the **Real-time Control** stage performs control strategies from the table that match actual islands.

Reference [43] demonstrates such a WAMS-based CSS scheme on the WECC 179-bus power system.

In the **Offline Analysis** stage, planning engineers may study generator coherency to determine potential OOS conditions. The slow-coherency analysis method can help estimate generator coherency by the slowest modes [44]. The method assumes coherent generator groups to be almost independent of the size of the disturbance and the level of details of generator models so that the linearized, simplified model of the system having all generators in the classical model can be used [14]. That assumption is not true under large disturbances like cascading failures, so this method only provides an approximate, initial guess on generation coherency. It does provide insights on what generators tend to become coherent than the others in general, based on which the grouping can be refined after time-domain simulations on selected severe contingencies. This stage may identify a number of small, strongly coherent groups of generators. Each group has a high probability to preserve its synchronism and is merged to one equivalent generator together with neighboring lines and load buses. That largely simplifies the problem and now separation points are only chosen from the remaining lines between those groups. The stability of generators within each group is ensured by stabilization control after CSS.

For each OOS scenario, algorithms like the method proposed in [8, 9] can be applied to quickly find all separation strategies satisfying C-I (generation coherency constraint), C-II (power balance constraint), C-III (transmission capacity constraint), and the others as well. Alternatively, optimal strategies can be solved to minimize the generation-load imbalance (C-II) while satisfying the other constraints. When the system operating condition changes, the separation points may drift. There is a compromise between the number and applicability of the separation points since it may not be economically practical to install separation relays at all possible separation points.

Regarding stabilization control on each formed island, adequate real and reactive power reserves need to be planned to help arrest frequency and voltage excursions and stabilize generators after CSS. Spinning reserves allow fast increase of the generation in an island to balance the surplus of load and support the frequency; unimportant or dispatchable loads in the island can also be shed to match generation. Reactive power reserves include shunt capacitors and reactors and dynamic VAR sources such as SVCs and STATCOMs, which can control voltage profile in an island. For each potential island, a strategy table on stabilization control can be developed offline in this stage. The strategy table needs to be comprehensive to cover a wide variety of operating conditions and each strategy needs to be validated sufficiently by simulation studies. Once an island is formed, a real-time algorithm finding the strategy matching best the actual condition is also necessary.

In the **Online Monitoring** stage, separation boundaries are predicted by means of real-time WAMS data. Major disturbances can usually be reflected from real-time changes in amplitudes or shapes of inter-area oscillation modes whose frequencies typically range in 0.1–1 Hz. Therefore, this online stage monitors dominant inter-area oscillation modes for probable separation boundaries. Measurement-based modal analysis can help identify the dominant inter-area modes and their mode shapes. For instance, the spectral analysis based method in [45] and the phase-locked loop (PLL) based algorithm in [46] can be applied to a latest time window of PMU data to differentiate inter-area modes from local modes and estimate the mode shape of each dominant mode as an important indicator on how generators actually cluster regarding the mode. Accordingly, the most probable separation boundaries should be identified.

The **Real-time Control** stage takes the final action of CSS when the stability and integrity of the power system cannot be preserved. In general, real-time, accurate prediction of angular instability for a large-scale multi-machine power system is a difficult problem, so this stage requires the most future efforts for research and development among all three stages. In operations, power companies usually adopt an approximate, often conservative approach that compares the angle distance between two selected substations with a preset threshold in, e.g., 120°–180°. Thus, before a more accurate real-time method for prediction of transient instability is available, an engineering compromise is to monitor the real-time angle distance across each probable separation boundary by PMUs and compare it with a threshold determined by offline simulation studies. Alternatively, the angle distance can be extrapolated utilizing its measured waveforms from the latest time window to foresee a violation of the threshold before it happens [43]. Moreover, when two potential islands oscillate about a dominant mode in opposite directions, i.e., having phasing difference close to 180°, they have increased risk to separate. Paper [47] suggests also considering phasing differences across potential separation boundaries in predicting the final separation boundary.

5 Conclusions

Controlled system separation (CSS) is considered the final resort against power blackouts in smart grid. Although there have been works respectively addressing three key questions "where", "when", and "how" on CSS, these questions are in fact coupled into a single complex problem on CSS, which has rarely been studied in literature. The main technical gaps are in the lack of a fast and reliable method for real-time prediction of transient instability for large power systems, which is critical for answering "where" and "when", and in the lack of powerful algorithms and computing resources to overcome the inherent computational complexities with the CSS problem. Breakthroughs in those two areas will expedite the research on CSS in smart grid. For instance, both technical gaps are being reduced with increasing applications of PMU-based WAMS and high-performance parallel computers to the real-time operations of power grids. Finally, this article also suggests a unified framework for research and development of a practical WAMS-based CSS scheme that addresses all three key questions in a systematic way.

References

1. Prevention of Power Failures. U.S. Federal Power Commission, Government Printing Office, Washington, D.C., U.S. (1967) http://chnm.gmu.edu/blackout/archive/a_1965.html
2. U.S.-Canada Power System Outage Task Force, "Final Report on the August 14th Blackout in the United States and Canada" https://reports.energy.gov
3. FERC/NERC, Arizona-Southern California Outages on September 8, 2011: Causes and Recommendations, April 2012
4. Final Report—System Disturbance on 4 November 2006, Union for the Co-ordination of Transmission of Electricity
5. Report on the Grid Disturbances on 30th July and 31 July 2012, Central Electricity Regulatory Commission, August 8 (2012)
6. IEEE PES CAMS Task Force on Understanding, Prediction, Mitigation and Restoration of Cascading Failures), *Initial Review of Methods for Cascading Failure Analysis in Electric Power Transmission Systems* (IEEE PES General Meeting, Pittsburgh, 2008)

7. NASPI Task Force on Synchrophasor Protection Applications, Integrating Synchrophasor Technology into Power System Protection Applications, NASPI Engineering Analysis Task Team White Paper No. NAPSI-2016-TR-007, September 2016
8. K. Sun, D. Zheng, Q. Lu, Splitting strategies for islanding operation of large-scale power systems using OBDD-based methods. IEEE Trans. Power Syst. **18**(2), 912–923 (2003)
9. Q. Zhao, K. Sun, D. Zheng, A study of system splitting strategies for island operation of power system: a two-phase method based on OBDDs. IEEE Trans. on Power Syst. **18**(4), 1556–1565 (2003)
10. K. Sun, D. Zheng, Q. Lu, Searching for feasible splitting strategies of controlled system islanding. IEE Proc. Generat. Trans. Distribut. **153**(1), 89–98 (2006)
11. K. Sun, D. Zheng, Q. Lu, A simulation study of OBDD-based proper splitting strategies for power systems under consideration of transient stability. IEEE Trans. Power Syst. **20**(1), 389–399 (2005)
12. X. Li, Q. Zhao, Parallel implementation of OBDD-based splitting surface search for power system. IEEE Trans. Power Syst. **22**(4), 1558–1593 (2007)
13. H. You, V. Vittal, Z. Yang, Self-healing in power systems: an approach using islanding and rate of frequency decline-based load shedding. IEEE Trans. Power Syst. **18**, 174–181 (2003)
14. H. You, V. Vittal, X. Wang, Slow coherency-based islanding. IEEE Trans. Power Syst. **19**(1), 483–491 (2004)
15. X. Wang, V. Vittal, System islanding using minimal cutsets with minimum net flow. IEEE PES Power Systems Conference and Exposition, Oct 10–13, 2004
16. B. Yang, V. Vittal, Slow-coherency-based controlled islanding—a demonstration of the approach on the August 14, 2003 blackout scenario. IEEE Trans. Power Syst. **21**(4), 1840–1847 (2006)
17. G. Xu, V. Vittal, Slow coherency based cutset determination algorithm for large power systems. IEEE Trans. Power Syst. **25**(2), 877–884 (2010)
18. G. Xu, V. Vittal, A. Meklin, J.E. Thalman, Controlled islanding demonstrations on the WECC system. IEEE Trans. Power Syst. **26**(1), 334–343 (2011)
19. L. Ding, F.M. Gonzalez-Longatt, P. Wall, V. Terzija, Two-step spectral clustering controlled islanding algorithm. IEEE Trans. Power Syst. **28**(1), 75–84 (2013)
20. L. Ding, P. Wall, V. Terzija, Constrained spectral clustering based controlled islanding. Int. J. Elect. Power Energy Syst. **63**, 687–694 (2014)
21. R.J. Sánchez-García, M. Fennelly, S. Norris, N. Wright, G. Niblo, J. Brodzki, J.W. Bialek, Hierarchical spectral clustering of power grids. IEEE Trans. Power Syst. **29**(5), 2229–2237 (2014)
22. H. Song, J. Wu, K. Wu, A wide-area measurement systems-based adaptive strategy for controlled islanding in bulk power systems. Energies **7**, 2631–2657 (2014)
23. J. Quirós-Tortós, R. Sánchez-García, J. Brodzki, J. Bialek, V. Terzija, Constrained spectral clustering-based methodology for intentional controlled islanding of large-scale power systems. IET Generat. Trans. Distrib. **9**(1), 31–42 (2015)
24. L. Ding, Y. Guo, P. Wall, Performance and suitability assessment of controlled islanding methods for online WAMPAC application. Int. J. Elect. Power Energy Syst. **84**, 252–260 (2017)
25. J. Li, C.-C. Liu, K.P. Schneider, Controlled partitioning of a power network considering real and reactive power balance. IEEE Trans. Smart Grid **1**(3), 261–269 (2010)
26. W. Liu, L. Liu, D.A. Cartes, Binary particle swarm optimization based defensive islanding of large scale power systems. Int. J. Comput. Sci. Appl. **4**, 69–83 (2007)
27. A.Y. Abdelaziz, W. El-Khattam, Mohammed, Application of angle-modulated particle swarm optimization technique in power system controlled separation WAP. Int. J. Intell. Syst. Appl. Eng. **2**(3), 51–57 (2014)
28. M.R. Aghamohammadi, A. Shahmohammadi, Intentional islanding using a new algorithm based on ant search mechanism. Int. J. Elect. Power Energy Syst. **35**, 138–147 (2012)
29. F. Tang, H. Zhou, Q. Wu, H. Qin, J. Jia, K. Guo, A tabu search algorithm for the power system islanding problem. Energies **8**, 11315–11341 (2015)

30. F. Jabari, H. Seyedi, S.N. Ravadanegh, Large-scale power system controlled islanding based on backward elimination method and primary maximum expansion areas considering static voltage stability. Int. J. Elect. Power Energy Syst. **67**, 368–380 (2015)
31. Y. Wu, Y. tang, B. Han, M. Ni, A topology analysis and genetic algorithm combined approach for power network intentional islanding. Int. J. Elect. Power Energy Syst. **71**, 174–183 (2015)
32. P.A. Trodden, W.A. Bukhsh, A. Grothey, K.I.M. McKinnon, MILP formulation for controlled islanding of power networks. Int. J. Elect. Power Energy Syst. **45**, 501–508 (2013)
33. P.A. Trodden, W.A. Bukhsh, A. Grothey, K.I.M. McKinnon, Optimization-based islanding of power networks using piecewise linear AC power flow. IEEE Trans. Power Syst. **29**(3), 1212–1220 (2014)
34. T. Ding, K. Sun, C. Huang, Z. Bie, F. Li, Mixed-integer linear programming-based splitting strategies for power system islanding operation considering network connectivity, IEEE Syst. J. https://doi.org/10.1109/JSYST.2015.2493880
35. T. Ding, K. Sun, Q. Yang, A.W. Khan, Z. Bie, Mixed integer second order cone relaxation with dynamic simulation for proper power system islanding operations. IEEE J. Emerg. Selected Topics Circ. Syst. **7**(2), 295–306 (2017)
36. C.G. Wang, B.H. Zhang, Z.G. Hao, J. Shu, P. Li, Z.Q. Bo, A novel real time searching method for power system splitting boundary. IEEE Trans. Power Syst. **25**(4), 1902–1909 (2010)
37. F. Raak, Y. Susuki, T. Hikihara, Data-driven partitioning of power networks via Koopman mode analysis. IEEE Trans. Power Syst. **31**(4), 2799–2808 (2016)
38. M. Jin, T.S. Sidhu, K. Sun, A new system splitting scheme based on the unified stability control framework. IEEE Trans. Power Syst. **22**(1), 433–441 (2007)
39. K. Sun, T.S. Sidhu, M. Jin, Online pre-analysis and real-time matching for controlled splitting of large-scale power networks, in *IEEE International Conference on Future Power Systems* (Amsterdam, Nov 16–18, 2005)
40. R. Diao, V. Vittal, K. Sun, S. Kolluri, S. Mandal, F. Galvan, Decision tree assisted controlled islanding for preventing cascading events, in *IEEE PES Power Systems Conference and Exposition* (Seattle, WA, March 15–18, 2009)
41. N. Senroy, G.T. Heydt, V. Vittal, Decision tree assisted controlled islanding. IEEE Trans. Power Syst. **21**(4), 1790–1797 (2006)
42. MdQ Ahsan, A.H. Chowdhury, S.S. Ahmed, I.H. Bhuyan, M.A. Haque, H. Rahman, Technique to develop auto load shedding and islanding scheme to prevent power system blackout. IEEE Trans. Power Syst. **27**(1), 198–205 (2012)
43. K. Sun, K. Hur, P. Zhang, A new unified scheme for controlled power system separation using synchronized phasor measurements. IEEE Trans. Power Syst. **26**(3), 1544–1554 (2011)
44. J.H. Chow, R. Galarza, P. Accari et al., Inertial and slow coherency aggregation algorithms for power system dynamic model reduction. IEEE Trans. Power Syst. **10**(2), 680–685 (1995)
45. D.J. Trudnowski, Estimating electromechanical mode shape from synchrophasor measurements. IEEE Trans. Power Syst. **23**(3), 1188–1195 (2008)
46. K. Sun, Q. Zhou, Y. Liu, A phase locked loop-based approach to real-time modal analysis on synchrophasor measurements. IEEE Trans. Smart Grid **5**(1), 260–269 (2014)
47. K. Sun, X. Luo, J. Wong, Early Warning of Wide-Area Angular Stability Problems Using Synchrophasors, in *IEEE PES General Meeting* (23–26 July 2012, San Diego, 2012)

Part IV
Cyber-Physical Security

Smart Grid Security: Attacks and Defenses

Azwirman Gusrialdi and Zhihua Qu

Abstract Electric grids in the future will be highly integrated with information and communications technology. The increase in use the of information technology is expected to enhance reliability, efficiency, and sustainability of the future electric grid through the implementation of sophisticated monitoring and control strategies. However, it also comes at a price that the grid becomes more vulnerable to cyber-intrusions which may damage the physical system. This chapter provides an overview of cyberattacks on power systems from a system theoretical perspective by focusing on the tight coupling between the physical system and the communication network. It is demonstrated via several attack scenarios how the adversary may cause significant impacts on the power system by intercepting the communication channel and without possibly being detected. The attack strategies and the corresponding countermeasures are formulated and analyzed using tools from optimization, dynamical systems, and control theory.

1 Introduction

Electric grids (physical systems) in the future will be highly integrated with information and communications technology (cyber-layer) resulting in a complex cyber-physical system (CPS). The communication technology has been mainly used by the supervisory control and data acquisition (SCADA) systems for the purpose of sensing, monitoring, and control of the power systems. The increase in use of information technology is expected to enhance reliability, efficiency, and sustainability of the future electric grid through the implementation of sophisticated monitoring and control strategies such as advanced demand side management system. On the other hand, information and communication technology of the power grids have started evolving from isolated structures into a more open and networked environ-

A. Gusrialdi · Z. Qu (✉)
Department of ECE, University of Central Florida, Orlando, FL 32816, USA
e-mail: qu@ucf.edu

A. Gusrialdi
e-mail: azwirman.gusrialdi@ucf.edu

© Springer Nature Switzerland AG 2019
J. Stoustrup et al. (eds.), *Smart Grid Control*, Power Electronics
and Power Systems, https://doi.org/10.1007/978-3-319-98310-3_13

ment via TCP/IP and Ethernet. Since the information and communication technology is known to be vulnerable to cyber-intrusions and cyberattacks, potential network intrusion may cause physical damage to the power network due to the tight coupling between the power system and the cyber-layer.

An example of the most recent cyberattack on power systems is the attack on the Ukraine power grid in December 2015, which is a synchronized and coordinated cyberattack, causing a 6-h blackout and affecting hundreds of thousands of customers [48]. Investigations revealed that the attack was initiated by the BlackEnergy3 malware delivered via phishing emails and activated by the employee. Specifically, the attack was a hijack of SCADA network by targeting field devices with malicious firmware which facilitates foreign attacker to remotely open the substation breakers [30]. The attack consists of multiple stages [12] including gaining a foothold into the IT networks and harvesting credentials to access the industrial control system (ICS) network, using existing remote access tools to issue commands, using telephone systems in generating denial-of-service attack to deny access to customers report for outages, and delaying restoration efforts by erasing master boot records on workstations via a modified KillDisk firmware attack.

1.1 Adversary Models

Adversary model in general is composed of attack strategy and the adversary resources, that is the model (system) knowledge, disclosure, and disruption resources as discussed in [62]. Moreover, many well-known attack schemes can be conveniently categorized based on the adversary resources as depicted in Fig. 1. Model knowledge is the most important component of the adversary model since it can be used by the adversary to construct complex and undetectable attacks with more significant impacts on the physical systems. The disclosure resources such as a set of actuator,

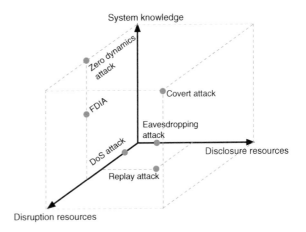

Fig. 1 Attack space for CPS [62]

sensor channels, and communication channel that can be accessed during the attack enable the adversary to gather sensitive information about the system by violating data confidentiality. It is worth noting that disclosure attack alone cannot damage the physical system but it can be used to construct more complex attacks which could affect the physical dynamics of the systems. Disruption resources are used by the adversary to disturb the system operation, that is by violating data integrity of the systems. The attacker can affect the system operation via physical resources or data deception resources such as by modifying control action or sensor measurement.

In order to discuss the adversary model in more details, let us take the false data injection attack (FDIA) as an example. As will be shown later in the chapter, in order to launch an undetectable FDIA, an adversary requires [30, 31]: (i) information of power system operation and features of target system, (ii) capability of manipulating meter measurements, (iii) system knowledge such as network topology, electrical parameters, and bad data detection scheme. These requirements might raise question on the practical feasibility for launching FDIA. However, the previously mentioned Ukraine power grid attack illustrates the plausibility of the above requirements as discussed in [30]. To be more precise, the adversary can obtain substantial collection of knowledge on the targeted power system via the Internet which provides access to power system and vendor-specific information, research publications to update the knowledge of power system innovation, and industrial control standards and network protocols. Knowledge of network topology and system parameters can also be gathered via one-off observation of components and market data. In addition, meter measurements can be manipulated by compromising meters locally, modifying control center database, or intercepting and forging data sent to the control center. The adversary can also take advantage of the vulnerabilities in network protocols, firewalls, encryption, and VPN connections. For example: (i) intercepting and forging communication messages of SCADA is relatively easy since its communication network is not equipped with secure communication protocol; (ii) password-protected access can be obtained via keystroke loggers as in the case of the Ukraine blackout; (iii) database manipulation is also possible given that the adversary successfully obtains credentialed access via VPNs.

1.2 Cybersecurity Countermeasures

In addition to ensuring grid's reliability against random failures, smart grid security objectives in a traditional security include: (i) availability, that is to ensure reliable access to information, (ii) integrity, that is to ensure information authenticity against information modification, and (iii) confidentiality, that is to protect personal privacy and proprietary information. Availability and integrity are the most important objectives from the perspective of system reliability. On the other hand, confidentiality becomes more important for operation involving interactions with customers, such as demand response. The existing solutions to network security can be applied to detect and mitigate network threats from the smart grid. For example, packet-based detection can be used to detect denial-of-service attack by identifying a significant increase in packet transmission failure. Moreover, cryptographic based approaches such as encryption, authentication, and key management can be utilized to deal with attacks targeting integrity and confidentiality. We refer the interested reader to a survey of results in [64] for the details.

However, the tools from network security alone are not sufficient to address challenges in smart grid security since they do not take into account the physical attacks through direct interaction with the components in the grid, including the stability and control performance of the physical systems. For example, by placing a shunt around a meter, the integrity of a meter can be violated without the need of breaking the cybersecurity countermeasure implemented to protect the data sent through the network. Since it is prohibitive to protect all sensors in the grid, an innovative approach is required to address the security challenges in a smart grid. As demonstrated in [41] and will be demonstrated through the chapter, the combination of network security countermeasures (to deal with the network level) and system theoretic approaches (to consider the physical aspects of the grid) show a great promise to address cyber-physical security of the future grid.

1.3 Objective of the Chapter

The chapter provides an overview of (cyber) attacks on power systems together with their countermeasures. It should be noted that there already exist several survey papers on cyberattacks and security of smart grid, for example [19, 31, 61, 64]. The work [64] presents a comprehensive survey of cybersecurity issues for smart grid from a communication network perspective by focusing on security requirements, network vulnerabilities, attack countermeasures, and secure communication protocols and architecture. In addition, the authors in [61] provide a survey of security issues for smart grid based on a data-driven approach including data generation, acquisition, storage, and processing. However, the above survey papers do not discuss in details the coupling between the cyber-layer (e.g., communication network) and the physical system, including the impact of cyberattacks on the physical systems. On the other hand, the work [19, 31] presents a survey of smart grid security from the cyber-physical system perspective, namely by considering the coupling between the communication network and physical system. In particular, the survey paper [19] provides a comprehensive review of attack scenarios with significant impact on the smart grid operation together with the corresponding defense mechanisms. The attack schemes are grouped based on the infrastructure of the power grid, namely generation, transmission, distribution, and electricity market. However, the papers lack technical details and analysis which are of interest for researchers, in particular, from control systems society. In contrast, this chapter presents a sample of attack schemes on smart grid and their countermeasures from a system theoretic perspective. Each attack scheme and its associated defense mechanism (when applicable) are systematically formulated and analyzed using tools from optimization, dynamical systems, and control theory.

2 Insider Attacks

An insider attack occurs when an authorized operator, for example of the supervisory control and data acquisition (SCADA) system, misuses the permissions and injects malicious commands including changing data values, manipulating control signals, and opening breakers in order to bring damages to the power grid. Since the malicious operator has detailed knowledge on the system operation and his/her commands are legal, he/she could then easily design an attack with a high success rate and great impact without being detected. Moreover, complicit insider may also provide critical information to outsider attackers who use the information to increase the effectiveness of their attacks. Hence, insider attack is considered as one of the most dangerous threats on security of critical infrastructures [4]. One of the most well-known real-world examples of insider attack on SCADA system is the attack on Maroochy water services in Australia in 2000 [56] where a disgruntled ex-employee sent control signals to various pumps in the system resulting in the release of thousands of gallons of sewage into the surrounding area.

In SCADA system, an alarm will be triggered when abnormal changes or failures occur in the power system. The alarms in SCADA system can be categorized into transient alarms which are short term and can be easily remotely resolved by the operator, and permanent alarms which are long term and cannot be resolved remotely [45]. Disgruntled operator can cause damage to the power system through the following actions [45]: (i) not responding to the alarms resulting in cascading failure, (ii) sending incorrect responses to the incoming alarm, (iii) creating new alarms by modifying the topology or load transfer resulting in power failure, opening output feeders or improper equipment settings. In addition to the previously mentioned scenarios, insider attacks can also be launched by physically accessing the power system device located in the remote substations. Even though some of the devices are equipped with basic password authentication, a trusted insider attack could easily bypass this basic protection technique.

Since the command from the authorized but disgruntled operator is legal, traditional security technique including intrusion detection system becomes ineffective against insider attacks. Moreover, operator with authorized access to the control systems could still launch the insider attacks even though the networks are isolated from the public networks. Since the operation of SCADA system in general has a regular pattern, one possible strategy to counteract the insider attack is by detecting anomalous behavior. Anomaly detection in SCADA system is based on determining whether the current behavior deviates from the normal behavior. This can be achieved by periodically monitoring the logs of the SCADA and applying knowledge-based method, case-based reasoning [3], statistical method [7, 45], or machine learning based methods [15]. However, the previously mentioned countermeasure is only valid for detecting attacker at the network level. As previously mentioned, in some cases, the attacker may have physical access to power system device such as remote substations and gain complete control of the device. Basic password authentication may not be able to protect the device if the attacker is a trusted one. Moreover, it is not practical for the power system company to protect their devices by deploying physical locks and tracking all physical keys. To address this issue, a software-based solution, namely an overlay network of gateway devices, is developed in [14] which provides authenticated access control and security monitoring for the vulnerable interfaces.

3 Outsider Attacks: Non-stealthy Case

Outsider attack refers to attackers who inject false data or modify control signals by (remotely) intercepting the communication channel and modifying the transmitted data accordingly, hence by taking advantage of the deployment of information and communication topology in the smart grid. The objective is in general to drive the system to an unsafe state. In the following section, we provide several scenarios of outsider attacks together with the defense strategies (if applicable).

3.1 Attacks on Load Frequency Control

Load frequency control (LFC) is one of the automatic control loops in SCADA power systems whose goal is to keep the frequency of power system at a nominal value (i.e., 60 Hz) by adjusting power generation set-point. In the following, we illustrate how the attacker could destabilize the frequency of the power system away from its nominal value by launching denial-of-service attack based on the results presented in [35]. First, the power system can be decomposed into several areas where each area consists of generators and load and is interconnected with the neighboring areas through the tie lines. Representing all generators in an area by one single machine, the dynamics of LFC for the i-th area is given by

$$\dot{\omega}_i = -\frac{D_i^G}{M_i^G}\omega_i + \frac{\Delta P_i^M}{M_i^G} - \frac{\Delta P_i^{tie}}{M_i^G} - \frac{\Delta P_i^L}{M_i^G}$$

$$\dot{\Delta P}_i^M = -\frac{1}{T_i^t}\Delta P_i^M + \frac{1}{T_i^t}\Delta P_i^v$$

$$\dot{\Delta P}_i^v = -\frac{1}{R_i T_i^G}\omega_i - \frac{1}{T_i^G}\Delta P_i^v + \frac{1}{T_i^G}\Delta P_i^c \qquad (1)$$

$$\dot{\Delta P}_i^{tie} = \sum_{j=1, j\neq i}^{N} 2\pi T_{ij}(\omega_i - \omega_j)$$

$$\dot{E}r_i = \beta_i \omega_i + \Delta P_i^{tie}$$

where ω_i, ΔP_i^M, ΔP_i^v, ΔP_i^c, ΔP_i^{tie}, ΔP_i^L denote the deviation of frequency, generator mechanical power, turbine valve position, load reference set-point, tie-line power flow in area i, and load, respectively. Moreover, M_i^G, D_i^G, T_i^G, T_i^t, T_{ij}, β_i, Er_i, R_i, N are moment of inertia of generator i, damping coefficient of generator i, time constant of governor i, time constant of turbine i, stiffness constant, frequency bias factor of area i, i-th area control error, speed droop coefficient, and number of areas respectively. The state-space model of (1) can then be written as

$$\dot{x}_i = A_{ii}x_i + B_i u_i + \sum_{j=1, j \neq i}^{N} A_{ij}x_j + F_i \Delta P_i^L \tag{2}$$

where $x_i = [\omega_i, \Delta P_i^M, \Delta P_i^v, \Delta P_i^{tie}, Er_i]^T$ and $u_i = \Delta P_i^c$. For simplicity, it is assumed that ΔP_i^L is constant. The overall interconnected system can then be compactly written as $\dot{x} = A_c x + B_c u$ and its discretized model is given by

$$x(k + 1) = Ax(k) + Bu(k) \tag{3}$$

where $x = [x_1^T, x_2^T, \ldots, x_N^T]^T$ and $u = [u_1, u_2, \ldots, u_N]^T$. After receiving the measurements telemetered in remote terminal units, the control center then computes the optimal control input given by

$$u(k) = -Kx(k). \tag{4}$$

However, the communication channels in the sensing loop may be compromised by the adversary. The adversary may launch denial-of-service (DoS) attack by jamming the communication channel, attacking networking protocol, or flooding the network traffic which results in that the measurement packets sent from the sensor through this channel will be lost [47]. Assuming that the controller in (4) is equipped with zero-order hold, the DoS attack can then be modeled as

$$\begin{cases} x(k) = x(k) & \text{no attack} \\ x(k) = x(k-1) & \text{under DoS attack} \end{cases}$$

Defining augmented state $\bar{x}(k) = [x^T(k), x^T(k-1)]$, the overall system under DoS attacks can then be modeled as the following switched system:

$$\bar{x}(k+1) = \Phi_{\sigma_i} \bar{x}(k) \tag{5}$$

where $\sigma_i = 1$, $\sigma_i = 2$ denote the no attack mode and under DoS attack mode, respectively, and

$$\Phi_1 = \begin{bmatrix} A - BK & 0 \\ I & 0 \end{bmatrix}; \quad \Phi_2 = \begin{bmatrix} A & -BK \\ 0 & I \end{bmatrix}.$$

It is known that the switching among stable or/and unstable subsystems may result in that the whole system becomes stable or unstable [32]. Hence, the adversary can destabilize switched system (5) by choosing a switching rule using the well-known stability result of linear switched system [34]. Alternatively, the attacker can choose a constant $0 < \zeta < 1$ (if any) such that the average system $\Phi_\zeta = \zeta\Phi_1 + (1 - \zeta)\Phi_2$ has an eigenvalue with magnitude outside the unity circle. One countermeasure to Dos attacks is by reconfiguring the routing topology of the network so that malicious nodes can be isolated [58].

Note that in addition to DoS attacks, other types of attacks on the LFC and Automatic Generation Control which may yield instability also include time-delay switch attack [54, 55], resonance attack [68], and integrity attack [57].

3.2   Interdiction Attack

An interdiction refers to the tripping of lines, transformers, generators, buses, and substations in the transmission grid by an attacker under limited budget in order to cause the largest possible disruption to the grid. As discussed in [51], the problem in general can be formulated as the following bi-level optimization:

$$\max_{\mu\in\Delta} \min_p \; C^T p \quad \text{subject to} \quad g(p, \mu) \le b, \quad p \ge 0, \tag{6}$$

where the binary vector μ denotes the interdiction plan, namely, $\mu_k = 1$ if component k of the system is attacked and zero otherwise and discrete set Δ represents the attacks that may be carried out. The inner optimization in (6) is the standard optimal power flow whose goal is to minimize the total generation plus total load shedding costs. The vector p denotes the generation outputs, power flow, phase angles, and load shedding while vector C is the generation and load shedding cost. The set of inequality constraints in (6) represent constraints on thermal limit, generating unit output, power balance at each bus, and interdiction attack budget. Optimization problem (6) can also be used to study power system's resilience against natural disasters by identifying critical sets of power grid's components and can be solved using mixed-integer bi-level programming [2], greedy search [6], or generalized Benders decomposition [52].

One strategy to minimize the risk of disruption due to interdiction attacks is by hardening targets, acquisition of spare critical components, and surveillance [63]. The challenge is that how to allocate the available and constrained countermeasure resources to the power system such that their effectiveness is maximized. The strategy should also consider the attacker's response after the countermeasure resources are in place and the mitigation measures that can be performed after the attack. As discussed in [71], the defense strategy can be formulated as the following trilevel optimization problem:

$$\min_{y} \ \max_{\mu \in \{\Delta \setminus y\}} \ \min_{p} \ C^T p \quad \text{subject to} \quad \bar{g}(p, \mu, y) \le \bar{b}, \quad p \ge 0, \tag{7}$$

where it is assumed that a protected network element cannot be attacked. The elements of binary vector y, i.e., $y_k = 1$ if element k need to be defended and zero otherwise. Similar to (6), the inequality constraints include thermal limit, generating unit output limit, power balance at each bus, interdiction resource (budget), and defense budget.

3.3 Attack on Circuit Breaker

A circuit breaker is an electrical switch designed to protect electrical circuit in power system from damage typically caused by overload or short circuit. Modern circuit breakers are remotely controlled and connected to the control center via the communication network. Hence, another strategy of the adversary to destabilize the power grid is by compromising communication channel and the control signal in order to manipulate the circuit breaker. In the following, we summarize the results on attack of circuit breaker presented in [36].

The action of a circuit breaker in power system can be modeled as the following variable structure system:

$$\dot{x} = \begin{cases} f_1(x, t), & s(x) > 0 \\ f_2(x, t), & s(x) \le 0 \end{cases}, \tag{8}$$

where x denotes the state (e.g., rotor angle and generator frequency) and $s(x)$ is a state-dependent switching signal which is also called *switching surface*. According to the sign of $s(x)$, the system switches between $f_1(x, t)$, i.e., when the load is connected, and $f_2(x, t)$, when the load is not connected. One interesting property of variable structure system (8) is the so-called sliding mode in which $s(x) = 0$ is attractive so that the trajectories within the subset of state space will converge, confine to the surface, and then slide along the surface in the direction or away from the equilibrium point. Even though the power system is stable when the breaker is static (either open or close), since the switching between stable subsystems may induce instability, the attacker can then destabilize the power system by controlling a circuit breaker, i.e., determining the switching surface $s(x)$ so that the whole switched system is unstable.

3.4 Load Altering Attacks

While the attacks described previously target generation and power distribution and control, the attacker can also aim the consumption sector in power system by altering the loads at certain grid locations. The goal of the attacker is to alter the loads in order to damage the grid through circuit overflow, disturbing the balance between power demand and supply or destabilizing the frequency of power system away from its nominal value. Demand side management (i.e., demand-response program) has been widely adopted to minimize peak demand and shift this load to off-peak hours and also to improve system operation. Due to the increase in use of information technology in demand side management which makes the loads to be remotely accessible via the Internet, the adversary can modify the load by compromising the communication network [44]. In the following, we demonstrate how the adversary can alter the loads by corrupting the control signals in demand-response program based on the results presented in [1, 16].

3.4.1 Direct Load Control

One of the commonly used demand side management method is direct load control in which some loads such as water heating and air conditioning are under direct control of the utility. To this end, the load control systems are equipped with two-way communications in order to send command signals (via power line carrier [66] or Internet [40]), such as switch on or switch off commands or operational power level to the appliances being controlled. The attacker can then compromise the command signals by changing the volume of certain vulnerable loads in an *abrupt* manner which could potentially yield a large spike in the aggregate demand. In addition, an adversary can also launch a more sophisticated load altering attack by changing the load over time which is called *dynamic* load altering attack (D-LAA) [1].

A D-LAA can be launched in both open-loop and closed-loop fashion. In an open-loop D-LAA, the adversary does not monitor the grid condition and the impact of the load manipulated while implementing the attack through a feedforward controller. On the other hand, in a closed-loop D-LAA, the attacker monitors the grid condition, such as measuring the voltage magnitude or frequency, by hacking into the existing monitoring system and use these information to control the load trajectory via a feedback controller. Moreover, the adversary can compromise the vulnerable load at one victim bus or multiple vulnerable loads at several victim buses in a coordinated fashion.

One of the objectives of closed-loop D-LAA is to deviate the frequency of the power system from its nominal value (60 Hz in North America) which affects the power system stability by changing the load based on the frequency measurement through the frequency sensor co-located with the victim bus or located at some bus on the same interconnection network called as *sensor bus*. It should be noted that it is sufficient for the adversary to hack into the remote load control systems in launching the attack. Hence, he/she does not need to have access to the transmission-level SCADA system where the impact of area frequency is taken place. Specifically, the adversary first monitors frequency at sensor bus and sends the measurements to the controller. Based on these measurements, the adversary calculates the amount of vulnerable load P_{LV} and remotely controls the load at that amount by compromising the command signals in direct load control program [1].

3.4.2 Indirect Load Control

Indirect load control is an alternative to direct load control in demand-response programs which allows the utility to control the load indirectly via incentives such as real-time pricing sent through the Internet. This type of load control can also be automated with the use of energy consumption scheduling (ECS) [43] in smart meters which schedule the timing and amount of energy consumption for each controllable household appliances based on the price information such that the cost of energy is minimized.

Let us assume that the aggregated demand of consumers at time instant k can be written as $d_k(\lambda_k) = b_k + w_k(\lambda_k)$, where λ_k denotes the clearing price determined by the independent system operator (ISO), b_k is the power required to satisfy the main consumer needs at each instant k and is independent of the pricing mechanism, and $w_k \geq 0$ denotes the price-responsive demand which is the amount of demand that can be controlled by pricing signal λ_k. In general, a decrease in price values yields an increase in the load demand. The following constant elasticity of own-price model has been widely utilized to characterize the total price-responsive demand [33] and is given by $w(\lambda_k) = D\lambda_k^\varepsilon$, where $D > 0$ denotes a scaling constant and $\varepsilon \in (-1, 0)$ captures how the price λ_k affects the demand. Moreover, a model of the supply of electricity is proposed in [60] based on linear regression between supply and cost and is given by $su(\lambda_k) = p\lambda_k + q$, where parameters p, q are estimated using historical market data. Based on the above models, define the unbalance (error) between the electric power supply and demand as $er_k = su(\lambda_k) - d_k(\lambda_k)$. The ISO aims at adjusting the price signal λ_k (which must be carefully designed so that it will not yield instability [50]) in order to keep er_k close to zero by measuring the error er_k. To this end, the following price-setting algorithm which guarantees the stability of the feedback system is proposed in [60]:

$$\lambda_k = \lambda_{k-1} - \frac{2\eta}{p - D\varepsilon(\lambda_0)^{\varepsilon-1}} er_{k-1} \tag{9}$$

where $\eta \in (0, 1)$ influences the convergence rate of the price, λ_0 is the operational point, and $p = \dot{s}u(\lambda_0)$. Briefly speaking, the price will increase if er_{k-1} is negative to influence consumers to reduce their consumption.

Since the calculated price signal λ_k is sent through a communication channel such as Internet, the adversary can then compromise a portion of the communication channels to send a falsified price signal. For example, by sending a price which is lower than the actual price values, the advisory can cause a large spike in the total load demand which can further lead to economical losses and unstable behavior of the system. Let ρ denote the amount of communication channels which is compromised and each of these consumers received the modified price value [16]: $\hat{\lambda}_k = \lambda_k + \delta\lambda_k^a$, where $\delta\lambda_k^a$ is the false price information injection launched by the adversary which can take any value. The price-responsive demand for the set of compromised loads is then equal to $\rho w_k(\lambda_k, \delta\lambda_k^a) = \rho D(\lambda_k + \delta\lambda_k^a)^\varepsilon$. Moreover, in order to avoid detection, it is assumed that $|\delta\lambda_k^a| \ll \lambda_k$. The objective of the attacker is to maximize the potential damage of the system by maximizing the mismatch between power generated and consumed, i.e., the error er_k. Sensitivity analysis (based on sensitivity functions) is performed in [16] to quantify the impact of the attack by looking at the response of the system to a perturbation (i.e., $\delta\lambda_k^a$) of a specific frequency. It is observed that the impact of supply–demand mismatch is severe for most frequencies. On the other hand, it is also found that smaller values of gain η in (9), i.e., a slower control action, can attenuate the impact of high-frequency component of the attack time series.

An attack-resilient controller can be designed by estimating the parameters of the system, namely the disturbance using the historical data. To this end, first the discrete-time state-space model of feedback real-time pricing problem is given by [16] $er_k = \left(p - D\varepsilon(\lambda_0)^{\varepsilon-1}\right)\lambda_k - \left(\rho D\varepsilon(\lambda_0)^{\varepsilon-1}\right)\delta\lambda_k$, which can be written as the following generic linear discrete-time system: $\tilde{x}_{k+1} = A\tilde{x}_k + Bu_k + \Gamma d_k$, where $A = 0$, $B = p - D\varepsilon(\lambda_0)^{\varepsilon-1}$, $\Gamma = -\left(\rho D\varepsilon(\lambda_0)^{\varepsilon-1}\right)$, $\tilde{x}_{k+1} = er_k$, $u_k = \lambda_k$, and $d_k = \delta\lambda_k$. Next, consider the following observer [26]:

$$z_{k+1} = z_k + K\left((A - I)x_k + Bu_k + \Gamma\hat{d}_k\right), \quad \hat{d}_k = Kx_k - z_k \qquad (10)$$

where $\hat{d}_k$ is the estimated disturbance (price signal). Moreover, assume that the disturbance is slowly time-varying, that is, $|d_k - d_{k-1}| \leq T\mu$ for some constant μ and sampling period T. Note that the defender in reality does not know the amount of compromised nodes, that is Γ is unknown. However, using the approximate value of Γ given by $\hat{\Gamma}$ and setting $K = \hat{\Gamma}^{-1}(1 - \phi)$ where $\phi \in (-1, 1)$, it is shown in [16] that the estimation error $\hat{e}_k = \Gamma d_k - \hat{\Gamma}\hat{d}_k$ is bounded by $|\hat{e}_\infty| \leq \frac{|\Gamma|T\mu}{1-|\phi|}$.

4 Outsider Attacks: Stealthy Case

The objective of this type of outsider attacks is to drive the system to an unsafe state while not being detected, i.e., stealthy, by traditional anomaly detector (such as bad data detection test) designed to detect possible deviations from the nominal behavior. In general, more resources such as model knowledge and the feature of anomaly detector are required by the adversary to launch the stealthy attacks.

4.1 Attacks on State Estimation

State estimation (SE) is a key function in smart grid due to its wide applications such as for contingency analysis, load forecasting, and calculating locational marginal pricing for power markets. Control center typically receives two types of data from sensors deployed throughout the grid. The first one is digital data $s \in \{0, 1\}^d$ which indicates the on and off states of various switches and line breakers used to construct the network topology. The second one is analog meter data z consisting of a vector of bus injection and power flow measurements. SE aims at estimating the states of the power systems (e.g., bus voltage magnitudes and phase angles) denoted by $x \in \mathbb{R}^n$ from a limited set of measurements $z \in \mathbb{R}^m$ (with $m > n$) which satisfies

$$z = h(x) + e \tag{11}$$

where $h(x)$ is a nonlinear relation between the states x and measurements z and $e \in \mathbb{R}^m$ is Gaussian noise vector with zero mean and covariance matrix Σ_e. Equation (11) is commonly solved using the weighted least square method given by the following optimization problem:

$$\min_{\hat{x}} \frac{1}{2} (z - h(\hat{x}))^T \Sigma_e^{-1} (z - h(\hat{x})) \tag{12}$$

where $\hat{x}$ denotes the estimated state. If the phase difference is small, the linear approximation of (11) (also known as DC model) can then be written as

$$z = Hx + e \tag{13}$$

where H is the Jacobian matrix of $h(\cdot)$. Assume that system (13) is observable, namely rank$(H) = n$. It is known that for linearized model (13), the solution to (12) is analytically given by

$$\hat{x}(z) = (H^T \Sigma_e^{-1} H)^{-1} H^T \Sigma_e^{-1} z. \tag{14}$$

Next, based on (14), let us define the residue vector r given by the difference between the measurement and the calculated value from the estimated state

$$r = z - H\hat{x} = Rz \tag{15}$$

where $R = I - H(H^T \Sigma_e^{-1} H)^{-1} H^T \Sigma_e^{-1}$, also called residual sensitivity matrix. Substituting (13), (14) into (15), the residue vector can then be written as

$$r = Re. \tag{16}$$

Since the measurements may be corrupted due to, for example, random noise, faulty sensors, and topological errors, bad data detection techniques have been widely developed to detect such abnormality using a threshold test over the residue (15). One example is J test [18] which uses the weighted least square error

$$J = r^T \Sigma^{-1} r \tag{17}$$

in the following threshold test

$$\begin{cases} \text{good} & \text{if } J \le \gamma \\ \text{bad} & \text{if } J > \gamma \end{cases} \tag{18}$$

where γ is the prespecified detection threshold.

While detector (18) is relatively effective against random noise, it lacks the ability to detect highly structure bad data (coordinated attack). In order to demonstrate this, we first introduce the false data injection attack problem. Note that a comprehensive review of the state of the art in false data injection attack in power systems is provided in [31]. For linearized model (13), the attacks can be modeled as

$$z^a = Hx + e + a \tag{19}$$

where $a \in \mathbb{R}^m$ denotes the bad data injection vector launched by the adversary which may depend on z, i.e., $a(z)$ and z^a is the corrupted measurement. This type of attack is also known as false data injection attack (FDIA).

Under corrupted measurement (19), residual vector r in (15) can be computed as

$$r = Re + Ra. \tag{20}$$

The FDIA is called stealthy (i.e., undetectable) if the attacker can still pass the bad data detection test, i.e., under residue r in (20), we have $J \le \gamma$. For example, if the attacker chooses attack vector a so that following condition holds:

$$Ra = 0, \tag{21}$$

then the residual under attack (19) will be equal to the one without attack, given in (16). Hence, any residue-based detector test, including (17), will not be able to detect the attack. If the attacker has knowledge on the network topology, i.e., matrix H, he/she can then simply launch the following stealthy injections to become stealthy:

$$a = Hc \tag{22}$$

where c is an arbitrary vector. Hence, the control center will believe that the true state is equal to $x + c$. Moreover, in launching stealthy false data injection, in general, the attacker aims at minimizing the number of sensors it needs to compromise in order to reduce the probability of being detected and also to minimize the cost of launching the attack. This problem is also called as the least-effort attack problem which in general can be formulated as the following optimization problem [10]:

$$\min_{c} \| H^{\overline{\mathscr{I}}} c \|_0 \quad \text{subject to} \quad H^{\mathscr{I}} c = 0, \quad \| c \|_\infty \ge \tau, \tag{23}$$

where some of the sensors, denoted by the set $\mathscr{S}$, are assumed secure. Hence, the attacker can only modify the measurements of the sensors given by the complementary set $\overline{\mathscr{S}}$. Note that $H^{\mathscr{S}}$ is a matrix composed of the rows in Jacobian H indexed by $\mathscr{S}$. The last constraint in (23) shows that the attacker tries to guarantee a minimum distortion at at least one attack position where τ is a predefined threshold. Note that a similar problem formulation of finding the sparsest attack vector a has also been considered in [53, 70]. The above least-effort attack problem is in general NP-hard. Hence, one reasonable approach to solve it is to find the suboptimal solution to the original combinatorial problem. Several methods based on heuristic [38], graph theoretic [28], greedy [27], and sparse optimization [46] have been developed to obtain the suboptimal solution and reduce the computational complexity.

When matrix H is not known to the attacker, the attacker needs to first estimate the network topology before launching the stealthy false data injection attack. The authors in [23] demonstrate that the topology of the network in terms of Laplacian matrix can be estimated sufficiently accurate from the locational marginal prices using a regularized maximum likelihood estimator. Moreover, a method to estimate the topology from power flow measurements is proposed in [13]. Specifically, the idea is that when the system parameters such as active and passive loads vary in a small range, the topology information is embedded in the correlations among power flow measurements. Hence, using the power flow measurements and independent component analysis algorithm, the attacker can then estimate the network topology. While the work mentioned above deals with the estimation of the overall network topology, it is shown in [37] that it is possible for the attacker to launch stealthy false data injection attack by only estimating the topology of the local attacking region (and thus without knowing any information of the non-attacking region).

In addition to modifying measurement z by injecting attack vector a, the attacker can also simultaneously modify the network topology [8, 25], i.e., matrix H in (13) by corrupting the digital data s received by the control center according to

$$\overline{s} = s + b \pmod{2} \tag{24}$$

where $\overline{s}$ denotes the modified network data with corrupted Jacobian matrix $\overline{H}$ and attack vector $b \in \{0, 1\}^d$. The attacker's goal is to launch injections $(a(z), b)$ while satisfying the test (18). To illustrate this and for the sake of simplicity, let us consider the noiseless case which yields that the test (18) is equivalent to check if the received measurement data is in the column space of H. Based on this observation, it is shown in [25] that an attack to modify the topology (measurement matrix) H to $\overline{H}$ with injection vector a is undetectable if $z + a \in \text{Col}(\overline{H})$, $\forall z \in \text{Col}(H)$, where $\text{Col}(\cdot)$ denotes the column space. Hence, one strategy of the attacker in launching undetectable attack is simply by preserving the state. To this end, consider again the noiseless case. Given $z = Hx \in \text{Col}(H)$, the attacker can modify the measurement matrix H to $\overline{H}$ without being detected by injecting $a = (\overline{H} - H)x$ since $z + a = \overline{H}x \in \text{Col}(\overline{H})$. Furthermore, since $\overline{H}$ has a full column rank, the injection a can then be explicitly computed as $a = (\overline{H} - H)(H^T H)^{-1} H^T z$. Similar undetectable attack can also be constructed for noisy measurement case and by using only local information (the attacker does not have measurement of the entire network), see [25]. While the previously mentioned work consider FDIA for a linearized model, a method to construct a stealthy FDIA for AC state estimation in (11) is presented in [20] and guidelines for designing stealthy FDIA for nonlinear grid model without any need of information on the power grid topology and transmission line admittances are discussed in [9].

One of the strategies for the defender to mitigate FDIA is by securing or protecting the measurement or sensors so that they cannot be compromised [10, 11, 27]. One way to protect the measurements is by deploying advanced measurement units such as PMUs which are equipped with various security measures [27]. The goal of the defender is to find the smallest set $\mathscr{S}$ in (23) such that there is no stealthy and nonzero feasible a (note: an attack a is feasible if and only if $a_i = 0$ for all $i \in \mathscr{S}$). Before proceeding, let us define $\Gamma(\mathscr{S}) = \text{diag}(s_1, \ldots, s_m)$, where $s_i = 1$ if and only if $i \in \mathscr{S}$. A set $\mathscr{S}$ is called observable if and only if $\Gamma(\mathscr{S})H$ is of full column rank [41]. It can then be shown that the only feasible and stealthy attack is $a = 0$ if and only if $\mathscr{S}$ is observable [41]. Hence, the problem of finding the minimum number of sensors to be secured such that the control center can detect the compromised nodes is equivalent to the problem of computing the smallest observable set $\mathscr{S}$.

Another possible objective of the defender in securing the sensors is to meet a certain level of resilience with the minimum protection cost. To this end, let N_{Ai} be the minimum number of measurements that the attacker needs to control to inject bad data into the state of bus i without being detected. The defender then aims at solving the following optimization problem presented in [10]:

$$\min_{\mathscr{S}} |\mathscr{S}| \quad \text{subject to} \quad \min_{i \in \{1, \ldots, N\}} N_{Ai} \geq N_A,$$

where N_A is a predefined positive integer. Since the above optimization is also a combinatorial problem, one possible method to obtain a suboptimal solution is by adding one measurement into the set $\mathscr{S}$ at a time, until the constraint is met.

Alternative approaches in addition to protecting some of the measurements or sensors are summarized in [31] which include a system theoretic countermeasure (e.g., using noise as an authentication signal) to detect reply attack [42] and by dynamically changing the information structure of the grid [59].

4.2 Data Attacks in Power Market Operations

One application of the state estimation described previously is for the computation of the locational marginal price (performed by the independent system operator) in a wholesale electricity market that reflects the electricity price at each node in the network. Therefore, the attacker can make profitable market transactions by injecting false data to compromise several line sensors without being detected. To illustrate this, in the following, we summarize the results on data attacks in power market operations presented in [69].

The electric power market consists of several forward (ex-ante) and real-time (ex-post) markets. Given the expected load $L_{d_j}^*$, the goal of the ex-ante market (which takes place every 1 to 15 min prior to real-time spot market) is to determine the optimal power generation $P_{g_i}^*$ by solving the following security-constrained economic dispatch problem:

$$
\begin{aligned}
\min_{P_{g_i}^*} \quad & \sum_{i=1}^{G} C_i(P_{g_i}^*) \\
\text{subject to} \quad & \sum_{i=1}^{G} P_{g_i}^* = \sum_{j=1}^{D} L_{d_j}^*, \quad P_{g_i}^{min} \le P_{g_i}^* \le P_{g_i}^{max} \ \forall i = 1, \dots, G, \\
& P_{ij}^{min} \le P_{ij}^* \le P_{ij}^{max} \ \forall (i, j) \in E,
\end{aligned}
\tag{25}
$$

where G, D denote the number of generators and loads respectively, E is the set of lines in the power network, $P_{g_i}^{min}$, $P_{g_i}^{max}$ denote the minimum and maximum power available to the i-th generator, and P_{ij}^*, P_{ij}^{min}, P_{ij}^{max} are the power flow on line (i, j) and its minimum and maximum allowable power flow, respectively. However, the real-time values of P_g, L_d and P_{ij} may differ from the ones obtained from (25) due to the stochastic nature of real-time demand L_{d_j}. Hence, using the actual state estimation obtained from the SCADA system, the market management system calculates the ex-post locational marginal price which also corresponds to the settlement price for all market participants. Specifically, the market management system solves (25) in the small range of the actual system state which can be written as the following optimization problem:

$$
\begin{aligned}
\min_{\Delta P_{g_i}} \quad & \sum_{i=1}^{G} C_i(\Delta P_{g_i} + \hat{P}_{g_i}) \\
\text{subject to} \quad & \sum_{i=1}^{G} \Delta P_{g_i} = 0, \quad P_{g_i}^{min} \le \Delta P_{g_i} \le \Delta P_{g_i}^{max} \ \forall i = 1, \dots, G \\
& P_{ij}^{min} \le \hat{P}_{ij} + \Delta P_{ij} \le P_{ij}^{max} \ \forall (i, j) \in E
\end{aligned}
\tag{26}
$$

where $\hat{P}_{g_i}$, $\hat{P}_{ij}$ denote the estimated power generation and power flow, respectively, and $\Delta P_{g_i}^{min}$, $\Delta P_{g_i}^{max}$ are chosen to be sufficiently small.

Since the market management system uses the state estimation to solve (26), a malicious third party can then make a profit by compromising the sensors and thus injecting false data to measurement data used for the state estimation as discussed in Sect. 4.1. Specifically, the attacker exploits the virtual bidding mechanism which is a legitimate financial instruments in many regional transmission organizations. In the virtual bidding, when a market participant purchases/sells a certain amount of virtual power at location i in day-ahead forward market, he/she is obliged to sell/purchase the same amount in the real-time market. If the attacker knows the network topology and also the optimal states P_g^*, L_d^*, P_{ij}^* from the ex-ante market, he/she can then execute the following strategy:

- During day-ahead forward market, the malicious third party buys and sells virtual power P_v at locations i_1 and i_2 at price $\lambda_{i_1}^f$, $\lambda_{i_2}^f$ respectively.
- Injects a in (19) to manipulate the locational marginal price of ex-post (real-time) market.
- During ex-post market, sells and buys virtual power P_v at locations i_1 and i_2 at the price of λ_{i_1}, λ_{i_2}.

The profit obtained by the attacker from the above virtual trading is given by

$$pf = (\lambda_{i_1} - \lambda_{i_1}^f)P_v + (\lambda_{i_2}^f - \lambda_{i_2})P_v = (\lambda_{i_1} - \lambda_{i_2} + \lambda_{i_2}^f - \lambda_{i_1}^f)P_v.$$

Note that the difference in price of two nodes $\lambda_{i_1} - \lambda_{i_2}$ is obtained by computing the Lagrangian multipliers in (26) which also depends on the compromised measurement. The goal of the attacker is to maximize the profit pf while satisfying the security constraint and without being detected. However, since the system is stochastic, the attacker can only guarantee the expected profit. Moreover, the problem is still hard due to the implicit relation between the compromised measurement and the price given by the Lagrangian multipliers. To address this issue, several heuristics are proposed in [69] by exploiting the structure of the ex-post formulation. To conclude this section, it is also possible for the attacker to launch more sophisticated attack while being stealthy, namely by combining a generalized FDIA and cyber topology attack with the goal to mislead customers so that they pay higher electricity bills in the wholesale market as presented in [29].

4.3   Spoofing Attack on GPS Receiver of a PMU

Phasor measurement unit (PMU) is one of the most important devices in a smart grid which uses the global positioning system (GPS) in order to obtain synchronized positive sequence phasor voltage and current measurements and thus enables real-time monitoring and control. Using signals transmitted by the GPS satellites, the PMU computes its own position and the offset of its clock with respect to the GPS time measured by the satellite clocks which enables all PMUs to synchronize their clocks and further derive a Coordinated Universal Time time stamp reference.

   To this end, first the receiver of PMU computes an estimate of its distance, called pseudorange ρ_i, from the i-th satellite by taking the difference between the time of signal transmission and reception and multiplying it by the propagation speed, which is equal to the speed of light l. If the satellite clock and receiver clock were perfectly synchronized, the pseudorange ρ_i is equal to the true satellite-to-receiver range and thus it would be sufficient for the receiver to compute its Earth-centered Earth-fixed

(ECEF) coordinate by using the information received from three satellites. However, in reality, the receiver clock has an offset t_u (not known a priori) which yields that the pseudorange is not equal to the true range d_i. The true range between the i-th satellite and the receiver can then be computed according to

$$\rho_i = d_i - lt_u \tag{27}$$

with $d_i = \sqrt{(x_i - x_u)^2 + (y_i - y_u)^2 + (z_i - z_u)^2}$, where x_i, y_i, z_i denote the i-th satellite's ECEF coordinates and x_u, y_u, z_u be the receiver's ECEF coordinates. Since the receiver needs to compute its position and the clock offset, it then requires to track at least four satellites. Note that the receiver can compute the satellite's position using the set of values broadcast by the GPS satellite known as the *ephemerides*. Let $\bar{\delta}_i = [\delta_i(1), \ldots, \delta_i(m)]^T$ be the vector containing the ephemerides broadcast by the i-th satellite. The satellite's ECEF coordinate can then be expressed as $x_i = f(\bar{\delta}_i, t)$, $y_i = g(\bar{\delta}_i, t)$, $z_i = h(\bar{\delta}_i, t)$ [22].

Errors in the GPS clock offset will affect PMU time synchronization which is critical for obtaining accurate phasor angle measurements. Hence, the attacker may aim at maximizing the error of the clock offset by spoofing the GPS signal, namely the ephemerides, pseudoranges, and the receiver position, and by taking into account the possibility that the GPS receiver may implement certain spoofing detection scheme. For the case of four satellites, the spoofing attack can be formulated as the following optimization problem [22]:

$$\max_{\bar{p}_u, \bar{\delta}_i, \rho_i} \quad \left(t_u(\bar{p}_u, \bar{\delta}_i, \rho_i) - t_u^* \right)^2$$

$$\text{subject to} \quad \rho_i = d_i(\bar{\delta}_i, \bar{p}_u) - ct_u(\bar{p}_u, \bar{\delta}_i, \rho_i), i = 1, 2, 3, 4$$
$$|\bar{p}_u(l) - \bar{p}_u^*(l)| \le \varepsilon_{\bar{p}_u}(l), l = 1, 2, 3, \tag{28}$$
$$|\bar{\delta}_i(j) - \bar{\delta}_i^*(j)| \le \varepsilon_{\bar{\delta}_i}(j), j = 1, \ldots, m, \; \forall i,$$
$$|\bar{p}_i(k) - \bar{p}_i^*(k)| \le \varepsilon_{\bar{p}_i}(k), k = 1, 2, 3, \; \forall i, \quad |\rho_i - \rho_i^*| \le \varepsilon_{\rho_i}, \forall i$$

where $\bar{p}_i = [x_i, y_i, z_i]^T$, $\bar{p}_u = [x_u, y_u, z_u]^T$, and constants $\varepsilon_{\bar{p}_u}(l), \varepsilon_{\bar{\delta}_i}(j), \varepsilon_{\bar{p}_i}(k), \varepsilon_{\rho_i}$ are some thresholds. Note that the notation $(\cdot)^*$ denotes the pre-attack value and $\bar{y}(k)$ refers to the kth entry of vector $\bar{y}$. The inequality constraints are added to demonstrate that the spoofing can still be stealthy even if the receiver applies a simple form of countermeasures by detecting large deviations in the variables in comparison to their pre-attack values.

A brief review on existing techniques to detect and mitigate spoofing threats is presented in [21]. The following countermeasures are recommended in [24, 65] to detect suspicious GPS signal activity: (1) amplitude discrimination, (2) time-of-arrival discrimination, (3) consistency of navigation inertial measurement unit (IMU) cross-check, (4) polarization discrimination, (5) angle-of-arrival discrimination, (6) cryptographic authentication. While the first two methods can be implemented in software on GPS receivers, their effectiveness are limited only to detect simple attacks. The next three techniques are effective against some sophisticated attacks. However, they require additional hardware to implement, such as multiple antenna or a high-grade IMU, and they are still ineffective against coordinated spoofing attack, for example, the one involving multiple GPS transmitters. Finally, the crypto-graphic authentication calls for significant changes to the current GPS signal coding scheme, specifically when implemented to the civilian GPS signal. Note that the attack scenario formulated in (28) can be detected using, e.g., the angle-of-arrival discrimination or cryptographic authentication methods [22].

4.4   *Coordinated Attacks and Systematic Defense*

As more functions of forecast, estimation, optimization, and control get enabled at the grid edge, the potential of coordinated attacks will inevitably increase. A holistic framework is needed to achieve resilience of wide-area monitoring [72] and control [49], distributed cooperative controls [17, 39], market and energy management [5, 67]. New analytic tools and systematic designs will have to be developed in the future research.

5   Summary and Challenges

In the chapter, we demonstrate through several scenarios how adversary can destabi-lize the power system and cause economic loss by intercepting the communication channel and further modifying the measurement and/or control signals. It is also shown that cybersecurity and widely used bad data detector are not sufficient to defend the power system against such attacks. Moreover, in the future, the adver-sary may launch more sophisticated attacks, for example, by simply combining the attacks on different layers of power systems. It is also demonstrated the potential of dynamical systems and control theory as analytical and design tools to defend the power grid against cyberattacks. Hence, it is promising to combine cybersecurity approach and system theoretic approach to improve the resilience of smart grid.

Following this chapter on the survey of smart grid security are four papers which discuss in more details and comprehensive manner the research challenges in smart grid security described above. The highlight of each research challenge paper is provided below.

1. The paper by Zhu provides a layered perspective of the smart grid security. Specifically, motivated by the OSI for the Internet and PRM models for enterprise and control systems, the author proposes to organize the smart grid into six layers, namely, physical layer, control layer, data communication layer, network layer,

supervisory layer, and management layer. Security problems at each layer are then identified and it is demonstrated that the solutions require a cross-layer viewpoint. It is also discussed that game and decision theory are promising tools to model the interactions among system components and the interaction between attackers and the system.

2. The paper by Chakhchoukh and Ishii considers attacks on state estimation. After describing attacks on the measurements and topology of the grid together with their impacts, the authors discuss cybersecurity solutions such as offline and online methods and propose necessary research directions including distributed algorithm by considering trade-off between reduction of execution time and maximization of detection, randomization the set of measurements, state estimation at distribution level, and development of cybersecurity testbeds.

3. The paper by Weerakkody and Sinopoli first identifies attack surfaces and also describes the potential strategies attackers may carry out to obtain benefits or cause damage. It is also discussed that the existing mechanisms in cybersecurity are insufficient for addressing smart grid security. Therefore, the authors propose the development of cyber-physical system security which augments cybersecurity with tools from system theory, namely by leveraging fundamental understanding of the dynamics of power systems to ensure security. Specifically, the authors envision research directions comprised of three thrusts namely attack detection (including passive and active methods), attack identification, and design of resilient control algorithm that can guide operators in performing corrective control actions.

4. The paper by Gusrialdi and Qu considers resilient control design for smart grid. The authors start by discussing the concept of passivity-short and demonstrate its potential to deal with heterogeneous dynamics and enable plug-and-play operation of the networked system. Next, the authors introduce two distributed control strategies to make the smart grid resilient against unknown attacks. The first strategy is based on constructing an observation graph which assigns a confidence level to the neighbors of each node in the network and gradually decreases the confidence level of the misbehaving nodes until they are isolated. The second strategy is by introducing an additional information flow, interconnected with the original networked system, which maintains the stability of the overall system by competitively interacting with the original networked system.

Acknowledgements This work is supported in part by U.S. Department of Transportation (award DTRT13GUTC51), by U.S. National Science Foundation (grant ECCS-1308928), by U.S. Department of Energy (awards DE-EE0006340 and DE-EE0007327), by L-3 Communication Coleman Aerospace (contract 1101312034), by Texas Instruments' awards, and by Leidos (contract P010161530).

References

1. S. Amini, F. Pasqualetti, H. Mohsenian-Rad, Dynamic load altering attacks against power system stability: attack models and protection schemes. IEEE Trans. Smart Grid **9**(4), 2862–2872 (2018)
2. J.M. Arroyo, F.D. Galiana, On the solution of the bilevel programming formulation of the terrorist threat problem. IEEE Trans. Power Syst. **20**(2), 789–797 (2005)
3. C. Balducelli, S. Bologna, L. Lavalle, G. Vicoli, Safeguarding information intensive critical infrastructures against novel types of emerging failures. Reliab. Eng. Syst. Saf. **92**(9), 1218–1229 (2007)
4. N. Baracaldo, J. Joshi, An adaptive risk management and access control framework to mitigate insider threats. Comput. Secur. **39**, 237–254 (2013)
5. S. Baros, D. Shiltz, P. Jaipuria, A. Hussain, A.M. Annaswamy, Resilient control of cyber-physical energy systems for security against cyber-attacks, in *DSpace @MIT*, http://hdl.handle.net/1721.1/107408
6. V.M. Bier, E.R. Gratz, N.J. Haphuriwat, W. Magua, K.R. Wierzbicki, Methodology for identifying near-optimal interdiction strategies for a power transmission system. Reliab. Eng. Syst. Saf. **92**(9), 1155–1161 (2007)
7. J. Bigham, D, Gamez, N. Lu, *Safeguarding SCADA Systems with Anomaly Detection* (Springer, Berlin, Heidelberg, 2003), pp. 171–182
8. Y. Chakhchoukh, H. Ishii, Coordinated cyber-attacks on the measurement function in hybrid state estimation. IEEE Trans. Power Syst. **30**(5), 2487–2497 (2015)
9. W.L. Chin, C.H. Lee, T. Jiang, Blind false data attacks against ac state estimation based on geometric approach in smart grid communications. IEEE Trans. Smart Grid (99), 1–1 (2017)
10. S. Cui, Z. Han, S. Kar, T.T. Kim, H.V. Poor, A. Tajer, Coordinated data-injection attack and detection in the smart grid: a detailed look at enriching detection solutions. IEEE Signal Process. Mag. **29**(5), 106–115 (2012)
11. R. Deng, G. Xiao, R. Lu, Defending against false data injection attacks on power system state estimation. IEEE Trans. Ind. Inf. **13**(1), 198–207 (2017)
12. E-ISAC, SANS: Analysis of the cyber attack on the ukrainian power grid: Defense use case, https://ics.sans.org/duc5. Accessed 20 June 2017
13. M. Esmalifalak, H. Nguyen, R. Zheng, Z. Han, Stealth false data injection using independent component analysis in smart grid, in *IEEE International Conference on Smart Grid Communications (SmartGridComm)* (2011), pp. 244–248
14. D. Formby, S.S. Jung, S. Walters, R. Beyah, A physical overlay framework for insider threat mitigation of power system devices, in *IEEE International Conference on Smart Grid Communications* (2014), pp. 970–975
15. I. Garitano, R. Uribeetxeberria, U. Zurutuza, A review of scada anomaly detection systems, in *6th International Conference SOCO Soft Computing Models in Industrial and Environmental Applications* (2011), pp. 357–366
16. J. Giraldo, A. Cárdenas, N. Quijano, Integrity attacks on real-time pricing in smart grids: impact and countermeasures. IEEE Trans. Smart Grid **8**(5), 2249–2257 (2017)
17. A. Gusrialdi, Z. Qu, M.A. Simaan, Competitive interaction design of cooperative systems against attacks. IEEE Trans. Autom. Control (2018)
18. E. Handschin, F.C. Schweppe, J. Kohlas, A. Fiechter, Bad data analysis for power system state estimation. IEEE Trans. Power Appar. Syst. **94**(2), 329–337 (1975)
19. H. He, J. Yan, Cyber-physical attacks and defences in the smart grid: a survey. IET Cyber-Phys. Syst. Theory Appl. **1**(1), 13–27 (2016)
20. G. Hug, J.A. Giampapa, Vulnerability assessment of ac state estimation with respect to false data injection cyber-attacks. IEEE Trans. Smart Grid **3**(3), 1362–1370 (2012)
21. A. Jafarnia-Jahromi, A. Broumandan, J. Nielsen, G. Lachapelle, GPS vulnerability to spoofing threats and a review of antispoofing techniques. Int. J. Navig. Obs. **2012** (2012)
22. X. Jiang, J. Zhang, B.J. Harding, J.J. Makela, A.D. Dominguez-Garcia, Spoofing GPS receiver clock offset of phasor measurement units. IEEE Trans. Power Syst. **28**(3), 3253–3262 (2013)

23. V. Kekatos, G.B. Giannakis, R. Baldick, Grid topology identification using electricity prices, in *IEEE PES General Meeting* (2014) pp. 1–5
24. E. Key, *Techniques to Counter GPS Spoofing*. Internal memorandum (1995)
25. J. Kim, L. Tong, On topology attack of a smart grid: undetectable attacks and countermeasures. IEEE J. Sel. Areas Commun. **31**(7), 1294–1305 (2013)
26. K.S. Kim, K.H. Rew, Reduced order disturbance observer for discrete-time linear systems. Automatica **49**(4), 968–975 (2013)
27. T.T. Kim, H.V. Poor, Strategic protection against data injection attacks on power grids. IEEE Trans. Smart Grid **2**(2), 326–333 (2011)
28. O. Kosut, L. Jia, R.J. Thomas, L. Tong, Malicious data attacks on the smart grid. IEEE Trans. Smart Grid **2**(4), 645–658 (2011)
29. G. Liang, S.R. Weller, F. Luo, J. Zhao, Z.Y. Dong, Generalized fdia-based cyber topology attack with application to the Australian electricity market trading mechanism. IEEE Trans. Smart Grid **9**(4), 3820–3829 (2018)
30. G. Liang, S.R. Weller, J. Zhao, F. Luo, Z.Y. Dong, The 2015 ukraine blackout: implications for false data injection attacks. IEEE Trans. Power Syst. **32**(4), 3317–3318 (2017)
31. G. Liang, J. Zhao, F. Luo, S. Weller, Z.Y. Dong, A review of false data injection attacks against modern power systems. IEEE Trans. Smart Grid **8**(4), 1630–1638 (2017)
32. D. Liberzon, *Switching in systems and control* (Springer, Science & Business Media, 2012)
33. M.G. Lijesen, The real-time price elasticity of electricity. Energy Econ. **29**(2), 249–258 (2007)
34. H. Lin, P.J. Antsaklis, Stability and stabilizability of switched linear systems: a survey of recent results. IEEE Trans. Autom. Control **54**(2), 308–322 (2009)
35. S. Liu, X.P. Liu, A. El Saddik, Denial-of-service (DoS) attacks on load frequency control in smart grids, in *2013 IEEE PES Innovative Smart Grid Technologies (ISGT)* (IEEE, 2013)
36. S. Liu, S. Mashayekh, D. Kundur, T. Zournts, K. Butler-Purry, A framework for modeling cyber-physical switching attacks in smart grid. IEEE Trans. Emerg. Topics Comput. **1**(2), 273–285 (2013)
37. X. Liu, Z. Bao, D. Lu, Z. Li, Modeling of local false data injection attacks with reduced network information. IEEE Trans. Smart Grid **6**(4), 1686–1696 (2015)
38. Y.Liu, P. Ning, M.K. Reiter, False data injection attacks against state estimation in electric power grids, in *Proceedings of the 16th ACM Conference on Computer and Communications Security* (2009), pp. 21–32
39. Y. Liu, H. Xin, Z. Qu, D. Gan, An attack-resilient cooperative control strategy of multiple distributed generators in distribution networks. IEEE Trans. Smart Grid **7**(6), 2923–2932 (2016)
40. J.T.K. Ma, T.M. Liu, L.F. Wu, New energy management system architectural design and intranet/internet applications to power systems in *1998 Proceedings of EMPD'98 International Conference on Energy Management and Power Delivery*, vol. 1 (1998), pp. 207–212
41. Y. Mo, T. Hyun-Jin Kim, K. Brancik, D. Dickinson, H. Lee, A. Perrig, B. Sinopoli, Cyber-physical security of a smart grid infrastructure. Proc. IEEE **100**(1), 195–209 (2012)
42. Y. Mo, B. Sinopoli, Secure control against replay attacks, in *2009 47th Annual Allerton Conference on Communication, Control, and Computing* (Allerton, 2009), pp. 911–918
43. A.H. Mohsenian-Rad, A. Leon-Garcia, Optimal residential load control with price prediction in real-time electricity pricing environments. IEEE Trans. Smart Grid **1**(2), 120–133 (2010)
44. A.H. Mohsenian-Rad, A. Leon-Garcia, Distributed internet-based load altering attacks against smart power grids. IEEE Trans. Smart Grid **2**(4), 667–674 (2011)
45. P.M. Nasr, A.Y. Varjani, Alarm based anomaly detection of insider attacks in SCADA system, in *Smart Grid Conference (SGC)* (2014), pp. 1–6
46. M. Ozay, I. Esnaola, F.T.Y. Vural, S.R. Kulkarni, H.V. Poor, Sparse attack construction and state estimation in the smart grid: centralized and distributed models. IEEE J. Sel. Areas Commun. **31**(7), 1306–1318 (2013)
47. K. Pelechrinis, M. Iliofotou, S.V. Krishnamurthy, Denial of service attacks in wireless networks: the case of jammers. IEEE Commun. Surv. Tutor. **13**(2), 245–257 (2011)
48. T. Pultarova, Cyber security—ukraine grid hack is wake-up call for network operators [news briefing]. Eng. Technol. **11**(1), 12–13 (2016)

49. A. Rajabi, R.B. Bobba, A resilient algorithm for power system mode estimation using synchrophasors, in *Proceeding of ICSS'16 Proceedings of the 2nd Annual Industrial Control System Security Workshop* (2016), pp. 23–29
50. M. Roozbehanit, M. Rinehart, M. Dahleh, S. Mitter, D. Obradovic, H. Mangesius, Analysis of competitive electricity markets under a new model of real-time retail pricing, in *2011 8th International Conference on the European Energy Market (EEM)* (2011), pp. 250–255
51. J. Salmeron, K. Wood, R. Baldick, Analysis of electric grid security under terrorist threat. IEEE Trans. Power Syst. **19**(2), 905–912 (2004)
52. J. Salmeron, K. Wood, R. Baldick, Worst-case interdiction analysis of large-scale electric power grids. IEEE Trans. Power Syst. **24**(1), 96–104 (2009)
53. H. Sandberg, A. Teixeira, K.H. Johansson, On security indices for state estimators in power networks, in *First Workshop on Secure Control Systems (SCS)* (2010)
54. A. Sargolzaei, K. Yen, M. Abdelghani, Delayed inputs attack on load frequency control in smart grid, in *2014 IEEE PES Innovative Smart Grid Technologies Conference (ISGT)* (IEEE, 2014), pp. 1–5
55. A. Sargolzaei, K.K. Yen, M.N. Abdelghani, Preventing time-delay switch attack on load frequency control in distributed power systems. IEEE Trans. Smart Grid **7**(2), 1176–1185 (2016)
56. J. Slay, M. Miller, Lessons learned from the maroochy water breach, in *International Conference on Critical Infrastructure Protection* (Springer, 2007), pp. 73–82
57. S. Sridhar, M. Govindarasu, Model-based attack detection and mitigation for automatic generation control. IEEE Trans. Smart Grid **5**(2), 580–591 (2014)
58. P. Srikantha, D. Kundur, Denial of service attacks and mitigation for stability in cyber-enabled power grid, in *2015 IEEE Power & Energy Society Innovative Smart Grid Technologies Conference (ISGT)* (2015), pp. 1–5
59. M. Talebi, C. Li, Z. Qu, Enhanced protection against false data injection by dynamically changing information structure of microgrids, in *2012 IEEE 7th Sensor Array and Multichannel Signal Processing Workshop (SAM)* (2012), pp. 393–396
60. R. Tan, V. Badrinath Krishna, D.K. Yau, Z. Kalbarczyk, Impact of integrity attacks on real-time pricing in smart grids, in *Proceedings of the 2013 ACM SIGSAC conference on Computer & communications security* (2013), pp. 439–450
61. S. Tan, D. De, W.Z. Song, J. Yang, S.K. Das, Survey of security advances in smart grid: a data driven approach. IEEE Commun. Surv. Tutor. **19**(1), 397–422 (2017)
62. A. Teixeira, I. Shames, H. Sandberg, K.H. Johansson, A secure control framework for resource-limited adversaries. Automatica **51**, 135–148 (2015)
63. (U.S.)., N.R.C.: Making the nation safer : the role of science and technology in countering terrorism / Committee on Science and Technology for Countering Terrorism, National Research Council. National Academy Press Washington, D.C (2002)
64. W. Wang, Z. Lu, Cyber security in the smart grid: survey and challenges. Comput. Netw. **57**(5), 1344–1371 (2013)
65. J. Warner, R. Johnston, *GPS spoofing countermeasures* (Los Alamos Research Paper LAUR-03-6163, 2003)
66. D. Weers, M. Shamsedin, Testing a new direct load control power line communication system. IEEE Trans. Power Deliv. **2**(3), 657–660 (1987)
67. P. Wood, D. Shiltz, T. Nudell, A. Hussain, A. Annaswamy, A framework for evaluating the resiliency of dynamic real-time market mechanisms. IEEE Trans. Smart Grid **7**(6), 2904–2912 (2016)
68. Y. Wu, Z. Wei, J. Weng, X. Li, R.H. Deng, Resonance attacks on load frequency control of smart grids. IEEE Transactions on Smart Grid **PP**(99), 1–1 (2017)
69. L. Xie, Y. Mo, B. Sinopoli, Integrity data attacks in power market operations. IEEE Trans. Smart Grid **2**(4), 659–666 (2011)
70. Q. Yang, J. Yang, W. Yu, D. An, N. Zhang, W. Zhao, On false data-injection attacks against power system state estimation: modeling and countermeasures. IEEE Trans. Parallel Distrib. Syst. **25**(3), 717–729 (2014)

71. Y. Yao, T. Edmunds, D. Papageorgiou, R. Alvarez, Trilevel optimization in power network defense. IEEE Trans. Syst. Man Cybern. Part C (Applications and Reviews) **37**(4), 712–718 (2007)
72. J. Zhang, P. Jaipuria, A. Hussain, A. Chakrabortty, Attack-resilient estimation of power system oscillation modes using distributed and parallel optimization: theoretical and experimental methods, in *Conference on Decision and Game Theory for Security (GameSec)* (2014)

Multilayer Cyber-Physical Security and Resilience for Smart Grid

Quanyan Zhu

Abstract The smart grid is a large-scale complex system that integrates communication technologies with the physical layer operation of the energy systems. Security and resilience mechanisms by design are important to provide guarantee operations for the system. This chapter provides a layered perspective of the smart grid security and discusses game and decision theory as a tool to model the interactions among system components and the interaction between attackers and the system. We discuss game-theoretic applications and challenges in the design of cross-layer robust and resilient controller, secure network routing protocol at the data communication and networking layers, and the challenges of the information security at the management layer of the grid. The chapter will discuss the future directions of using game-theoretic tools in addressing multilayer security issues in the smart grid.

1 Introduction

The smart grid aims to provide reliable, efficient, secure, and quality energy generation/distribution/consumption using modern information, communications, and electronics technologies. The integration with modern IT technology moves the power grid from an outdated proprietary technology to more common ones such as personal computers, Microsoft Windows, TCP/IP/Ethernet, etc. It can provide the power grid with the capability of supporting two-way energy and information flow, isolate and restore power outages more quickly, facilitate the integration of renewable energy resources into the grid, and empower the consumer with tools for optimizing energy consumption. However, in the meantime, it poses security challenges on power systems as the integration exposes the system to public networks.

Many power grid incidents in the past have been related to software vulnerabilities. In [1], it is reported that hackers have inserted software into the USA power grid, potentially allowing the grid to be disrupted at a later date from a remote location. As reported in [2], it is believed that an inappropriate software update has led to a

Q. Zhu (✉)
Tandon School of Engineering, New York University, Brooklyn, NY 11201, USA
e-mail: quanyan.zhu@nyu.edu; qz494@nyu.edu

© Springer Nature Switzerland AG 2019
J. Stoustrup et al. (eds.), *Smart Grid Control*, Power Electronics
and Power Systems, https://doi.org/10.1007/978-3-319-98310-3_14

225

recent emergency shutdown of a nuclear power plant in Georgia, which lasted for 48 hours. In [3], it has been reported that a computer worm, Stuxnet, has been spread to target Siemens SCADA systems that are configured to control and monitor specific industrial processes. On November 29, 2010, Iran confirmed that its nuclear program had indeed been damaged by Stuxnet [4, 5]. The infestation by this worm may have damaged Iran's nuclear facilities in Natanz and eventually delayed the start-up of Iran's nuclear power plant.

Modern power systems do not have built-in security functionalities, and the security solutions in regular IT systems may not always apply to systems in critical infrastructures. This is because critical infrastructures have different goals and assumptions concerning what needs to be protected, and have specific applications that are not originally designed for a general IT environment. Hence, it is necessary to develop unique security solutions to fill the gap where IT solutions do not apply.

In this chapter, we describe a layered architecture to address the security issues in power grids, which facilitates identifying research problems and challenges at each layer and building models for designing security measures for control systems in critical infrastructures. We also emphasize a cross-layer viewpoint toward security issues in power grids in that each layer can have security dependence on the other layers. We need to understand the trade-off between the information assurance and the physical layer system performance before designing defense strategies against potential cyber threats and attacks. As examples, we address three security issues of smart grid at different layers, namely, the resilient control design problem at the physical power plant, the data-routing problem at the network and communication layer, and the information security management at the application layers.

The rest of the chapter is organized as follows. In Sect. 2, we first describe the general multilayer architecture of cyber-physical systems and the related security issues associated with each layer. In Sect. 3, we focus on the cyber and physical layers of the smart grid and propose a general cross-layer framework for robust and resilient controller design. In Sect. 4, we discuss secure network routing problem at the data communication and networking layers of the smart grid. In addition, we discuss the centralized versus decentralized routing protocols and propose a hybrid architecture as a result of the trade-off between robustness and resilience in the smart grid. In Sect. 5, we present the challenges of the information security at the management layer of the grid. We conclude finally in Sect. 6 and discuss future research directions that can follow from the multilayer model using game-theoretic tools.

2 Multilayer Architecture

Smart grid comprises of physical power systems and cyber information systems. The integration of the physical systems with the cyberspace allows new degrees of automation and human–machine interactions. The uncertainties and hostilities existing in the cyber environment have brought emerging concerns for modern power systems. It is of supreme importance to have a system that maintains state awareness and an acceptable level of operational normalcy in response to disturbances, including threats of an unexpected and malicious nature [6].

The physical systems of the power grid can be made to be resilient by incorporating features such as robustness and reliability [7], while the cyber components can be enhanced by many cybersecurity measures to ensure dependability, security, and privacy. However, the integration of cyber and physical components does not necessarily ensure overall reliability, robustness, security, and resilience of the power system. The interaction between the two environments can create new challenges in addition to the existing ones. To address these challenges, we first need to understand the architecture of smart grids. The smart grid can be hierarchically organized into six layers, namely, physical layer, control layer, data communication layer, network layer, supervisory layer, and management layer. The first two layers, physical layer and control layer, can be jointly seen as physical environment of the system. The data communication layer and network layer comprise the cyber environment of the power grid. The supervisory layer together with the management layer constitute the higher level application layer where services and human–machine interactions take place.

The power plant is at the physical layer, and the communication network and security devices are at the network and communication layers. The controller interacts with the communication layer and the physical layer. An administrator is at the supervisory layer to monitor and control the network and the system. Security management is at the highest layer where security policies are made against potential threats from attackers. SCADA is the fundamental monitoring and control architecture at the control area level. The control center of all major U.S. utilities have implemented a supporting SCADA for processing data and coordinating commands to manage power generation and delivery within the EHV and HV (bulk) portion of their own electric power system [8].

To further describe the functions at each layer, we resort to Fig. 1, which conceptually describes a smart grid system with a layering architecture. The lowest level is the physical layer where the physical/chemical processes we need to control or monitor reside. The control layer includes control devices that are encoded with control algorithms that have robust, reliable, secure, and fault-tolerant features. The communication layer passes data between devices and different layers. The network layer includes the data packet routing and topological features of control systems. The supervisory layer offers human–machine interactions and capability of centralized decision-making. The management layer makes economic and high-level operational decisions.

In the following, we identify problems and challenges at each layer and propose problems whose resolution requires a cross-layer viewpoint.

Physical layer: The physical layer comprises of the physical plant to be controlled. It is often described by an ordinary differential equation (ODE) model from physical or chemical laws. It can also be described by difference equations, Markov models, or model-free statistics. We have the following challenges that pertain to the security

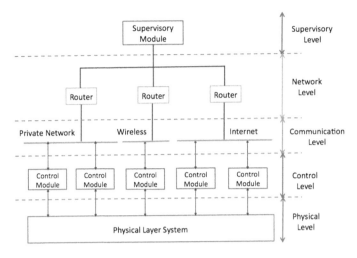

Fig. 1 A conceptual control system with layering

and reliability of the physical infrastructure. First, it is important to find appropriate measures to protect the physical infrastructure against vandalism, environmental change, unexpected events, etc. Such measures often need a cost-and-benefit analysis involving the value assessment of a particular infrastructure. Second, it is also essential for engineers to build the physical systems with more dependable components and more reliable architecture. It brings the concern on the physical maintenance of the control system infrastructures that demands cross-layer decision-making between the management and physical layers [9].

Control layer: The control layer consists of multiple control components, including observers/sensors, intrusion-detection systems (IDSs), actuators, and other intelligent control components. An observer has the sensing capability that collects data from the physical layer and may estimate the physical state of the current system. Sensors may need to have redundancies to ensure correct reading of the states. The sensor data can be fused locally or sent to the supervisor level for global fusion. A reliable architecture of sensor data fusion will be a critical concern. An IDS protects the physical layer as well as the communication layer by performing anomaly-based or signature-based intrusion detection. An anomaly-based ID is more common for physical layer whereas a signature-based ID is more common for the packets or traffic at the communication layer. If an intrusion or an anomaly occurs, an IDS raises an alert to the supervisor or work hand-in-hand with built-in intrusion-prevention systems (related to emergency responses, e.g., control reconfiguration) to take immediate action. There lies a fundamental trade-off between local decisions versus a centralized decision when intrusions are detected. A local decision, for example, made by a prevention system, can react in time to unanticipated events; however, it may incur a high packet drop rate if the local decision suffers high false negative rates due to incomplete information. Hence, it is an important architectural concern on whether the diagnosis and control modules need to operate locally with IDS or globally with a supervisor.

Communication layer: Communication layer is where we have a communication channel between control layer components or network-layer routers. The communication channel can take multiple forms: wireless, physical cable, blue-tooth, etc. The communication layer handles the data communication between devices or layers. It is an important vehicle that runs between different layers and devices. It can often be vulnerable to attacks such as jamming and eavesdropping. There are also privacy concerns of the data at this layer. Such problems have been studied within the context of wireless communication networks [10]. However, the goal of critical infrastructure may distinguish themselves from the conventional studies of these issues.

Network layer: The network layer concerns the topology of the architecture. It comprises of two major components: one is network formation, and the other one is routing. We can randomize the routes to disguise or confuse the attackers to achieve certain security or secrecy or minimum delay. Moreover, once a route is chosen, how much data should be sent on that route has long been a concern for researchers in communications [11–13]. In control systems, many specifics of the data form and rates may allow us to reconsider this problem in a control domain.

Supervisory layer: The supervisory layer coordinates all layers by designing and sending appropriate commands. It can be viewed as the brain of the system. Its main function is to perform critical data analysis or fusion to provide an immediate and precise assessment of the situation. It is also a holistic policy maker that distributes resources in an efficient way. The resources include communication resources, maintenance budget as well as control efforts. In centralized control, we have one supervisory module that collects and stores all historical data and serves as a powerful data fusion and signal-processing center [14, 15]. One key challenge at this layer is to defend against advanced persistent threats which behave stealthily, leverage social engineering, and exploit the vulnerabilities of the computer networks to obtain unauthorized credentials to access the control system networks [16, 17]. Hence, it is critical to implement security mechanism at this layer to detect intrusive, stealthy and deceptive behaviors, and ensure the integrity of information processing and the availability of critical services.

Management layer: The management layer is a higher level decision-making engine, where the decision-makers take an economic perspective toward the resource allocation problems in control systems. At this layer, we deal with problems such as (i) how to budget resources to different systems to accomplish a goal; (ii) how to develop policies to maintain data security and privacy; and (iii) how to manage patches for control systems, e.g., disclosure of vulnerabilities to vendors, development and release of patches [18].

Addressing the security challenges at the multiple layers of the smart grid requires a holistic and integrable framework that can capture different system features of the multilayer cyber-physical security problems. Game theory is a versatile quantitative tool which can be used to model different types of adversarial interactions between a defender and an attacker. For example, at the physical and the control layers, game-theoretic methods can be used to used to design robust and resilient controllers for dynamical systems in an uncertain or adversarial environment [19–22]. At the supervisory layer, game-theoretic methods can be used to understand spear-phishing attacks [23], insider threats [24], and the advanced persistent threats [17, 25, 26]. At the management layer, game theory serves as a primary tool to design strategies for security investment and information disclosure policies. At the network layer, game theory has been used as a quantitative method for analyzing network security policies and designing defense mechanisms [10, 19, 27, 28]. The wide range of application domains of game theory have made it an ideal tool for developing a unifying framework for a holistic and fundamental understanding of cyber-physical security across different layers of functionalities. In the following section, we will discuss the applications of game-theoretic methods for addressing control, network and management layer problems.

3   Robust and Resilient Control

The layered architecture in Fig. 1 can facilitate the understanding of the cross-layer interactions between the physical world and the cyber world. In this section, we aim to establish a framework for designing a resilient controller for the physical power systems. In Fig. 2, we describe a hybrid system model that interconnects the cyber and physical environments. We use $x(t)$ and $\theta(t)$ to denote the continuous physical state and the discrete cyber state of the system, which are governed by the laws f and Λ, respectively. The physical state $x(t)$ is subject to disturbances w and can be controlled by u. The cyber state $\theta(t)$ is controlled by the defense mechanism l used by the network administrator as well as the attacker's action a.

We view resilient control as a cross-layer control design, which takes into account the known range of unknown deterministic uncertainties at each state as well as the

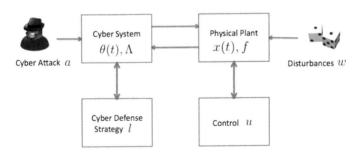

Fig. 2 The interactions between the cyber and physical systems are captured by their dynamics governed by the transition law Λ and the dynamical system f. The physical system state $x(t)$ is controlled by u with the presence of disturbances and noises. The cyber state $\theta(t)$ is controlled by the defense mechanism l used by the network administrator as well as the attacker's action a

random unanticipated events that trigger the transition from one system state to another. Hence, it has the property of disturbance attenuation or rejection to physical uncertainties as well as damage mitigation or resilience to sudden cyber attacks. It would be possible to derive resilient control for the closed-loop perfect-state measurement information structure in a general setting with the transition law depending on the control action, which can further be simplified to the special case of the linear quadratic problem.

The framework depicted in Fig. 2 can be used to describe the voltage regulation problem of a power generator subject to sudden faults or attacks. A power system has multiple generators interconnected through a large dynamic network. There are three types of attacks that can be considered, which are given as follows:

- Sensor Attacks: Attackers can launch a man-in-the-middle attack to introduce a bias to the measured parameters or multiply the sensed value by a constant.
- Actuator Attacks: Attackers can intrude the power control system and disrupt the physical control loops. The attack can cause an error on the generators' output torque, and consequently system dynamics are modified.
- Controller Attacks: An attacker can change the control signal sent through the SCADA system to an extent without being noticed by the system administrator. Consequently, the output of the controller is modified.

The framework can incorporate networked control system models to capture different aspects of network effects, for example, sampled-data systems, systems with delayed measurements, and model predictive control systems. The optimal design of the cyber and physical system can be made jointly by viewing each design process as a game-theoretic problem. For example, a zero-sum differential game problem can be used to design a robust controller while a stochastic game model can be used to design a defense strategy. With the joint game design, the framework yields control and defense strategies depend on both cyber states and physical states, and there is the need for the development of advanced computational tools to compute such joint control and defense strategies.

4 Secure Network Routing

One of the challenging issues at the data communication and networking layers of the smart grid in Fig. 1 is the assurance of secure routing of phasor measurement unit (PMU) and smart meter (SM) data in the open network, which is enabled by the adoption of IP-based network technologies. It is forecasted that 276 million smart grid communication nodes will be shipped worldwide during the period from 2010 to 2020, with annual shipments increasing dramatically from 15 million in 2009 to 55 million by 2020 [29]. The current dedicated network or leased-line communication methods are not cost-effective to connect large numbers of PMUs and SMs. Thus, it is foreseen that IP-based network technologies are widely adopted since they enable data to be exchanged in a routable fashion over an open network, such as the

Internet [30–34]. This will bring benefits such as efficiency and reliability, and risks of cyber attacks as well. Without a doubt, smart grid applications based on PMUs and SMs will change the current fundamental architecture of communication network of the power grid, and bring new requirements for communication security. Delay, incompleteness, and loss of PMU and SM data will adversely impact smart grid operation in terms of efficiency and reliability. Therefore, it is important to guarantee integrity and availability of those PMUs and SMs data. To meet the QoS requirements in terms of delay, bandwidth, and packet loss rate, QoS-based routing technologies have been studied in both academia and the telecommunications industry [35–38]. Unlike video and voice, data communications of PMUs and SMs have different meanings of real-time and security, especially in terms of timely availability [30, 39–43]. Therefore, QoS-based and security-based routing schemes for smart grid communications should be studied and developed to meet smart grid application requirements in terms of delay, bandwidth, packet loss, and data integrity.

We can leverage the hierarchical structure of power grids and investigate a routing protocol that maximizes the QoS along the routing path to the control room. In addition, the data communication rates between the super data concentrator can be optimized at the penultimate level with the control center. A hybrid structure of routing architecture is also highly desirable to enable the resilience, robustness, and efficiency of the smart grid.

Hierarchical routing: The smart grid has a multilayer structure that is built upon the current hierarchical power grid architecture. The end-users, such as households, communicate their power usage and pricing data with a local area substation which collects and processes data from SMs and PMUs. In the smart grid, the path for the measurement data may not be predetermined. The data can be relayed from smaller scale data concentrators (DCs) to some super data concentrators (SDCs) and then to the control room. With the widely adopted IP-based network technologies, the communications between households and DCs can be in a multi-hop fashion through routers and relay devices. The goal of each household is to find a path with minimum delay and maximum security to reach DCs and then substations. This optimal decision can be enabled by the automated energy management systems built-in SMs. Figure 3 illustrates the physical structure of the smart grid communication network. The PMUs and SMs send data to DCs through a public network. DCs process the collected data and send the processed data to SDCs through (possibly) another public network.

In the depicted smart grid, the data from a PMU or an SM has to make several hops to reach the control room. The decision for a meter to choose a router depends on the communication delay, security enhancement level, and packet loss rate. In addition, the decisions for a DC to choose an SDC also depends on the same criteria. The communication security at a node is measured by the number of security devices such as firewalls, intrusion-detection systems (IDSs), and intrusion-prevention systems (IPSs) deployed to reinforce the security level at that node. We can assign a higher utility to network routers and DCs that are protected by a larger number of firewalls, IDSs/IPSs and dedicated private networks in contrast to public networks.

Fig. 3 An example of the
physical structure of the
multilayer smart grid
communication network

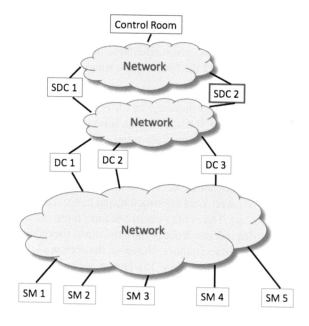

This relatively simple metric only considers one aspect of the control system cyber security. It can be further extended to include more security aspects by considering the authorization mechanisms, the number of exploitable vulnerabilities, potential damages as well as recovery time after successful attacks. The readers can refer to [6, 44–46] for more comprehensive metrics.

A trade-off with higher security is the latency and packet loss rate incurred in data transmission. A secure network inevitably incurs delays in terms of processing (encrypting/decrypting) and examining data packets. We can model the process of security inspection by a tandem queueing network. Since the arriving packets are inspected by IDS using signature-based or anomaly-based methods to detect malicious behaviors, each security device can be modeled with a queueing model. One simple example is the M/M/1 queue whose external arrival rate follows a Poisson process and the service time follows an exponential distribution. The latency caused by the security devices such as IDSs/IPSs is due to the number of predefined attack signatures and patterns to be examined [9, 27, 47]. In addition, devices such as IPSs can also lead to high packet loss due to their false negative rates in the detection.

Furthermore, a node with a higher level of security may be preferred by many meters or routers, eventually leading to a high volume of received data and hence higher level of congestion delay. Hence it leads to a distributed decision-making problem in which each device determines its route by assessing the trade-offs between security risks, the congestion delay and the quality of services. This problem can be analyzed using a game-theoretical approach to yield distributed routing decisions in the smart grid [12, 48]. The solution concept of mixed Nash equilibrium [49] as a solution outcome is desirable for two reasons. First, in theory, mixed Nash equilibrium always exists for a finite matrix game [49] and many learning algorithms such as fictitious play and replicator dynamics can lead to mixed Nash equilibrium [28, 50, 51]. Second, the randomness in the choice of routes makes it harder for an attacker to map out the routes in the smart grid.

Centralized versus decentralized architectures: A centralized routing architecture ensures the global efficiency, and it is robust to small disturbances from SMs and individual DCs or SDCs. However, it is costly to implement centralized planning on a daily basis for a large-scale smart grid. In addition, global solutions can be less resilient to unexpected failures and attacks as they are less nimble for changes in routes and it takes time for the centralized planner to respond in a timely manner.

On the other hand, decentralized decision-making can be more computationally friendly based on local information, and hence the response time to the emergency is relatively fast. The entire system becomes more resilient to local faults and failures, thanks to the independence of the players and the reduced overhead on the response to unanticipated uncertainties. However, the decentralized solution can suffer from high loss due to inefficiency [14, 15]. Hence, we need to assess the trade-off between efficiency, reliability, and resilience for designing the communication protocol between the control stations and the SDCs.

5   Management of Information Security

The use of technologies with known vulnerabilities exposes power systems to potential exploits. In this section, we discuss information security management which is a crucial issue for power systems at the management layer in Fig. 1. The timing between the discovery of new vulnerabilities and their patch availabilities is crucial for the assessment of the security risk exposure of software users [52, 53]. The security focus in power systems is different from the one in computer or communication networks. The application of patches for control systems needs to take into account the system functionality, avoiding the loss of service due to unexpected interruptions. The disclosure of software vulnerabilities for control systems is also a critical responsibility. Disclosure policy indirectly affects the speed and quality of the patch development. Government agencies such as CERT/CC (Computer Emergency Response Team/Coordination Center) currently act as a third party in the public interest to set an optimal disclosure policy to influence the behavior of vendors [54].

The decisions involving vulnerability disclosure, patch development, and patching are intricately interdependent. In Fig. 4, we illustrate the relationship between these decision processes. A control system vulnerability starts with its discovery. It can be discovered by multiple parties, for example, individual users, government agencies, software vendors or attackers, and hence can incur different responses. The discoverer may choose not to disclose it to anyone, may choose to fully disclose through a forum such as Bugtraq [55], may report to the vendor, or may provide to an attacker. Vulnerability disclosure is a decision process that can be initiated by those who have

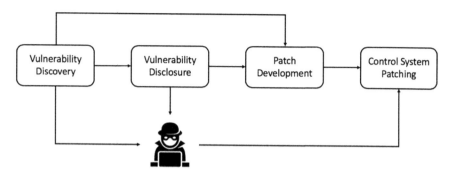

Fig. 4 A holistic viewpoint toward vulnerability discovery, disclosure, development, and patching. An attacker can discover a vulnerability or learn it from a disclosure process, eventually influencing the speed of patch application. A discoverer can choose to fully disclose through a forum or report to the vendor or may provide to an attacker. A vulnerability can be disclosed to a vendor for patch development or leaked to the attacker

discovered the vulnerability. Patch development starts when the disclosure process reaches the vendor and finally a control system user decides on the application of the patches once they become available. An attacker can launch a successful attack once it acquires the knowledge of vulnerability before a control system patches its corresponding vulnerabilities. The entire process illustrated in Fig. 4 involves many agents or players, for example, system users, software vendors, government agencies, attackers. Their state of knowledge has a direct impact on the state of vulnerability management.

We can compartmentalize the task of vulnerability management into different sub-modules: discovery, disclosure, development, and patching. The last two submodules are relatively convenient to deal with since the agents involved in the decision-making are very specific to the process. The models for discovery and disclosure can be more intricate in that these processes can be performed by many agents and hence specific models should be used for different agents to capture their incentives, utility, resources, and budgets. In [18], a dynamic model for control system patching is established to assist users in making optimal patching decisions. It has been shown that the optimal patching intervals are much shorter when risks of potential attacks are taken into account in the system. A dynamic game problem can thus be formulated to study the optimal frequency of patching to minimize the risk of an unpatched control system while an attacker aims to determine the time to launch the attack.

6   Discussions and Challenges

Security issues that arise in the smart grid constitute a pivotal concern in modern power-system infrastructures. In this chapter, we have discussed a six-layer security architecture for the smart grid, motivated by the OSI for the Internet and PRM models for enterprise and control systems. We have identified the security challenges present at each layer and pinpointed a holistic viewpoint for security solutions in the smart grid. The layered architecture facilitates the understanding of the trade-off between the information assurance at the cyber-related layers and the physical layer system performance.

We have presented security issues at three different layers. The resilient control design at the physical system is pivotal for modern power systems. We need a hybrid framework in which the occurrence of unanticipated events is modeled by a stochastic switching, and deterministic uncertainties are represented by the known range of disturbances. It is important to develop new methodologies to take the resilience of physical systems into consideration and enable a cross-layer control design for modern power grids.

At the data communication and network layers, we need to investigate the secure routing problem in the smart grid, which arises from the adoption of IP-based network technologies due to the wide use of PMUs and smart meters. It is important to leverage the multilayer structure of power grids and discuss a routing protocol that is based on distributed optimization of the quality-of-service along individual routing paths. The hybrid structure of the routing protocol is desirable to incorporate the desirable features of the centralized and decentralized architectures.

The use of information technologies in power systems poses additional potential threats due to the frequent disclosure of software vulnerabilities. At the higher level of the information security management layer, we have discussed a series of policy-making decisions on vulnerability discovery, disclosure, patch development and patching. We can use a system approach to understand the interdependencies of these decision processes.

Game-theoretic methods have provided formal approaches to model the adversarial interactions at multiple layers of the cyber-physical energy system. The game model has taken different forms to design resilient control systems, secure routing, and patching mechanisms. Understanding the security of multilayer energy system requires a holistic model that integrates the game models to provide an integrated framework for designing cross-layer security strategies. Mitigation of one security threat at one layer can be sometimes more effective than achieving it at other layers. In addition, the success of mitigation of certain attacks can rely on the strategies implemented at other layers. Hence cross-layer game-theoretic models are essential for developing an effective defense under budget constraints for the multilayer system as a whole.

More challenges as a result of the multilayer architecture of the smart grid security involves the integration of game theory, machine learning, control theory, and data-driven approaches for detection, automation, and reconfiguration in the smart grid. In addition to the security problems illustrated in the chapter, there are other security and privacy issues existing at each layer, for example, the jamming and eavesdropping problems at the data communication layer, the user data privacy problem at the management layer, and the system reliability problem at the network layer. Furthermore, the multilayer framework can be extended to study multi-agent systems. The interactions between subsystems in the smart grid can reside at the network, communication, and physical layers. It will be interesting to investigate the competition and cooperation for resources at multiple layers.

References

1. S. Gorman, Electricity grid in U.S. penetrated by spies. Wall Str. J., http://online.wsj.com/article/SB123914805204099085.html
2. B. Krebs, Cyber incident blamed for nuclear power plant shutdown. Wash. Post, http://www.washingtonpost.com/wp-dyn/content/article/2008/06/05/AR2008060501958.html
3. R. McMillan, Siemens: stuxnet worm hit industrial systems. Computerworld (2010), http://www.computerworld.com/s/article/print/9185419. Accessed 16 Sept 2010
4. Iran confirms Stuxnet worm halted centrifuges. CBS News, http://www.cbsnews.com/stories/2010/11/29/world/main7100197.shtml. 29 Nov 2010
5. S. Greengard, The new face of war. Commun. ACM **53**(12), 20–22 (2010)
6. D. Wei, K. Ji, Resilient industrial control system (RICS): concepts, formulation, metrics, and insights, in *Proceedings of International Symposium on Resilient Control Systems (ISRCS)*, Idaho Falls, ID, USA, Aug 2010
7. K. Moslehi, R. Kumar, Smart grid—a reliability perspective, in *Proceedings of Innovative Smart Grid Technologies (ISGT)*, January 2010, pp. 1–8
8. F.F. Wu, K. Moslehi, A. Bose, Power system control centers: past, present, and future. Proc. IEEE **93**(11), 1890–1909 (2005)
9. Q. Zhu, T. Başar, Towards a unifying security framework for cyber-physical systems, in *Proceedings of Workshop on the Foundations of Dependable and Secure Cyber-Physical Systems (FDSCPS-11)* (CPSWeek, Chicago, 2011)
10. M. Manshaei, Q. Zhu, T. Alpcan, T. Başar, J.-P. Hubaux, Game theory meets network security and privacy. ACM Comput. Surv. (CSUR) **45**(3), 25
11. W. Saad, Q. Zhu, T. Başar, Z. Han, A. Hjorungnes, Hierarchical network formation games in the uplink of multi-hop wireless networks, in *Proceedings of 28th IEEE Conference on Global Telecommunications (GLOBECOM)*, December 2009, Honolulu, Hawaii, USA (2009), pp. 2390–2395
12. Q. Zhu, D. Wei, T. Başar, Secure routing in smart grids, in *Proceedings of Workshop on the Foundations of Dependable and Secure Cyber-Physical Systems (FDSCPS-11)* (CPSWeek, Chicago, 2011)
13. Q. Zhu, Z. Yuan, J.B. Song, Z. Han, T. Başar, Dynamic interference minimization routing game for on-demand cognitive pilot channel, in *Proceedings of 28th IEEE Conference on Global Telecommunications (GLOBECOM)* (Miami, FL, USA, 2010)
14. T. Başar, Q. Zhu, Prices of anarchy, information, and cooperation in differential games. Dyn. Games Appl. **1**(1), 50–73 (2010)
15. Q. Zhu, T. Başar, Price of anarchy and price of information in linear quadratic differential games, in *Proceedings of American Control Conference (ACC)* (Baltimore, MD, USA, 2010)
16. S. Rass, A. Alshawish, M.A. Abid, S. Schauer, Q. Zhu, H. de Meer, Physical intrusion games—optimizing surveillance by simulation and game theory. IEEE Access **5**, 8394–8407 (2017)
17. S. Rass, Q. Zhu, GADAPT: A sequential game-theoretic framework for designing defense-in-depth strategies against advanced persistent threats, in *Proceedings of the 27th Conference on Decision and Game Theory for Security (GameSec)*, NY, USA, on 2–4 November 2016
18. Q. Zhu, M. McQueen, C. Rieger, T. Başar, Management of control system information security: control system patch management, in *Proceedings of Workshop on the Foundations of Dependable and Secure Cyber-Physical Systems (FDSCPS-11)* (CPSWeek, Chicago, 2011)
19. Z. Xu, Q. Zhu, Cross-Layer Secure Cyber-Physical Control System Design for Networked 3D Printers, in *Proceedings of 2016 American Control Conference*, Boston, USA, 6–8 July 2016

20. Z. Xu, Q. Zhu, A game-theoretic approach to secure control of communication- based train control systems under jamming attacks, in *Proceedings of the 1st International Workshop on Safe Control of Connected and Autonomous Vehicles* (SCAV, 2017) CPS Week, April 18–21, 2017 (Pittsburgh, PA, USA, 2017)

21. Z. Xu, Q. Zhu, A Cyber-Physical Game Framework for Secure and Resilient Multi-Agent Autonomous Systems, in *Proceedings of 54th IEEE Conference on Decision and Control (CDC)*, Osaka, Japan, 15–18 December 2015

22. Q. Zhu, T. Başar, Game-theoretic methods for robustness, security, and resilience of cyber-physical control systems: games-in-games principle for optimal cross-layer resilient control systems. IEEE Control Syst. **35**(1), 46–65 (2015)

23. J. Pawlick, Q. Zhu, Crowd-Sourcing Deception as a Mechanism for Phishing Defense, in *International Conference on Deceptive Behavior (Decepticon)*, University of Cambridge, UK, 24–26 Aug 2015

24. J. Casey, A. Morales, E. Wright, Q. Zhu, B. Mishra, Compliance signaling games: modeling deterrence of unintentional insider threats. J. Comput. Math. Organ. Theory **22**(3), 318–349 (2016)

25. Z. Xu, Q. Zhu, A Secure Data Assimilation for Large-Scale Sensing Networks Using an Untrusted Cloud, in *Proceedings of 20th IFAC World Congress*, Toulouse, France, 9-14 July 2017

26. J. Pawlick, S. Farhang, Q. Zhu, Flip the Cloud: cyber-physical signaling games in the presence of advanced persistent threats, in Proceedings of Conference on Decision and Game Theory for Security (GameSec), London, UK, 4–5 November 2015

27. Q. Zhu, H. Tembine, T. Başar, Network security configuration: a nonzero-sum stochastic game approach, in *Proceedings of of American Control Conference (ACC)* (Baltimore, MD, USA, 2010)

28. Q. Zhu, H. Tembine, T. Başar, Distributed strategic learning with application to network security, in *Proceedings of American Control Conference (ACC)* (San Francisco, CA, 2011)

29. Pike Research's report (2010), Smart grid networking and communications

30. K. Tomsovic, D. Bakken, V. Venkatasubramanian, A. Bose, Designing the next generation of real-time control, communication, and computations for large power systems. Proc. IEEE **93**(5), 965–979 (2005)

31. G.N. Ericsson, On requirements specifications for a power system communications system. IEEE Trans. Power Deliv. **20**(2), 1357–1362 (2005)

32. M.S. Amin, B.F. Wollenberg, Toward a smart grid: power delivery for the 21st century. IEEE Power Energy Mag. **3**(5), 34–41 (2005)

33. E. Santacana, G. Rackliffe, L. Tang, X. Feng, Getting smart. IEEE Power Energy Mag. **8**(2), 41–48 (2010)

34. US Department of Energy, "Grid 2030–a national vision for electricity's second 100 years," Technical Report, July 2003

35. D.H. Lorenz, A. Orda, QoS routing in networks with uncertain parameters. IEEE/ACM Trans. Netw. **6**(6), 768–778 (1998)

36. P. Van Mieghem, F.A. Kuipers, Concepts of exact QoS routing algorithms. IEEE/ACM Trans. Netw. **12**(5), 851–864 (2004)

37. G. Xue, S.K. Makki, Multi-constrained QoS routing: a norm approach. IEEE Trans. Comput. **56**(6), 859–863 (2007)

38. F. Kuipers, P. Van Mieghem, T. Korkmaz, M. Krunz, An overview of constraint-based path selection algorithms for QoS routing. IEEE Commun. Mag. **40**(12), 50–55 (2002)

39. US Department of Energy, "Communications requirement of smart grid technologies," 5 Oct 2010

40. G.N. Ericsson, On requirements specifications for a power system communications system. IEEE Trans. Power Deliv. **20**(2), 1357–1362 (2005)

41. P. McDaniel, S. McLaughlin, Security and privacy challenges in the smart grid. IEEE Secur. Priv. **7**(3), 75–77 (2009)

42. National Institute of Standards and Technology, "NIST framework and roadmap for smart grid interoperability standards," Release 1.0, Jan 2010
43. C.H. Hauser, D.E. Bakken, A. Bose, A failure to communicate: next generation communication requirements, technologies, and architecture for the electric power grid. IEEE Power Energy Mag. **3**(214), 47–55 (2005)
44. Department of Homeland Security, "Primer control systems cyber security framework and technical metrics," Technical Report, June 2009
45. W. Boyer, M. McQueen, Ideal based cyber security technical metrics for control systems, in *Proceedings of 2nd International Workshop on Critical Information Infrastructures Security*, October 2007
46. A. McIntyre, B. Becker, R. Halbgewachs, Security metrics for process control systems, SANDIA report, SAND2007-2070P, September 2007
47. Q. Zhu, T. Başar, Dynamic policy-based IDS configuration, in *Proceedings of 48th IEEE Conference on Decision and Control (CDC)* (Shanghai, P. R. China, 2009) December 2009
48. R. Hou, C. Wang, Q. Zhu, J. Li, Interference-aware QoS multicast routing for smart grid. Ad Hoc Netw. **22**, 13–26 (2014)
49. T. Başar, G.J. Olsder, *Dynamic Noncooperative Game Theory*, 2nd edn. Classics in Applied Mathematics (SIAM, Philadelphia, 1999)
50. D. Fudenberg, D. K. Levine, *The Theory of Learning in Games*, May 1998 (MIT Press, 1998)
51. Q. Zhu, H. Tembine, T. Başar, Heterogeneous learning in zero-sum stochastic games with incomplete information, in *Proceedings of 49th IEEE Conference on Decision and Control (CDC)* (Atlanta, GA, USA, 2010)
52. S. Frei, B, Tellenbach, B. Plattner, 0-Day patch: exposing vendors (in)security performance, *BlackHat* (2008)
53. M.A. McQueen, W.F. Boyer, T.A. McQueen, S. McBride, Empirical estimates of 0-day vulnerabilities in control systems, in *SCADA Security Scientific Symposium*, 21–22 Jan 2009
54. Symantec Inc, "Symantec Internet Security Threat Report" (2003), http://www.symantec.com
55. Bugtraq, http://www.securityfocus.com/

Cyber Security for Power System State Estimation

Yacine Chakhchoukh and Hideaki Ishii

Abstract State estimation is a critical application that provides situational awareness and permits efficient operation of the smart grid. The secure, accurate, and fast computation of the state estimates is crucial to execute the complex decisions and diverse control actions needed in real time to provide reliable, economic, and safe power systems that integrate distributed and intermittent renewable generation. This chapter discusses research directions to evaluate the cyber security and develop novel algorithms for securing today and tomorrow's power state estimation and grid operation.

1 Introduction

Power systems are essential in the functioning and development of our modern society. Unfortunately, the modern power systems are vulnerable to cyber attacks that could degrade their performance and cause blackouts [1, 2]. Indeed, the power grid is becoming increasingly complex and the need for implementing sophisticated cyber systems for its automatic operation raises serious concerns regarding its safety. Recently, the US administration warned power companies against cyber attacks such as the ones that targeted Ukraine's power grid in December 2015 [3].

The power systems are evolving toward the so-called smart grid, which enables increased integration of intermittent generation from renewable energy sources such as solar and wind with classical sources such as coal, nuclear, natural gas, and hydroelectric [2]. For example, renewable generation is forecasted to englobe more than two-thirds of all installed generation capacity between now and 2030 [4].

Y. Chakhchoukh
Department of Electrical and Computer Engineering, University of Idaho,
875 Perimeter Drive, MS 1023, Moscow, ID 83844-1023, USA
e-mail: yacinec@uidaho.edu

H. Ishii (✉)
Department of Computer Science, Tokyo Institute of Technology,
4259-J2-54, Nagatsuta-cho, Midori-ku, Yokohama 226-8502, Japan
e-mail: ishii@c.titech.ac.jp

© Springer Nature Switzerland AG 2019
J. Stoustrup et al. (eds.), *Smart Grid Control*, Power Electronics
and Power Systems, https://doi.org/10.1007/978-3-319-98310-3_15

The rapid development of energy storage, sensing, communication, computing technologies, and distributed automatic control will aid this transition. The resulting power grid can be viewed as a large interconnected cyber-physical system [2]. The cyber part will contain all the communication, data analysis, computation, and control needed by the power systems. Developing the cyber part and its applications helps improve the operation of the future power systems but increases vulnerabilities to cyber attacks introduced by malicious agents and hackers. Cyber attackers can be individuals, groups, organizations, or even nations and could be motivated by inducing financial gains or creating nuisance by targeting the power grid. Such attacks can be possible due to the development of complex communication networks, the Internet, and different viruses and malwares. Modern power system control networks are interconnected at certain points with traditional Information Technology (IT) enterprise networks and the Internet. Intruders will have the possibility to access the power systems and modify the normal operation of the system. Actually, manipulations or attacks committed by malicious intruders can result in tremendous adverse effects in both the cyber and the physical worlds.

The vulnerabilities of the smart grid toward cyber attacks are not fully understood and cyber attack impacts could range from a modified electricity market and degraded operation to a threatened integrity of the grid causing material loss and destruction, and even cascading blackouts. The power grid is considered to be an important critical infrastructure and today's economy and society depends on its stable, efficient, and secure operation [5]. The amount of cyber security threats and the success rate of cyber attacks on current Information Technology and Operational Technology systems pose a currently immeasurable amount of risk to this critical infrastructure on which our society and economy depend. Since the reliability and costs can be affected by attacks, it is vital to insure the security and safety of the cyber system against malicious intruders [6, 7]. Research and development of techniques and algorithms for securing critical control systems in the power grid is imperative.

The focus of this chapter is on the cyber security issues that arise in the context of state estimation (SE) in power systems. Real-time operation of the power systems uses the SE results, which consists of the evaluation of voltage magnitudes and phase angles at chosen buses or substations [8, 9]. Several grid operation tools and power market tasks such as contingency analysis, unit commitment, economic dispatch, and locational marginal prices (LMPs) computation rely on an available and accurate state estimation. The SE is also needed by the operator in order to picture the power system condition clearly, which permits situational awareness in order to take the optimal corrective control actions. The results from SE are useful to run a Security Analysis (SA), or the so-called Operational Reliability Analysis (ORA). In the ORA, a contingency analysis is executed to check if the system is (N-1) secure. The term (N-1) secure means that the power system is still stable and in an acceptable state region after any single major contingency. A contingency could be, for example, a loss of a major generator, a transmission line or a large load change. The results from the contingency analysis will determine the need of the operator to intervene in the operation. If the system is safe, then the optimized electricity markets will determine the grid operation and the different power flows in the grid. Otherwise, the operator will take the appropriate actions such as generation rescheduling to insure the safety of the system which is fundamental.

In order to implement the future smart grid technology, it is necessary to identify the cyber-threats and the cyber-vulnerabilities of the real-time operation of realistic power systems and propose novel theoretical and practical solutions. This is a multidisciplinary area which requires complementary research expertise from systems control, signal processing, and power systems. The process is in general composed of three stages: (1) To assess the cyber-vulnerabilities and their consequences on the power grid. The emphasis here consists in providing a study of attacks targeting the power state and topology of the grid and their impacts on the real-time operation, control and power markets. (2) To propose novel procedures to detect and isolate cyber attacks. The techniques studied consist of adapting new robust signal processing methods for the linear and nonlinear regression and time series contexts, filtering and forecasting in the presence of cyber attacks and non-stationarity. (3) To counter or mitigate the impact resulting from cyber attacks and take the convenient correcting operation actions to ensure a resilient, reliable, efficient, and economical operation of the whole power system.

The organization of this chapter is as follows: In Sect. 2, we first provide an overview of the static state estimation problem and the bad data detection schemes. In Sect. 3, we discuss cyber attack models, configuration, and consequences on the power systems. Section 4 provides interesting and necessary future research directions. Finally, Sect. 5 concludes the chapter.

2   Static Power State Estimation

The static state estimation is run after collecting measurements from the supervisory control and data acquisition units (SCADAs) at remote terminal units (RTUs), and the results are communicated to the control center every 2–5 s. One important objective of the state estimation is to detect accidental bad data, i.e., bad measurements, topology errors, and line parameter inaccuracies and to correct this erroneous sensed data using the power model and available redundant measurements. To fulfill this objective, SE modules from different energy management systems (EMS) vendors are equipped with bad data detectors (BDD).

The objective consists in estimating the vector $x \in \mathbb{R}^n$ obeying a linear regression (DC) or nonlinear regression model (AC). The vector $z \in \mathbb{R}^m$ contains communicated readings from SCADAs. The number n of states is estimated from a larger number of measurements m. The AC model considers reactive power measurements and permits to estimate voltage magnitudes as well as phase angles at different buses or substations. The DC model assumes the voltages to be equal to 1 per unit (p.u.) at all buses and estimates only the phase angles. Obtaining the phase angles gives a clear picture about the power flow paths in the grid. The obtained models are linear regression for DC and nonlinear regression for AC.

In the AC formulation, the state x follows the equation:

$$z = h(x) + e. \tag{1}$$

The vector x provides the power states, i.e., voltage magnitudes and phase angles at the buses of interest. The error vector e is random and assumed Gaussian with zero mean and covariance matrix R. The nonlinear vector function $h(\cdot)$ is known as the measurement function and reflects the grid topology and transmission line parameters. The grid topology is estimated in the topology processor module and is updated continuously by collecting readings of the circuit breakers' binary states (i.e., 0 or 1 that can be obtained from node breaker models). The binary states provide the information about whether the different transmission lines are open or closed [8]. The different line parameters are available in the operators' database and are exploited to reconstruct the nonlinear function $h(\cdot)$. The line parameters are also estimated or updated when needed [8, 9].

In practice, the SE is solved by running an iterative algorithm based on the weighted least squares (WLS) [8], i.e., at the $k + 1$ iteration, the state estimate $\hat{x}^{k+1}$ is related to the gain matrix $G(\hat{x}^k) = \left(H^{(k)}(\hat{x}^k)\right)^T R^{-1} H^{(k)}(\hat{x}^k)$ as

$$\hat{x}^{k+1} = \hat{x}^k + \Delta x^k, \tag{2}$$

$$G(\hat{x}^k)\Delta x^k = \left(H^{(k)}\right)^T (\hat{x}^k) R^{-1} \left(z - h(\hat{x}^k)\right). \tag{3}$$

The matrix $H^{(k)}$ is the Jacobian of $h(\cdot)$ with respect to x at step k. The gain matrix $G(\hat{x}^k)$ is factorized following the LQ decomposition. The inverse matrix of $G(\hat{x}^k)$ can also be computed to evaluate Δx^k.

After the convergence of the algorithm [9], the obtained residuals, i.e., $r = z - h(\hat{x}^k)$ are analyzed in BDD modules to flag possible outliers or bad data. The bad data could be due to natural failures such as sensor, communication channels misbehavior or intrusions and cyber attacks. The most practical outlier detection rules are known as the chi-square (χ_2) test and the "3σ" rejection rule [9]. In the power systems literature, the largest normalized residual rejection (LNR) has been proposed as well. Basically, the largest normalized residual or element in r is rejected if it does not obey the Gaussian assumption (i.e., measurement is rejected if its normalized residual absolute value is larger than 3). The estimation is rerun after removing the detected measurement until no residual is flagged as outlying [9].

3   Cyber Attack Models in the State Estimation and Their Consequences

In this section, we introduce several classes of cyber attack models in the static SE problem. We provide an overview of the current state of research and discuss important future directions to enhance the security for the SE and critical systems affected by inaccuracies in SE.

In general, the security of SCADA systems is a real widespread practical concern since their use is pervasive [5]. In fact, SCADA systems require adapted security studies and newly developed tools that go beyond solutions available in Information Technology (IT). For example, SCADA systems installed in power systems are generally inexpensive, vulnerable, and have long life cycles. The life cycles of SCADA sensors are of a few decades, which means that their defense should be adaptable with possible continuous updates.

The existing bad data detectors implemented at control centers are useful for accidental or random sensor and communication channel errors, but are not adapted to counter sophisticated cyber attacks [10–14]. Cyber attacks targeting SE can be classified into different types such as denial-of-service (DoS) where data is not available or missing which can result in certain states to be unobservable; eavesdroppers which analyze the communication traffic to gain private information [7, 15] and raise privacy concerns; and integrity attacks where the data communicated is modified by a "man-in-the-middle" access where attackers are intermediate nodes in the communication. This latter type of attacks is also known as false data injection (FDI) cyber attacks. They can result in intentionally modified measurements communicated to the control center that could change the state in a stealthy fashion where the attacks could, under certain conditions, escape bad data detectors (BDDs) integrated into existing SE modules at energy management systems (EMS). FDI attacks are invisible and hence raise a lot of concern about the operation of the power grid. We will concentrate on this last type of attacks in the following sections.

3.1 False Data Injection Attacks on Measurements

Analyzing (2), we notice that at each step, a linear regression problem is solved where the state increment Δx^k is evaluated from the residual $r = z - h(x^k)$ regressed on $H^{(k)}$. This means that the estimation is run iteratively after linearizing the regression in each step. With slight abuse of notation, the problem can be reformulated as

$$z = Hx + e. \tag{4}$$

The above linear equation or regression represents also the DC formulation problem where the WLS solution is given by $\hat{x} = (H^T H)^{-1} H^T z$. The covariance of the error vector e is assumed to be equal to the identity matrix for simplicity. Notice that the WLS algorithm corresponds to the maximum likelihood estimator under the assumption that the errors are Gaussian [16].

In this context and as proposed in [10, 11], a man-in-the-middle attack could be generated, for example, in the communication between RTUs and the control center. This attack could create a contamination in the measurements by adding the vector equal to a as $z_a = z + a$. In particular, if the attacker has knowledge of the system topology (i.e., the Jacobian matrix), he can generate an attack with $a = Hc$ [11], that is,

$$z_a = z + a = z + Hc. \tag{5}$$

In this case, the attacker is able to change the state estimate to $\hat{x}_a = \hat{x} + c$ where he controls the state vector bias $c \in \mathbb{R}^n$. The residuals are unchanged, i.e., $r_a = z_a - H\hat{x}_a = z - H\hat{x} = r$, which means that the attack is stealthy to the classical bad data detectors (BDD) based on analyzing the residuals. In other words, a bad data detector that analyzes the vector r_a will not detect any change due to the stealthy attack. Stealthy attacks could be generated on both the DC and AC formulations of state estimation as shown in [11, 13]. Due to the sparsity of the power systems and the matrix H, the attacker does not need to target all sensors or have a global knowledge of the topology when targeting a few buses by a stealthy attack. The intrusion would mislead the operator at the control center because he obtains a modified result that does not reflect the actual state of the grid. All consequent actions at the EMS are contaminated by false data injection (FDI) cyber attacks. The impact of bad data and attacks on SE impacting power markets is discussed in [14, 17].

3.2 FDI Attacks Targetting the Topology of the Grid

False data injection (FDI) type attacks could also target the topology of the grid [18–20]. The topology represents the connectivity of the power system and is updated constantly over time in the topology processor. The binary readings from the circuit breakers representing the transmission line states (i.e., open or closed) are communicated to the control center. An intruder can modify these readings as well as the SCADA analog measurements reflecting, for example, power flows on neighboring lines in a coordinated fashion confirming the false state of the line. Such an attack allows a malicious update of the topology undetected by BDD. This type of FDI attacks clearly requires more knowledge and skills of the system by the intruder. The attacker needs the knowledge of the topology and the measurements or the actual grid state. He needs access to the circuit breaker states and the SCADA analog measurements communicated to the control center [18]. The consequences can be more dangerous and complex for the operation than those caused by attacks only on the measurements as considered in the previous subsection. In [18, 20], stealthy cyber attack strategies on both the power state and the topology of the grid are discussed.

Figure 1 illustrates the 14 bus system where bus 5 is targeted by a cyber attack. The system could be decomposed in subsystems as proposed in [22]. For example, subsystem 1 includes buses 1, 2, and 5 and their connecting lines. The other subsystems are cyclic {2, 4, 5}, {2, 3, 4}, {4, 7, 9}, {6, 9, 10, 11, 13, 14}, {6, 12, 13}, {4, 5, 6, 9, 10, 11}, and radial {7, 8} [20, 22]. The decomposition maximizes the number of bad measurements detected while insuring the observability of the whole system [23].

Figure 2a, b illustrate the vulnerability of several estimators [21, 24] toward stealthy attacks targeting the topology and the state through simulations for the IEEE 30 bus system. In both figures, the final state errors and estimate $\hat{x}$ are shown after

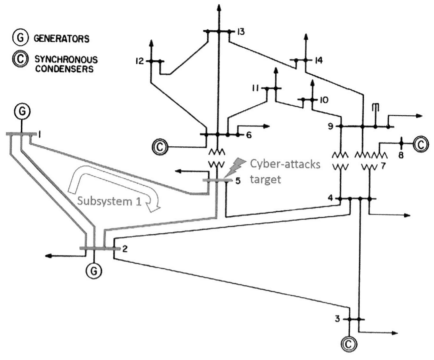

Fig. 1 IEEE 14 bus system (© 2016 IEEE. Reprinted with permission from [39])

cyber-intrusions escaped detection. In this simulation, the true state is obtained from solving a power flow. In practice, the real state is not known by the operator. The figures illustrate the errors in the absence (clean) and presence of stealthy attacks (RA and LTS) assuming the ideal case where the true state is known. In Fig. 2a, the phase angle of a single state at one bus in the system is targeted with a large error (i.e., phase angle at bus 5). In Fig. 2b, the attacker manages to manipulate the value of the final estimated phase at bus 6 because he has enough access to the grid information. If the attacker can target a large number of sensors, then all estimators illustrated in the figure will become vulnerable. In Fig. 2a, RA represents the popular "3σ" rejection rule applied to normalized WLS residuals. The curve labeled "clean" gives the estimation errors in the absence of cyber attacks. In Fig. 2b, WLSc denotes the WLS applied to the clean non-attacked topology and the curve labeled "true" gives the value obtained with the power flow solution, which represents the real state. LNR denotes the popular largest normalized residual rejection [8]. LTS represents the diagnostic of the measurements using the least trimmed squares estimator (LTS). The LTS is a robust estimator that is adapted to handle false data in the topology [8, 16, 23, 25, 26]. It was shown in [20, 22, 23] that the classical commercial BDD

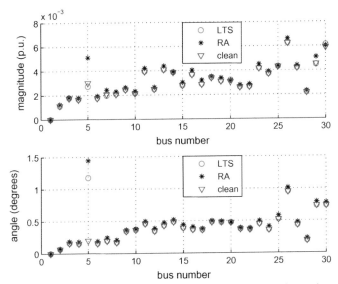

(a) Monte Carlo average absolute error of the SE estimate in the IEEE 30 bus system.

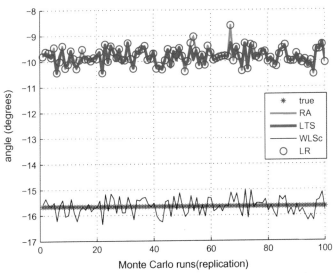

(b) Monte Carlo replications of the SE estimate of the voltage phase angle (deg) at bus 6 in the IEEE 14 bus system.

Fig. 2 Errors of AC power phase angle estimation caused by stealthy cyber attacks targeting the topology (©2015 IEEE. Reprinted with permission from [20, 21])

based on analyzing the residual of the WLS is not effective against random errors and attacks on the topology. Application of robust estimation techniques needs further verification.

3.3 Cyber Security of the PMU-Based State Estimation

Recently, phasor measurement units (PMUs) are being deployed due to the incentives provided by the US Department of Energy [27–29]. PMUs have higher reporting rate (30–120 measurements every second) and are better synchronized than SCADAs due to their use of the Global Positioning System (GPS) clocks [29, 30]. According to the North American SynchroPhasor Initiative (NASPI), around 1800 synchrophasor units or PMUs were available across North America in 2015 [31].

PMUs measure directly the states, i.e., voltage magnitudes and phase angles. The number of available PMUs is, however, still limited in practice because of their associated high costs. Power companies are interested in combining both their existing SCADAs and the newly installed PMUs in estimating the system state using the so-called hybrid state estimator. The state of the grid is estimated at regular intervals, i.e., every several seconds to a few minutes. Novel state estimation algorithms exploiting PMUs are gaining a lot of interest in the recent literature [29, 32–35]. These PMU-based state estimators are important to control the grid using the wide area measurement system (WAMS) technologies [27]. In [33], the authors proposed to buffer the data from PMUs to resolve the disparity in the reporting frequency between SCADAs (every few seconds) and PMUs (every 1/30 s). An optimal buffer length could be derived to ensure a good trade-off between tracking the fast changes in the grid states versus maximizing the time interval of the data exploited from PMUs to increase the accuracy of the estimates [33, 36].

Monitoring the grid with PMUs that are capable of delivering large amounts of real-time data creates cyber-vulnerabilities. Indeed, PMUs are vulnerable to both random bad data and cyber attacks [27, 35, 37, 38]. For example, intruders could create attacks by spoofing the GPS clocks of PMUs. However, the delivered data could be exploited in a clever way to improve both the cyber security and the operation of the grid. Practical and novel algorithms could be exploited to notify the operator at the control center when abnormal measurements are detected [39]. Furthermore, these techniques could correct the bad data introduced by cyber attacks automatically and secure the control in the power grid. One important research direction consists of proposing data-driven algorithms and power system model-based approaches that improve the cyber security of the whole power system operation.

Modeling time and space dependencies in multiple PMUs and estimating the existing correlation could be used in order to detect outliers or cyber attacks in PMU signals [38, 40, 41]. This approach improves also the accuracy of the hybrid state estimation. The technique proposed in [38] provides a sophisticated defense mechanism against stealthy cyber attacks and was shown to make the task of cyber attackers extremely complicated and tedious. The preliminary simulation results have considered an SE resolution of 2 s. This implies that every 2 s, a new estimate of the grid state is obtained, which might be considered to be a low resolution in the future power grid. Several authors have recently proposed to reconstruct the power states from PMU measurements to increase the SE resolution in order to track the rapid changes expected in the future smart grid [34, 42]. The SE will be refreshed very frequently (i.e., every few fractions of a second), which will completely change the real-time control of the power systems. This is an important emerging area for future studies.

Furthermore, the dynamic state estimation, which has gained interest thanks to the available PMU measurements, is executed in a nonlinear context using extended Kalman filters (EKF) [43] or unscented Kalman filters (UKF) [44]. Using a dynamic SE in the WAMS context enables the anticipation of power system dynamics and necessary fast controls. To improve the practical implementation of the dynamic SE, the authors in [45] proposed a decentralized algorithm that uses the UKF. Both the EKF and the UKF are vulnerable to bad data [25, 46] and cyber attacks [47]. The reference [48] shows the possibility of generating stealthy attacks in the general case of networked control systems obeying a dynamic linear state space representation. The theoretical results could be adapted to the case of power state estimation problem. Recently, some authors have implemented robust versions of both EKF and UKF using robust methods such as the Generalized M-estimator [46] and the least absolute value [35]. Offline PMU-based diagnostic techniques that improve the detection of cyber attacks [49] and errors on the topology or parameters are also being developed [50]. These techniques consist generally in identifying vulnerable sensors to be secured. Exploiting forecasts of PMUs and loads could improve the robustness and cyber security of the SE as proposed recently in [51]. Effective cyber security solutions considering realistic large power systems and hybrid state estimation for both static and dynamic approaches that handle attacks on PMU signals remain necessary for future grid operation security.

3.4 Assessing SE Cyber-Vulnerabilities and Their Consequences on the Power Grid Operation

So far, our discussion on security has been limited to SE on its own. However, assessing the impacts of cyber attacks targeting the power state estimation is very important in order to evaluate the danger of the different cyber attack types and configurations. That is, we must go beyond quantifying the power state modification to analyze the real cascading consequences of cyber attacks on contingency analysis, control actions, power markets, and power flows. While this issue is very vast and is outside the scope of this chapter, we would like to mention a few important directions for future research.

Developing and implementing metrics to assess the cyber security of practical power systems is necessary. Such metrics could be developed considering not only the impacts on the security of the operation but also the economic impacts on power markets. Some authors are starting to study cyber attacks consequence on power markets [14, 17], system operation [19] and tools to quantify cyber security [52]. Cyber attacks consequences will be quantified by static and dynamic studies on realistic systems. To reach this goal several universities are developing cyber security testbeds for the smart grid that emulate the real-time behavior of a large power system [53–56]. Testbeds are necessary to research the cyber-vulnerabilities that account for the complexity and the different interactions between the system part and the cyber part of the smart grid in real-time conditions. Fluctuations in frequency and power flows, oscillation modes, and voltage magnitudes will be considered in deriving these metrics for cyber security quantification. This will allow to rank the proposed cyber security methods and enable combining different solutions in an optimal and cost-effective way in order to secure the grid operation.

4   Future Directions

Here, we outline several future research directions related to the SE problem from a slightly broader perspective.

The cyber security solutions discussed so far can be classified into two main categories, online and offline solutions [51]. The offline solutions are remedial actions developed offline with no time constraints on the computation. An example of such methods consists of finding the minimum number of sensors to be secured and their positions in order to make a stealthy attack unfeasible [11, 17, 49]. On the other hand, online solutions are techniques that update their capabilities and models in real time using available sensed data. An example of the latter solutions consists of implementing robust estimation tools that detect an attack by comparing the modified measurement to a majority of clean data collected and analyzed in real time.

Sophisticated robust estimation tools have been developed recently in the signal processing and statistics literature [16, 25]. These methods are becoming practical thanks to the fast-evolving computation power. Novel techniques could be developed or adapted from robust statistics and signal processing to improve the detection of cyber attacks. This has the advantage of providing online adaptable methods that could reduce investments in expensive secure sensors. Robust signal processing methods and machine learning techniques exploit newly available data to update their models and detection procedures [57, 58].

For example, to enhance the overall cyber security of the static AC SE, the work in [39] introduces an approach that runs several robust least trimmed squares (LTS) estimators with different breakdown points or rejection percentages in parallel to improve the detection of cyber attacks targeting both the measurements and the topology of the grid. This approach allows us to not only robustly estimate but also accurately identify the presence of attacks. Also, as an alternative approach to detect the presence of stealthy cyber attacks, machine learning techniques are exploited in [57, 58]. The work in [59] introduces a statistical outlier detection approach using a recently proposed machine learning technique called density ratio estimation (DRE) [60]. Combining such different techniques should be further investigated.

While the proposed methods in the literature are effective and promising on computer simulated data and theoretically justified, their performance and implementation rely heavily on the power system complexity and collected data used during the estimation or the learning process. The collected data should be as realistic as possible to validate the effectiveness of the proposed methods in practical control centers allowing their real implementation. Cyber security testbeds are necessary to collect large amounts of data from SCADAs and PMUs to assess the performance of the proposed methods in real-life conditions. Investigating the sensitivity of the proposed machine learning methods, for example, when attacks are present in the learning process is of great interest. Furthermore, decomposing the grid in several subsystems (Fig. 1) and executing the proposed algorithms in a distributed fashion to reduce the computational burden allows the implementation of the online defense. Some authors proposed distributed state estimation methods [29, 61–63]. A trade-off should be ensured between a decomposition that maximizes detection versus a decomposition that reduces execution times. Finally, power systems are very sparse, i.e., each bus is connected only to a limited number of buses. Sparsity should be considered when adapting the robust techniques both to evaluate their robustness characteristics [26] and their implementation algorithms.

Moving target defense (MTD) is useful to defend the SE against cyber attacks. The objective of the MTD is to increase the complexity of the system so as to increase the attack cost for the intruder by reducing his knowledge of the system. This objective is achieved in [64] by randomizing the set of measurements and the topology of the grid. The topology of the grid change is reflected by a few changing line impedances thanks to distributed flexible AC transmission system (D-FACTs). The work [65] proposed to randomize the set of measurements as well and obtained an improved cyber security of the SE. This research direction could be explored further by improving the randomization and integrating PMUs as well.

For the case of an increased SE resolution to track fast changes in the future power systems, cyber security becomes even more challenging since the procedure will be fully automatic and the algorithms implemented to secure the operation need to converge very fast. Reference [34] proposed a method that provides robustness against random errors occurring in PMU sensors but sophisticated cyber attacks were not studied. Furthermore, the dynamic state estimation, which is also gaining interest and is vulnerable to both bad data and cyber attacks [46, 47], will allow even the anticipation of the control in the wide area measurement systems (WAMS).

The state estimation at distribution level is also a very interesting research topic for future investigation [66]. It may improve the monitoring of distribution systems especially with increased distributed generation and storage such as photovoltaic panels (PVs) and electric vehicles. The SE at the distribution level has been considered as more challenging because of the limited measurements redundancy and the imbalance at the distribution level requiring to consider the three phases separately [66]. This research could also enhance the control of the distribution systems which are vulnerable to cyber attacks [67, 68].

Finally, control is evolving with the development of wide area measurement system (WAMS) technologies [27, 69]. The authors in [70] have shown the vulnerability of the automatic generation control (AGC) module at the EMS toward cyber attacks. The AGC provides automatic frequency regulation of the power system while insuring that the scheduled power exchanges between adjacent power areas and utilities are met. In [37], the authors studied the effect of cyber attacks spoofing the GPS clocks of PMUs. They proposed algorithmic solutions to secure the damping of inter-area oscillation modes in the WAMs that will be deployed to control the future grid. The classical (N-1) secure operation of the grid criteria is not sufficient in the context of cyber-intrusions. Since the consequences of cyber attacks are tremendous on the grid, several layers of defense measures should provide resistance against the effects of cyber attacks. If a cyber attack is missed by the sophisticated data-analytic tools or a cyber attack is introduced directly in the control orders sent to controllers, relays, tap changers or Industrial Control Systems (ICS) at Remote Terminal Units (RTUs), the power system should be able to mitigate or limit the bad consequences using resilient and robust control.

5   Conclusions

The state estimation problem has significantly raised the concerns in the last decade about its vulnerability and security toward cyber attacks. The importance of state estimation is significant in the operation of the smart grid where it can be exploited not only in creating vulnerabilities and intrusions but also in implementing security measures. In this chapter, the critical current and future research applied to improve the state estimation safety against threats from the cyberspace has been discussed. Improving the cyber security of the state estimation combines multilayer defense systems. Novel robust signal processing and data-analytic methods could be very effective especially with the presence of synchrophasor measurements. Assessing the cyber security of the grid by evaluating the impact of undetected attacks is crucial. Proposing techniques for resilient control that limits and mitigates the impact of undetected attacks would complement the detection of attacks to ensure a secure and efficient operation of the grid. In general, the discussed research could be adapted to many engineering fields where industrial control systems are implemented. Hence, these directions should be explored further for the enhancement of the cyber security for the safety and well-being of the society.

Acknowledgements This work was supported by Japan Science and Technology Agency under the CREST Program, Grant No. JPMJCR15K3.

References

1. Terrorism and the electric power delivery system. Technical Report (National Academy of Engineering Press, U.S., 2012)
2. S.K. Khaitan, J.D. McCalley, C.C. Liu, *Cyber Physical Systems Approach to Smart Electric Power Grid* (Springer, Heidelberg, 2015)
3. IR-ALERT-H-16-056-01-Cyber-attack against Ukrainian critical infrastructure. Technical Report Industrial Control Systems Cyber Emergency Response Team, https://ics-cert.us-cert.gov/alerts/IR-ALERT-H-16-056-01, Feb 2016
4. Strengthening America's energy security with offshore wind. Technical Report (U.S. Department of Energy (DoE)), http://www.nrel.gov/docs/fy12osti/49222.pdf, Apr 2012
5. T.G. Lewis, *Critical Infrastructure Protection in Homeland Security: Defending a Networked Nation* (Wiley, New Jersey, second edition, 2015)
6. M. Govindarasu, P.W. Bauer (eds.), Special section on keeping the smart grid safe. IEEE Power Energy Mag. **10**(1) (2012)
7. Y. Mo, T.H.-H. Kim, K. Brancik, D. Dickinson, H. Lee, A. Perrig, B. Sinopoli, Cyber-physical security of a smart grid infrastructure. Proc. IEEE **100**(1), 195–209 (2012)
8. A. Abur, A. Gomez-Exposito, *Power System State Estimation: Theory and Implementation* (CRC Press, New York, 2004)
9. A. Monticelli, Electric power system state estimation. Proc. IEEE **88**(2), 262–282 (2000)
10. Y. Liu, M.K. Reiter, P. Ning, False data injection attacks against state estimation in electric power grids, in *Proceedings of 16th ACM Conference on Computer and Communications Security* (2009), pp. 21–32
11. Y. Liu, P. Ning, M.K. Reiter, False data injection attacks against state estimation in electric power grids. ACM Trans. Inf. Syst. Secur. **14**(1), 1–33 (2011)
12. A. Teixeira, S. Amin, H. Sandberg, K.H. Johansson, S.S. Sastry, Cyber security analysis of state estimators in electric power systems, in *Proceedings of 49th IEEE Conference on Decision and Control (CDC)* (2010), pp. 5991–5998
13. G. Hug, J.A. Giampapa, Vulnerability assessment of AC state estimation with respect to false data injection cyber-attacks. IEEE Trans. Smart Grid **3**(3), 1362–1370 (2012)
14. L. Jia, J. Kim, R.J. Thomas, L. Tong, Impact of data quality on real-time locational marginal price. IEEE Trans. Power Syst. **29**(2), 627–636 (2014)
15. W. Wang, Z. Lu, Cyber security in the smart grid: survey and challenges. Comput. Netw. **57**(5), 1344–1371 (2013)
16. R.A. Maronna, R.D. Martin, V.J. Yohai, *Robust Statistics: Theory and Methods* (Wiley Series in Probability and Statistics. Wiley, Chichester, 2006)
17. O. Kosut, L. Jia, R.J. Thomas, L. Tong, Malicious data attacks on the smart grid. IEEE Trans. Smart Grid **2**(4), 645–658 (2011)
18. J. Kim, L. Tong, On topology attack of a smart grid: undetectable attacks and countermeasures. IEEE J. Sel. Areas Commun. **31**(7), 1294–1305 (2013)
19. A. Ashok, M. Govindarasu, Cyber attacks on power system state estimation through topology errors, in *IEEE Power Energy Society General Meeting* (2012), pp. 1–8
20. Y. Chakhchoukh, H. Ishii, Coordinated cyber-attacks on the measurement function in hybrid state estimation. IEEE Trans. Power Syst. **30**(5), 2487–2497 (2015)
21. Y. Chakhchoukh, H. Ishii, Cyber attacks scenarios on the measurement function of power state estimation, in *Proceedings of American Control Conference (ACC)*, June 2015, pp. 3676–3681
22. M.G. Cheniae, L. Mili, P.J. Rousseeuw, Identification of multiple interacting bad data via power system decomposition. IEEE Trans. Power Syst. **11**(3), 1555–1563 (1996)
23. L. Mili, M.G. Cheniae, P.J. Rousseeuw, Robust state estimation of electric power systems. IEEE Trans. Circuits and Syst. I Fundam. Theory Appl. **41**(5), 349–358 (1994)
24. Y. Chakhchoukh, H. Ishii, Robust estimation for enhancing the cyber security of power state estimation, in *IEEE PES General Meeting*, July 2015, pp. 1–6

25. A.M. Zoubir, V. Koivunen, Y. Chakhchoukh, M. Muma, Robust estimation in signal processing: a tutorial-style treatment of fundamental concepts. IEEE Signal Process. Mag. **29**(4), 61–80 (2012)
26. L. Mili, C.W. Coakley, Robust estimation in structured linear regression. Ann. Statist. **24**(6), 2593–2607 (1996)
27. A. Chakrabortty, P.P. Khargonekar, Introduction to wide-area control of power systems, in *Proceedings of American Control Conference*, June 2013, pp. 6758–6770
28. F. Aminifar, M. Fotuhi-Firuzabad, A. Safdarian, A. Davoudi, M. Shahidehpour, Synchrophasor measurement technology in power systems: panorama and state-of-the-art. IEEE Access **2**, 1607–1628 (2014)
29. M. Kezunovic, S. Meliopoulos, V. Venkatasubramanian, V. Vittal, *Application of Time-Synchronized Measurements in Power System Transmission Networks* (Springer, Heidelberg, 2014)
30. A.G. Phadke, J.S. Thorp, *Synchronized Phasor Measurements and Their Applications, Power Electronics and Power Systems*, 2nd edn. (Springer, New York, 2008)
31. Map of PMUs in North America. Technical Report (North American Synchrophasor Initiative, March 2015)
32. B. Xu, A. Abur, Optimal placement of phasor measurement units for state estimation. Technical Report (Power Systems Engineering Research Center (PSERC), 2005)
33. Q. Zhang, Y. Chakhchoukh, V. Vittal, G.T. Heydt, N. Logic, S. Sturgill, Impact of PMU measurement buffer length on state estimation and its optimization. IEEE Trans. Power Syst. **28**(2), 1657–1665 (2013)
34. M. Göl, A. Abur, A hybrid state estimator for systems with limited number of PMUs. IEEE Trans. Power Syst. **30**(3), 1511–1517 (2015)
35. A. Rouhani, A. Abur, Linear phasor estimator assisted dynamic state estimation. IEEE Trans. Smart Grid **9**(1), 211–219 (2018)
36. V. Murugesan, Y. Chakhchoukh, V. Vittal, G.T. Heydt, N. Logic, S. Sturgill, PMU data buffering for power system state estimators. IEEE Power Energy Technol. Syst. J. **2**(3), 94–102 (2015)
37. Y. Wang, A. Chakrabortty, Distributed monitoring of wide-area oscillations in the presence of GPS spoofing attacks, in *Proceedings of IEEE Power and Energy Society General Meeting (PESGM)*, July 2016, pp. 1–5
38. Y. Chakhchoukh, V. Vittal, G.T. Heydt, H. Ishii, LTS-based robust hybrid SE integrating correlation. IEEE Trans. Power Syst. **32**(4), 3127–3135 (2017)
39. Y. Chakhchoukh, H. Ishii, Enhancing robustness to cyber-attacks in power systems through multiple least trimmed squares state estimations. IEEE Trans. Power Syst. **31**(6), 4395–4405 (2016)
40. Y. Chakhchoukh, V. Vittal, G.T. Heydt, PMU based state estimation by integrating correlation. IEEE Trans. Power Syst. **29**(2), 617–626 (2014)
41. M. Wu, L. Xie, Online identification of bad synchrophasor measurements via spatio-temporal correlations, in *Proceedings of Power Systems Computation Conference (PSCC)*, June 2016, pp. 1–7
42. M. Glavic, T. Van Cutsem, Reconstructing and tracking network state from a limited number of synchrophasor measurements. IEEE Trans. Power Syst. **28**(2), 1921–1929 (2013)
43. E. Ghahremani, I. Kamwa, Dynamic state estimation in power system by applying the extended Kalman filter with unknown inputs to phasor measurements. IEEE Trans. on Power Syst. **26**(4), 2556–2566 (2011)
44. S. Wang, W. Gao, A.P.S. Meliopoulos, An alternative method for power system dynamic state estimation based on unscented transform. IEEE Trans. Power Syst. **27**(2), 942–950 (2012)
45. A.K. Singh, B.C. Pal, Decentralized dynamic state estimation in power systems using unscented transformation. IEEE Transa. Power Syst. **29**(2), 794–804 (2014)
46. J. Zhao, M. Netto, L. Mili, A robust iterated extended Kalman filter for power system dynamic state estimation. IEEE Trans. Power Syst. **32**(4), 3205–3216 (2017)
47. O. Kosut, Malicious data attacks against dynamic state estimation in the presence of random noise, in *Proceedings of IEEE Global Conference on Signal and Information Processing*, Dec 2013, pp. 261–264

48. A. Teixeira, I. Shames, H. Sandberg, K.H. Johansson, A secure control framework for resource-limited adversaries. Automatica **51**, 135–148 (2015)
49. J. Kim, L. Tong, On phasor measurement unit placement against state and topology attacks, in *Proceedings of IEEE International Conference on Smart Grid Communications (SmartGrid-Comm)*, Oct 2013, pp. 396–401
50. Y. Lin, A. Abur, Strategic use of synchronized phasor measurements to improve network parameter error detection. IEEE Trans. Smart Grid **9**(5), 5281–5290 (2018)
51. A. Ashok, M. Govindarasu, V. Ajjarapu, Online detection of stealthy false data injection attacks in power system state estimation. IEEE Trans. Smart Grid **9**(3), 1636–1646 (2018)
52. A. Teixeira, K.C. Sou, H. Sandberg, K.H. Johansson, Secure control systems: a quantitative risk management approach. IEEE Control Syst. **35**(1), 24–45 (2015)
53. C.M. Davis, J.E. Tate, H. Okhravi, C. Grier, T.J. Overbye, D. Nicol, SCADA cyber security testbed development, in *Proceedings of 38th North American Power Symposium*, Sept 2006, pp. 483–488
54. A. Hahn, A. Ashok, S. Sridhar, M. Govindarasu, Cyber-physical security testbeds: Architecture, application, and evaluation for smart grid. IEEE Trans. Smart Grid **4**(2), 847–855 (2013)
55. T. Yardley, R. Berthier, D. Nicol, W.H. Sanders, Smart grid protocol testing through cyber-physical testbeds, in *Proceedings of IEEE PES Innovative Smart Grid Technologies Conference (ISGT)*, Feb 2013, pp. 1–6
56. V. Venkataramanan, A. Srivastava, A. Hahn, Real-time co-simulation testbed for microgrid cyber-physical analysis, in *Proceedings of Workshop on Modeling and Simulation of Cyber-Physical Energy Systems (MSCPES)*, April 2016, pp. 1–6
57. M. Esmalifalak, N.T. Nguyen, R. Zheng, Z. Han, Detecting stealthy false data injection using machine learning in smart grid, in *Proceedings of IEEE Global Communications Conference (GLOBECOM)*, Dec 2013, pp. 808–813
58. M. Esmalifalak, L. Liu, N. Nguyen, R. Zheng, Z. Han, Detecting stealthy false data injection using machine learning in smart grid. IEEE Syst. J. **11**(3), 1644–1652 (2017)
59. Y. Chakhchoukh, S. Liu, M. Sugiyama, H. Ishii, Statistical outlier detection for diagnosis of cyber attacks in power state estimation, in *Proceedings of IEEE PES General Meeting*, July 2016, pp. 1–5
60. M. Sugiyama, T. Suzuki, T. Kanamori, *Density Ratio Estimation in Machine Learning* (Cambridge University Press, 2012)
61. L. Xie, D.H. Choi, S. Kar, H.V. Poor, Fully distributed state estimation for wide-area monitoring systems. IEEE Trans. Smart Grid **3**(3), 1154–1169 (2012)
62. W. Jiang, V. Vittal, G.T. Heydt, A distributed state estimator utilizing synchronized phasor measurements. IEEE Trans. Power Syst. **22**(2), 563–571 (2007)
63. V. Kekatos, G.B. Giannakis, Distributed robust power system state estimation. IEEE Trans. Power Syst. **28**(2), 1617–1626 (2013)
64. M.A. Rahman, E. Al-Shaer, R.B. Bobba, Moving target defense for hardening the security of the power system state estimation, in *Proceedings of the First ACM Workshop on Moving Target Defense*, Nov 2014, pp. 59–68
65. Y. Yao, Z. Li, MTD-inspired state estimation based on random measurements selection, in *North American Power Symposium (NAPS)*, Sept 2016, pp. 1–6
66. R. Singh, B.C. Pal, R.A. Jabr, Choice of estimator for distribution system state estimation. IET Gener. Trans. Distrib. **3**(7), 666–678 (2009)
67. Y. Isozaki, S. Yoshizawa, Y. Fujimoto, H. Ishii, I. Ono, T. Onoda, Y. Hayashi, Detection of cyber attacks against voltage control in distribution power grids with PVs. IEEE Trans. Smart Grid **7**(4), 1824–1835 (2016)
68. A. Teixeira, G. Dan, H. Sandberg, R. Berthier, R.B. Bobba, A. Valdes, Security of smart distribution grids: Data integrity attacks on integrated volt/var control and countermeasures, in *Proceedings of American Control Conference*, June 2014, pp. 4372–4378
69. A. Chakrabortty, Co-designing communication and control systems for wide-area control of power systems, in *Proceedings of American Control Conference (ACC)*, July 2016, pp. 2667–2667
70. S. Sridhar, M. Govindarasu, Model-based attack detection and mitigation for automatic generation control. IEEE Trans. Smart Grid **5**(2), 580–591 (2014)

Challenges and Opportunities: Cyber-Physical Security in the Smart Grid

Sean Weerakkody and Bruno Sinopoli

Abstract In this chapter, we develop a vision to address challenges in securing the smart grid. Despite recent innovations, grid security remains a critical issue. The infrastructure is highly vulnerable due to its large scale, connectivity, and heterogeneity. Moreover, attacks on cyber-physical systems and the grid have been realized, most notably the attack on the Ukraine power system in 2015. While techniques in cyber security are useful, their implementation is not sufficient to secure the smart grid. Consequently, we advocate for research in cyber-physical system security, an interdisciplinary field which combines tools from both cyber security and system theory. Within this field, we argue that engineers need to develop a framework of accountability comprised of three main research thrusts: (1) the detection of attacks, (2) the attribution of attacks to particular malicious components and devices on the grid, and (3) the resilient design of systems and algorithms to ensure acceptable performance in the presence of malicious behavior. To close, we discuss the need for a unifying language and set of tools to address these problems, as we consider additional research in compositional security.

1 Introduction

Securing the smart grid, one of society's most crucial resources, is a significant problem. There exists ample motivation for attackers to target the smart grid. Economically, an adversary can tamper with smart meters to reduce bills or compromise sensors to elicit a profit in the electricity market. Attackers may also perturb the grid for more nefarious reasons including terrorism. Unfortunately, securing the grid is an open challenge [11, 14, 17, 19]. The electricity grid, due to developments in sensing and communication technologies, is becoming increasingly connected.

S. Weerakkody (✉) · B. Sinopoli
Department of Electrical and Computer Engineering, Carnegie Mellon University,
Pittsburgh, PA 15213, USA
e-mail: sweerakk@andrew.cmu.edu

B. Sinopoli
e-mail: brunos@ece.cmu.edu

© Springer Nature Switzerland AG 2019
J. Stoustrup et al. (eds.), *Smart Grid Control*, Power Electronics
and Power Systems, https://doi.org/10.1007/978-3-319-98310-3_16

257

Moreover, many heterogeneous components are being installed. The infrastructure is highly complex, connecting generation, transmission, and distribution subsystems that are often managed by multiple distinct parties. The scale of this infrastructure creates additional challenges as the system's size makes it impractical to guarantee physical security to any significant fraction of the grid. A decentralized, diverse, and connected system provides numerous attack surfaces for adversaries. Finally, there exists precedence for attacks on the grid, most notably the attack in Ukraine [25]. In this chapter, we argue that while existing tools in cyber security are necessary, these methods alone are insufficient because they do not account for the underlying dynamics of the grid. First, cyber security does not account for the impact of physical attacks. For example, the secrecy of encrypted sensor data can be compromised by placing an unencrypted sensor beside a legitimate sensor. Cyber security also fails to provide prescriptive tools to ensure resilient performance during an attack. Special system theoretic insight is required to ensure graceful degradation of performance under attack.

To address these challenges, we recommend the development of cyber-physical system security. The smart grid is a cyber-physical system (CPS), which is the embedding of communication, sensing, and computing technologies into physical systems to improve efficiency and reliability. Other CPSs include transportation networks, water distribution systems, smart buildings, and waste management systems. A science of CPS security will address the smart grid's role as a combination of a safety critical physical system and a complex information technology (IT) infrastructure. We expect that security researchers, with the aid of the controls community, will develop tools that ensure properties of secrecy, integrity, and availability while accounting for an attacker that can impact the grid's physical dynamics. It is likely impractical to remove all attack surfaces in the smart grid. We recommend that researchers instead focus on developing methods to respond to attacks once they occur. Specifically, we envision research developed around a framework of accountability comprised of three thrusts: (1) the detection of attacks, (2) the attribution of malicious behavior to specific entities, and (3) resilient system design to ensure graceful degradation of the grid when under attack.

The rest of the chapter is formulated as follows. In Sect. 2, we review case studies in CPS security to motivate research and help identify weaknesses in the smart grid. Next, in Sect. 3, we characterize potential malicious behavior on the grid, summarizing entry points and possible attack strategies. Then, in Sect. 4, we call for the development of CPS security to defend the smart grid. Here, we highlight weaknesses of traditional approaches and summarize the benefits CPS security can provide. Afterward, in Sect. 5, we develop a set of research goals centered around creating a framework of accountability. Finally, in Sect. 6, we bring attention to the burgeoning problem of compositional security and call for efforts to bridge the gaps that exist in CPS security.

2   Case Studies

We next review two case studies. The first, Stuxnet, while not an attack on the grid, did target supervisory control and data acquisition (SCADA) architectures. The second case study is the hack on the Ukraine power system.

2.1   Stuxnet

The Stuxnet worm was a malware that attacked uranium enrichment facilities in Iran, damaging 1,000 centrifuges [7]. It mainly spread through USB sticks and local networks. Stuxnet was designed to remain dormant unless it detected specific config-urations and model numbers [13]. Stuxnet used four unknown or zero-day exploits against Microsoft Windows. Two allowed the malware to spread across networks using USBs and shared printers. An additional two exploits elevated privileges of Stuxnet to execute code. Moreover, Stuxnet leveraged two stolen certificates [3] to install rootkit drivers. It also infected Siemens Programmable Logic Controllers (PLCs) used to control centrifuges at the plant. For stealthiness, Stuxnet used the first PLC rootkit.

As a result, Stuxnet compromised centrifuges by increasing the gas pressure of uranium hexafluoride and varying rotational speeds. While rotor speed was not a measured variable, gas pressure was likely monitored. In order to avoid detection, Stuxnet used a replay attack, sending prior measurements to the SCADA system. By avoiding detection, Stuxnet damaged centrifuges for long periods of time without defender interference. The worm demonstrated that mechanisms for cyber security can fail, here due to zero-day exploits and stolen certificates. A layer of physical secu-rity to actively detect a replay attack likely would have curtailed much of Stuxnet's impact. This will be discussed in greater detail when we examine attack detection.

2.2   Ukraine Power Grid Attack

We now examine the December 2015 attack on the Ukraine power grid [1, 25]. The hack caused blackouts over several hours, affecting thousands of customers. The attack was initiated months prior to the outage when the BlackEnergy3 malware was delivered to operators and employees through infected Microsoft Word documents via phishing emails. The BlackEnergy3 malware allowed hackers to establish a con-nection with the command and control server, harvest valid credentials, and perform reconnaissance. The adversaries likely used valid credentials to deliver the KillDisk malware and schedule uninterrupted power supply (UPS) outages. On December 23, 2015, the adversaries carried out an attack on the grid. Using remote access, the adver-saries hacked workstations and tripped breakers. Malicious firmware was delivered to converters, preventing operators from remotely operating breakers. Meanwhile, a telephone denial of service was carried out, cutting off communication between customers and providers. The UPS outages impeded backup power at control centers while the KillDisk malware destroyed data.

The Ukraine power attack provides ample lessons for researchers. It highlighted a need for operators to implement better security policies and educate employees. Network access may have been prevented if employees were better trained to recognize phishing emails. Additionally, the ability for employees to remotely access the system was a contributing factor to the attacker's success, allowing adversaries to carry out actions remotely after obtaining valid credentials. Restricting remote access and requiring two- factor authentication may have thwarted the adversary. A cyber-physical approach to security can improve resilience. Here, better automatic robust control mechanisms, implemented at the grid, might have prevented widespread blackouts.

3 Adversarial Models

To establish a vision for securing the grid, we must identify the threats that the system faces through an adversarial model. To do this, we first identify attack surfaces. Later, we investigate potential strategies attackers may carry out to obtain benefits or cause damage. The resulting attack model will help to inform cyber-physical strategies for providing security.

3.1 Attack Surfaces

An adversary can access the grid or SCADA systems by leveraging entry points. For example, an attacker can target improperly configured or nonexistent firewalls. Alternatively, attackers can leverage weaknesses in existing SCADA protocols such as DNP3 and Modbus. The Ukraine hack illustrated how VPN connections allow attackers to enter a network. An easy way for an adversary to enter the network is to target smaller heterogeneous devices. As an example, Stuxnet was able to spread across networks due to shared printers. In both Stuxnet and the Ukraine hack, employee actions likely allowed malware to enter the system. In addition to USB devices (Stuxnet) and phishing emails (Ukraine), laptops can be used to breach the system, especially if these devices are transferred in and out of the network. Attacks can also occur at the supply chain. In particular, if production is not performed securely, adversaries can secretly install backdoors in devices. An adversary may also physically access components on the grid. Unfortunately, due to the grid's size, it is impractical to guarantee the physical security of smart meters or even PMUs. Furthermore, substations are often unattended and are only monitored remotely. The actions of malicious insiders also cannot be underestimated. Malicious insiders can use grid knowledge and system access to attack the infrastructure. A notable example of this is the Maroochy Shire incident [31], where a disgruntled former employee hacked a SCADA system performing waste management. Thwarting malicious insiders is challenging. Thus, care should be given when granting permissions to employees.

3.2 Attack Strategies

Next, we consider the strategies an adversary chooses to violate properties of secrecy, integrity, and availability. Notably, we discuss the physical impact of an adversary's behavior, motivating the study of CPS security.

Secrecy Violations: Violating secrecy has diverse consequences in a power system. The privacy of smart meter data is essential as it can reveal sensitive information about consumers [26]. This data, however, can be used benignly to help operators forecast demand and increase efficiency. Fortunately, the confidentiality of meter data should not adversely affect the performance of a power system. As such, addressing this issue is out of scope in this work. Alternatively, an attacker can try to learn sensitive information about the grid infrastructure including its SCADA monitoring system. It is widely speculated that the hackers of the Ukraine power grid leveraged their stolen credentials and access to HMI workstations to learn more about the SCADA system and grid infrastructure, allowing them to maximize the impact of their attack. In Sect. 4.1, we will discuss mechanisms in cyber security that can protect sensitive information. As these countermeasures are not full-proof, we will detail recommendations for controls researchers in Sect. 5.1.

Integrity Violations: Integrity attacks, where information is modified, can significantly impact the smart grid. Individual consumers may perform isolated attacks to modify power consumption measurements in smart meters. This energy theft at a small scale should not affect the health of the infrastructure. Nonetheless, the integrity of other information can be safety critical. For instance, an adversary who modifies the price being sent by an independent system operator (ISO) can cause fluctuating demand and instability [34]. Here, low electricity prices can lead to heavy demand, that cannot be met with matching generation. This can lead to possible blackouts. Attackers can also violate the integrity of sensor measurements [2, 10, 12, 14, 27]. These attacks can be realized by targeting communication protocols. Alternatively, sensors may also be altered by attackers with access to the device.

Example **False Data Injection Attacks**—We consider the following model of transmission in a power system

$$y_k = H\theta_k + e_k + a_k. \tag{1}$$

Here, y_k consists of power flow measurements at each line and power injection measurements at each bus at time k. Also, θ_k consists of bus voltage angles, $e_k \sim \mathcal{N}(0, \Sigma_e)$ is Gaussian sensor noise, and a_k is the attacker's input. An estimator to obtain $\hat{\theta}_k$ and a residue detector can be constructed as follows:

$$\hat{\theta}_k = (H^T \Sigma_e^{-1} H)^{-1} H^T \Sigma_e^{-1} y_k, \quad z_k = y_k - H\hat{\theta}_k, \quad \|z_k\| \overset{\mathcal{H}_1}{\underset{\mathcal{H}_0}{\gtrless}} \tau. \qquad (2)$$

$\mathcal{H}_0$ is the hypothesis that the system is operating normally while $\mathcal{H}_1$ is the hypothesis that there is faulty or malicious behavior. If $a_k = Hc_k$, the attacker biases the state estimate by c_k, without changing the detection statistic. This attack can yield a profit in the market [46] and more importantly, cause SCADA operators to take incorrect actions, possibly leading to instability.

The integrity of control commands must be protected. In the Ukraine attack, the adversaries modified commands to trip breakers and schedule UPS outages. Similarly, attacks on software can have a profound effect on a CPS, potentially allowing attackers to manipulate any grid device. Stuxnet was able to modify software affecting PLCs, leading to centrifuge damage. Traditional countermeasures for detecting integrity attacks are considered in Sect. 4.1. As these methods are insufficient, we detail recommendations to detect and isolate integrity attacks in Sects. 5.1 and 5.2.

Availability Violations: Adversaries who carry out denial of service (DoS) attacks on the smart grid affect availability. While real-time data availability is not typically crucial in software systems, the real-time availability of information is often essential in CPSs due to the physical dynamics. The ability to deliver control inputs is often necessary to stabilize a CPS and achieve objectives. Likewise, sensor measurements on the grid are necessary to inform proper feedback control actions. Additionally, an absence of pricing information can affect the market and lead to undesirable changes in electricity demand. References to work in anti-jamming are provided in Sect. 4.1. As these anti-jamming tools are not prescriptive for ensuring grid performance, recommendations for resiliency are provided in Sect. 5.3.

4 A Case for Cyber-Physical System Security

Responding to the attacks will require multidisciplinary contributions from both security researchers as well as system scientists. In this section, we will describe existing tools in cyber security and then show that these existing mechanisms are insufficient for addressing smart grid security, necessitating our vision of CPS security.

4.1 Cyber Security Countermeasures

In this subsection, we summarize existing tools in cyber security, identifying strengths and weaknesses to inform future research.

Limiting Attack Surfaces: Well- configured firewalls can serve as a perimeter around SCADA systems and prevent access. Stringent policies can also limit an adversary's attack surface. For example, operators can prohibit USB sticks and laptops from being carried in and out of the SCADA system. The Ukraine attack would have been prevented if employees were trained to spot phishing emails or if two-factor authentication was used during VPN connections.

Secrecy, Integrity, and Availability: To help guarantee secrecy, operators can encrypt sensitive information. More generally, prior work in cyber security has examined mechanisms to stop sensitive information from leaking. Defenders can leverage information flow analysis to promote confidentiality [5]. Information flow analysis is a set of tools in software security designed to prevent illicit flows of information in computing systems. Various tools have been considered to prevent invalid information flows [37] and to quantify information leakage when it does occur [32]. Authentication protocols can be used to verify the identity of different devices, components, and operators and guarantee message integrity. Authenticated encryption, in particular, can provide message secrecy and detect integrity attacks. This ideally will enable secure communication for instance between sensors and the SCADA system, power distributors and smart meters, and equipment and field crews. The root of trust in such a system are cryptographic keys. Due to the sheer size of the grid, developing a key management system, which allows for key generation, sharing, replacement, and recovery is a nontrivial challenge [45]. Detecting software modification is a difficult problem. Typical antivirus software are ineffective against zero-day exploits and may be corrupted. Remote attestation can be a promising method to detect integrity attacks on software [29]. Finally, previous work has examined means to ensure the availability of data. These include preventing DoS and jamming attacks. We refer the reader to a survey of results found in [22, 24].

4.2 Toward Cyber-Physical System Security

Unfortunately, the tools available in cyber security are insufficient for the smart grid. First, the scale, connectivity, and heterogeneity of the grid make it impossible to remove all attack surfaces. Even established barriers such as firewalls can be broken. Next, the listed tools cannot account for physical attacks where an adversary physically interacts with the grid. For instance, to violate the secrecy of an encrypted sensor, a second unencrypted sensor can be placed in close proximity. Additionally, the integrity of a meter can be violated by placing a shunt, allowing electricity to bypass this device. Similarly, firewalls and authenticated encryption cannot prevent an attacker with physical access from modifying control commands. Availability can be compromised by physically shielding sensors and actuators. Such an attack cannot be mitigated by standard anti-jamming technologies. Cyber security also fails to provide prescriptive tools to deal with the physical layer of the smart grid as it fails to account for properties of stability and performance. For instance, upon detecting that a software system is compromised, one common solution is a system reboot. This, however, can be dangerous due to the inertia and dynamics of the grid. Instead, in a CPS, actions should be carefully taken to ensure graceful degradation.

We propose the development of cyber-physical system security which augments cyber security with tools from system theory. We recommend that researchers leverage a fundamental understanding of the dynamics of power systems to ensure security. A stochastic model of the physical dynamics on the grid can serve as a ground truth, allowing defenders to recognize and isolate anomalies by analyzing sensor measurements. Control inputs and other degrees of freedom can be used as active monitors, allowing the defender to differentiate observed behavior under attack and expected behavior under normal operation. This will be discussed further when we describe active detection. Finally, resilient algorithms from control theory can guide operators in performing corrective actions, which ensure the grid meets key performance objectives.

We stress that CPS security should consist of tools from both system theory and cyber security. Like cyber security, system theory on its own can be ineffective. One shortcoming is that detection is always probabilistic due to stochastic system modeling. Meanwhile, an attacker who breaks cryptographic primitives without key access will almost certainly be caught. Methods from system theory also often introduce high-level abstractions, which can be better captured via cyber security. A layered approach that considers cyber security and system theory will be imperative in securing a CPS.

5   Accountability in Cyber-Physical Systems

We now present a vision to guide progress in CPS security research. It is likely impossible to completely prevent attacks on the smart grid, first due to the grid's immense size and diversity and second due to weaknesses in the implementation of standard tools in cyber security. Consequently, systems researchers should focus on resiliently responding to attacks when they occur. We recommend developing a framework for accountability in CPSs [4], centered around responding to attacks, and consisting of three research thrusts:

1. The detection of attacks on the grid.
2. The attribution of attacks to malicious components.
3. Resilient design to ensure graceful degradation under attack.

We now delve deeper into these research areas. For generality, we consider the model of a generic control system.

$$x_{k+1} = Ax_k + Bu_k + w_k, \quad y_k = Cx_k + v_k. \tag{3}$$

Here, $x_k \in \mathbb{R}^n$ is the state at time k, which in power systems can correspond to bus voltage angles and magnitudes. Next, $u_k \in \mathbb{R}^p$ is a set of control inputs. Moreover, $y_k \in \mathbb{R}^m$ is a vector of measurements , for instance real power flows at transmission lines and power injections at each bus. Finally, $w_k \in \mathbb{R}^n$ and $v_k \in \mathbb{R}^m$ are process and sensor noise.

5.1 Detection

The first step to responding to an attack is detection. Both passive detection and active detection require investigation. *Passive detection* is essentially equivalent to standard detection theory [8, 44], where the defender attempts to determine whether the system is healthy or operating abnormally. One classical detector is the χ^2 Detector.

Example χ^2 **Detector**—One standard detector is a χ^2 detector. Given measurements y_k, and information $\mathcal{I}_k$, a χ^2 detector is given as follows

$$z_k \triangleq y_k - \mathbb{E}[y_k|\mathcal{I}_k], \quad \sum_{t=k-T+1}^{k} z_k^T (\mathrm{Cov}(z_k))^{-1} z_k \underset{\mathcal{H}_0}{\overset{\mathcal{H}_1}{\gtrless}} \tau. \tag{4}$$

Here, the residue z_k characterizes the difference between measurements y_k and their expected values. As mentioned, the null hypothesis $\mathcal{H}_0$ refers to normal operation, while the alternative hypothesis $\mathcal{H}_1$ refers to an attack.

The changing composition of the grid creates new opportunities to improve detection. PMUs, which enables faster, synchronized sensing, are changing detection from a static problem to a dynamic one. Dynamic detection creates challenges for adversaries, who must ensure their attacks generate outputs compatible with dynamical models. Data gathered from smart meters can also help operators model demand. Detectors should address tradeoffs between the probability of detection, probability of false alarm, and time to detection. Detection is time critical, especially if an unchecked adversary can disrupt service. To detect attackers quickly, the defender might have to sacrifice average detection performance. In the χ^2 detector, the threshold τ determines the tradeoff between the probability of false alarm and the probability of detection while the window size T balances the tradeoff between the probability of detection and detection time.

Passive detection can be ineffective against knowledgeable adversaries. Model-aware adversaries with channel access can potentially perform stealthy covert attacks [33], zero dynamics attacks [23, 35], false data injection attacks [18], and replay attacks [20]. Unfortunately, tactics in cyber security such as authenticated encryption are ineffective against physical attacks or might be broken (as in the Ukraine

Fig. 1 Active detection in CPS. To detect stealthy attacks a defender changes his control policy or system their parameters

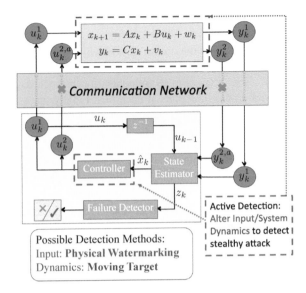

attack). The threat of stealthy attacks motivates research in *active detection*, where the defender alters aspects of the CPS to recognize adversarial behavior. Figure 1 illustrates active detection.

For instance, under certain scenarios, an attack is stealthy only if a defender uses a particular set of control policies. In this case, a defender can actively detect an adversary by changing his or her control strategy.

Example **Physical Watermarking**—One example of active detection is physical watermarking [21, 28, 39]. Here, the defender inserts a secret noisy control input Δu_k or watermark on top of the optimal input u_k^*

$$u_k = u_k^* + \Delta u_k. \tag{5}$$

The measurements y_k are ideally correlated to the physical watermark through the dynamics of the system. The absence of a watermark in the sensor measurements is indicative of faulty or malicious behavior. Physical watermarking is effective against replay attacks, where an adversary sends a sequence of prior outputs to the defender, the same strategy used in Stuxnet. The watermarks act like a cyber-physical nonce, verifying the freshness of the measurements being delivered to the operator. Figure 2 illustrates physical watermarking.

We recommend that researchers examine using control inputs as active monitors. Scientists should identify realistic attack strategies, whose stealthiness is linked to control policies on the grid. Alternative control policies can then be implemented to reveal these otherwise stealthy attacks. Detailed studies must be performed to address tradeoffs between control performance and security. The defender must have

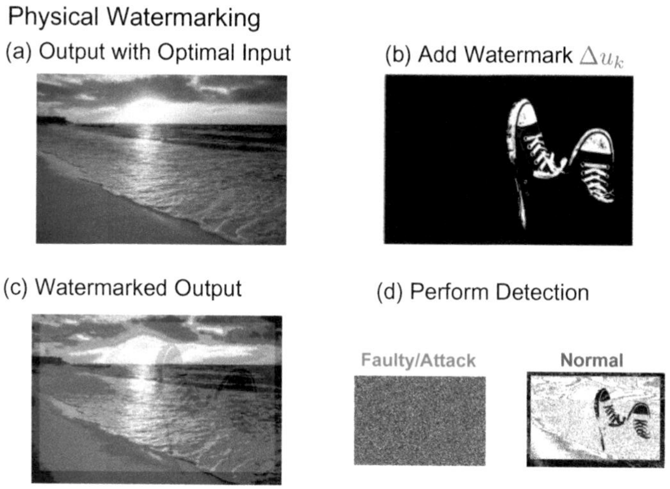

Physical Watermarking

(a) Output with Optimal Input

(b) Add Watermark $\triangle u_k$

(c) Watermarked Output

(d) Perform Detection

Faulty/Attack   Normal

Fig. 2 An illustration of physical watermarking. **a** Output with optimal control input, susceptible to replay attack. **b** Add watermark $\triangle u_k$ on top of optimal control input u_k. **c** Output is now correlated with watermark. **d** If we can not detect watermark in output the system is faulty/under attack

computable metrics to evaluate security and system performance. In the case of physical watermarking, the design of the watermark was determined by solving optimization problems that aim to maximize detection performance (as related to the expected detection statistic) subject to constraints on linear quadratic Gaussian costs. This active detection can modify the system's root of trust. For instance, control strategies may need to be private, necessitating encryption of inputs. Control inputs can also be used to preserve secrecy. For instance, a worthwhile task is to determine how changing a control policy can conceal the dynamics on the grid from an attacker [47], who passively observes input/output traffic.

There exist scenarios where active detection at the control input is ineffective, notably when an adversary uses model knowledge to construct stealthy attacks. We then envision a defender performing active detection by changing parameters of the system, in particular the dynamics itself.

Example **The Moving Target Approach**—Another active method of detection is the moving target [40]. Here, the defender introduces an authenticating subsystem on top of the original system. This subsystem is designed so it is causally affected by the normal dynamics of the system, but itself has no impact on the normal dynamics. The authenticating subsystem also has random time-varying dynamics, whose real-time realization is unknown to the attacker. The time-varying dynamics can, for instance, be determined by a cryptographically secure pseudorandom number generator (PRNG) whose seed is available to the defender and forms a root of trust. Existing dynamics or dedicated hardware with time-varying components can be used to realize the moving target. The hardware, for instance, can consist of circuits with controllable potentiometers and time-varying capacitors. These devices should be

Fig. 3 Moving target
approach in a CPS

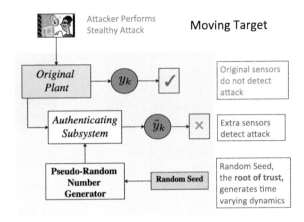

affected by the true state and thus might be affixed to generators, transformers, or power lines. An adversary who disrupts the normal dynamics of a system, may not be revealed by the sensors y_k. However, with proper design, an attack would be revealed by sensors $\tilde{y}_k$, which measure the extra states correlated to the normal dynamics. Moreover, because the dynamics are time varying, the attacker will be unable to use system identification to completely hide his their impact. If the defender does not care about controlling the extra dynamics, the moving target allows the system to remain at peak performance. Figure 3 illustrates the moving target approach.

We recommend that researchers evaluate changing system parameters on the grid to improve detection. To do this, a balance must be struck between maintaining performance and improving security. In the moving target approach, introducing an independent authenticating subsystem alleviates concerns about reducing control performance. However, if meaningful parameters on the grid are altered, a careful evaluation must occur. Interested parties will need to gauge which components on the grid can be modified and then analytically weigh the delicate performance–security tradeoff.

Additionally, practitioners must select a root of trust when changing parameters to inform subsequent design. For instance, if an attacker can not read information being sent across input/output channels, a single set of carefully chosen perturbations may be sufficient to detect attackers [36]. However, if this information is deemed available to attackers, researchers will need to consider the attacker's use of system identification techniques [15]. The defender may have to hide sensitive information about the dynamics by changing his policy or system parameters in a time-varying fashion. Researchers will then need to demonstrate that an attacker cannot glean sensitive information from public channels, perhaps using information theoretic bounds. To close, we argue the composition of detection tools must be considered. As only certain detection strategies are suitable for certain attacks, researchers must consider the problem of systematically combining detection methods in order to adequately account for realistic threats, while maintaining efficiency.

5.2 Attribution

While attack detection is necessary, it does not always prescribe a solution for system recovery. This motivates research in attack identification, which allows operators to attribute malicious behavior to a subset of malfunctioning devices and components and respond accordingly.

Example **Identifying Malicious Sensors and Actuators in Generic Control Systems**—Identifying malicious sensors and actuators in control systems has been well investigated. For instance, in [23], algebraic conditions were provided to determine when a system can identify q malicious sensors and actuators. The problem of identifying malicious sensors is related to resilient estimation. Qualitatively, by isolating a set of malicious sensors, resilient estimation can be performed by only using trusted sensors. It has been shown in deterministic systems that performing perfect state estimation under sensor attacks is equivalent to performing perfect identification [6].

The computational efficiency of attribution requires attention. If only a single failure is allowed (e.g., only one transmission line can fail [43]), the number of failure states to consider is the number of corruptible components. It might still be viable to leverage hypothesis testing to identify corrupt devices in this scenario. In the case of multiple failures, the number of possible sets of malicious nodes is combinatorial and practitioners must weigh the tradeoff between considering a large number of contingencies and efficiency. Relaxations should also be investigated. For instance, when attribution is posed as an optimization problem (as in [6]), we can look for convex relaxations. Satisfiable modulo theory (SMT) solvers [30] may also be promising.

There exist scenarios where intelligent attackers are able to generate unidentifiable attacks [6]. However, generating unidentifiable attacks, in many cases, relies on knowledge of the system parameters [41]. Consequently, we recommend the development of a theory of active identification to be applied in the smart grid. Researchers should prove that proposed active identification techniques accurately attribute malicious behavior to specific entities. The tradeoff between performance and security once again must be examined and operators must carefully select an appropriate root of trust for the CPS.

5.3 Resilient System and Control Design

The ultimate goal in securing the smart grid is to ensure that even when under attack, there exists graceful degradation, and essential services can be provided. The tools of cyber security are ill-equipped to deal with resilient control since they do not consider the underlying physical dynamics. Thus, significant contributions are required from the controls community to ensure that power grids remain functional in the presence of attacks.

After detection and attribution, operators can alter their control strategy to maintain performance. For example, if a subset of measurements is corrupted, resilient estimation on healthy measurements can still allow the grid to remain functional. Alternatively, if measurements are jammed, local robust control loops can be implemented at the plant level until connection is reestablished with SCADA operators. Additionally, if a transmission line fails, optimal power flow can be solved under the system's new configuration.

Because resilience is time sensitive, corrective actions should be automatic and designed offline. Since the possible set of contingencies is immense, risk assessment is needed to prioritize solutions. Engineers must identify the most high-risk failures and design control actions to respond to these failures. Due to a lack of trusted resources under attack, creative solutions may be required. In cases where online corrective action is analytically deemed impossible, researchers and involved parties must weigh the cost and benefits of resilient offline design [38]. As an example, ensuring a system is resilient to a fixed number of sensor attacks may require redundant sensing hardware.

Of course, identifying malicious devices is not instantaneous. Engineers must account for the gap in time between when detection and identification occurs. Robust controllers, which meet minimal performance standards in the presence of large classes of failures, can be used here. An interesting example of such a scenario is described in [16], where after detection, local set-theoretic control is performed until more countermeasures can be deployed. Developing similar robust controllers on the grid is a valuable research goal.

The scalability of the proposed approaches must be examined, due to the size and connectivity of the grid. While solutions can be implemented at the level of control areas for which there is often a designated SCADA system, operators should account for contingencies which require cooperation at a larger scale. Along these lines, efficient architectures must be developed to integrate solutions in detection, attribution, and resilient control.

We conclude by considering why the aforementioned three-step approach is necessary. If an operator can develop robust algorithms for estimation and control, why is detection/attribution necessary? The proposed approach results in better performance. Under normal operation, optimal estimators and controllers will outperform robust estimators and controllers. Upon detection, conservative countermeasures for control are likely similar to alternative robust algorithms. Finally, once attribution has occurred, specially designed countermeasures will outperform general robust techniques.

6   Bridging the Gap: Cyber-Physical System Security

In this chapter, we examined smart grid security, a necessity given the grid's crucial role in society and the precedence for attacks. We discussed standard techniques from cyber security and showed these methods are not sufficient for the security of the grid. We motivated the necessity of CPS security, which combines tools from both cyber security and system theory. We then proposed a vision for securing the grid centered around a framework of accountability consisting of detection, attribution, and resilient system design.

Nevertheless, despite the significant work in CPS security, the field as a whole is disjointed. While intelligent techniques have been proposed in both cyber security and system theory, most of these approaches fail to model the real interactions that exist between computing devices and physical components. A science of compositional security must be developed to address these concerns. Researchers should examine how changing the security properties of individual cyber and physical subsystems affect the security of the entire connected system. Proving security properties for the grid through analysis of the system as a whole is likely impractical. Consequently, a toolbox of analytical methods should be developed to answer when security properties hold after individual, disparate, subsystems are composed.

The lack of progress toward developing a science of compositional security is likely due to the fact that there are few individuals who are well versed in both cyber security and system theory. As such, interdisciplinary approaches will aid in addressing the existing gaps in CPS security. To help this effort, we recommend that a common language be developed to consider problems in both cyber security and physical system security. A unifying framework will allow researchers to better communicate ideas as well as implement tools across disciplines. A relevant direction is information flow analysis which, as mentioned earlier, has significant applications in developing security policies for confidentiality [9]. Recent work, has demonstrated that the tools of information flow can be applied to problems of detection and identification [4, 42] in CPS. Thus, developing a unified theory in terms of information flow could help to bridge the gaps that exist in smart grid security.

References

1. Analysis of the cyber attack on the Ukrainian power grid (2016)
2. R.B. Bobba, K.M. Rogers, Q. Wang, H. Khurana, K. Nahrstedt, T.J. Overbye, Detecting false data injection attacks on DC state estimation, in *First Workshop on Secure Control Systems, CPSWEEK* (2010)
3. T. Chen, Stuxnet, the real start of cyber warfare? [editor's note]. IEEE Netw. **24**(6), 2–3 (2010)
4. A. Datta, S. Kar, B. Sinopoli, S. Weerakkody, Accountability in cyber-physical systems, in *Science of Security for Cyber-Physical Systems Workshop (SOSCYPS)* (IEEE, 2016), pp. 1–3
5. D.E. Denning, A lattice model of secure information flow. Commun. ACM **19**(5), 236–243 (1976)
6. H. Fawzi, P. Tabuada, S. Diggavi, Secure estimation and control for cyber-physical systems under adversarial attacks. IEEE Trans. Auto. Control **59**(6), 1454–1467 (2014)
7. D.P. Fidler, Was stuxnet an act of war? decoding a cyberattack. IEEE Secur. Priv. **9**(4), 56–59 (2011)
8. P.M. Frank, Fault diagnosis in dynamic systems using analytical and knowledge-based redundancy: a survey and some new results. Automatica **26**(3), 459–474 (1990)
9. J.A. Goguen, J. Meseguer, Security policies and security models, in *1982 IEEE Symposium on Security and Privacy* (IEEE, 1982), pp. 11–20

10. G. Hug, J.A. Giampapa, Vulnerability assessment of AC state estimation with respect to false data injection cyber-attacks. IEEE Trans. Smart Grid **3**(3), 1362–1370 (2012)
11. H. Khurana, M. Hadley, N. Lu, D.A. Frincke, Smart-grid security issues. IEEE Secur. Priv. **8**(1) (2010)
12. O. Kosut, L. Jia, R.J. Thomas, L. Tong, Malicious data attacks on the smart grid. IEEE Trans. Smart Grid **2**(4), 645–658 (2011)
13. R. Langner, Stuxnet: dissecting a cyberwarfare weapon. IEEE Secur. Priv. **9**(3), 49–51 (2011)
14. Y. Liu, P. Ning, M.K. Reiter, False data injection attacks against state estimation in electric power grids. ACM Trans. Inf. Syst. Secur. (TISSEC) **14**(1), 13 (2011)
15. L. Ljung, System identification (Wiley Online Library, 1999)
16. W. Lucia, B. Sinopoli, G. Franze, Networked constrained cyber-physical systems subject to malicious attacks: a resilient set-theoretic control approach (2016). arXiv:1603.07984
17. P. McDaniel, S. McLaughlin, Security and privacy challenges in the smart grid. IEEE Secur. Priv. **7**(3) (2009)
18. Y. Mo, E. Garone, A. Casavola, B. Sinopoli, False data injection attacks against state estimation in wireless sensor networks, in *49th IEEE Conference on Decision and Control (CDC)* (IEEE, 2010), pp. 5967–5972
19. Y. Mo, T.H.J. Kim, K. Brancik, D. Dickinson, H. Lee, A. Perrig, B. Sinopoli, Cyber-physical security of a smart grid infrastructure. Proc. IEEE **100**(1), 195–209 (2012)
20. Y. Mo, B. Sinopoli, Secure control against replay attacks, in *47th Annual Allerton Conference on Communication, Control, and Computing (Allerton)* (IEEE, 2009), pp. 911–918
21. Y. Mo, S. Weerakkody, B. Sinopoli, Physical authentication of control systems: designing watermarked control inputs to detect counterfeit sensor outputs. IEEE Control Syst. Mag. **35**(1), 93–109 (2015)
22. A. Mpitziopoulos, D. Gavalas, C. Konstantopoulos, G. Pantziou, A survey on jamming attacks and countermeasures in WSNs. IEEE Commun. Surv. Tutor. **11**(4) (2009)
23. F. Pasqualetti, F. Dörfler, F. Bullo, Attack detection and identification in cyber-physical systems. IEEE Trans. Autom. Control **58**(11), 2715–2729 (2013)
24. T. Peng, C. Leckie, K. Ramamohanarao, Survey of network-based defense mechanisms countering the DoS and DDoS problems. ACM Comput. Surv. (CSUR) **39**(1), 3 (2007)
25. T. Pultarova, Cyber security-Ukraine grid hack is wake-up call for network operators [news briefing]. Eng. Technol. **11**(1), 12–13 (2016)
26. S.R. Rajagopalan, L. Sankar, S. Mohajer, H.V. Poor, Smart meter privacy: a utility-privacy framework, in *2011 IEEE International Conference on Smart Grid Communications (SmartGridComm)* (IEEE, 2011), pp. 190–195
27. H. Sandberg, A. Teixeira, K.H. Johansson, On security indices for state estimators in power networks, in *First Workshop on Secure Control Systems, CPSWEEK* (2010)
28. B. Satchidanandan, P. Kumar, Dynamic watermarking: active defense of networked cyber-physical systems. Proc. IEEE **105**(2), 219–240 (2017)
29. E. Shi, A. Perrig, Designing secure sensor networks. IEEE Wirel. Commun. **11**(6), 38–43 (2004)
30. Y. Shoukry, A. Puggelli, P. Nuzzo, A.L. Sangiovanni-Vincentelli, S.A. Seshia, P. Tabuada, Sound and complete state estimation for linear dynamical systems under sensor attacks using satisfiability modulo theory solving, in *American Control Conference (ACC), 2015* (IEEE, 2015), pp. 3818–3823
31. J. Slay, M. Miller, Lessons learned from the Maroochy water breach, in *International Conference on Critical Infrastructure Protection* (Springer, 2007), pp. 73–82
32. G. Smith, On the foundations of quantitative information flow, in *International Conference on Foundations of Software Science and Computational Structures* (Springer, 2009), pp. 288–302
33. R.S. Smith, Covert misappropriation of networked control systems: presenting a feedback structure. IEEE Control. Syst. Mag. **35**(1), 82–92 (2015)
34. R. Tan, V. Badrinath Krishna, D.K. Yau, Z. Kalbarczyk, Impact of integrity attacks on real-time pricing in smart grids, in *Proceedings of the 2013 ACM SIGSAC Conference on Computer & Communications Security* (ACM, 2013), pp. 439–450

35. A. Teixeira, D. Pérez, H. Sandberg, K.H. Johansson, Attack models and scenarios for networked control systems, in *Proceedings of the 1st international conference on High Confidence Networked Systems* (ACM, 2012), pp. 55–64

36. A. Teixeira, I. Shames, H. Sandberg, K.H. Johansson, Revealing stealthy attacks in control systems, in *50th Annual Allerton Conference on Communication, Control, and Computing (Allerton)* (IEEE, 2012), pp. 1806–1813

37. D. Volpano, C. Irvine, G. Smith, A sound type system for secure flow analysis. J. Comput. Sec. **4**(2–3), 167–187 (1996)

38. S. Weerakkody, X. Liu, S.H. Son, B. Sinopoli, A graph-theoretic characterization of perfect attackability for secure design of distributed control systems. IEEE Trans. Control Netw. Syst. **4**(1), 60–70 (2017)

39. S. Weerakkody, Y. Mo, B. Sinopoli, Detecting integrity attacks on control systems using robust physical watermarking, in *53rd Annual Conference on Decision and Control (CDC)* (IEEE, 2014), pp. 3757–3764

40. S. Weerakkody, B. Sinopoli, Detecting integrity attacks on control systems using a moving target approach, in *54th Annual Conference on Decision and Control (CDC)* (IEEE, 2015), pp. 5820–5826

41. S. Weerakkody, B. Sinopoli, A moving target approach for identifying malicious sensors in control systems, in *54th Annual Allerton Conference on Communication, Control, and Computing* (IEEE, 2016), pp. 1149–1156. https://arxiv.org/pdf/1609.09043.pdf

42. S. Weerakkody, B. Sinopoli, S. Kar, A. Datta, Information flow for security in control systems, in *55th Conference on Decision and Control (CDC)* (IEEE, 2016), pp. 5065–5072

43. J. Weimer, S. Kar, K.H. Johansson, Distributed detection and isolation of topology attacks in power networks, in *Proceedings of the 1st International Conference on High Confidence Networked Systems* (ACM, 2012), pp. 65–72

44. A.S. Willsky, A survey of design methods for failure detection in dynamic systems. Automatica **12**(6), 601–611 (1976)

45. D. Wu, C. Zhou, Fault-tolerant and scalable key management for smart grid. IEEE Trans. Smart Grid **2**(2), 375–381 (2011)

46. L. Xie, Y. Mo, B. Sinopoli, False data injection attacks in electricity markets, in *First IEEE International Conference on Smart Grid Communications (SmartGridComm)* (IEEE, 2010), pp. 226–231

47. Y. Yuan, Y. Mo, Security in cyber-physical systems: controller design against known-plaintext attack, in *54th IEEE Conference on Decision and Control* (IEEE, 2015), pp. 5814–5819

Toward Resilient Operation
of Smart Grid

Azwirman Gusrialdi and Zhihua Qu

Abstract Electric grids in the future will be highly integrated with information and communications technology resulting in a complex cyber-physical system. The increase in the use of information technology is expected to enhance reliability, efficiency, and sustainability of the future electric grid through the implementation of sophisticated monitoring and control strategies. However, the information and communication technology is known to be vulnerable to cyber-intrusions, which may cause physical damage to the power network due to the tight coupling between the physical and cyber layers. This chapter first discusses potential strategies to detect stealthy attacks in a smart grid. Since attacks cannot be foreseen in advance, it is highly desirable to design control algorithms so that the networked system becomes resilient against unknown attacks. Moreover, the heterogeneity of the individual dynamics and the openness of the networked system introduce additional challenges in designing the resilient control algorithm. To this end, we discuss the concept of passivity-short and demonstrate its potential to deal with heterogeneous dynamics and enable plug-and-play operation of the networked system. Two distributed control strategies are then presented to guarantee the resilience of the networked system against unknown attacks.

1 Introduction

Electric grids (physical systems) in the future will be highly integrated with information and communications technology (cyber layer) resulting in a complex cyber-physical system. The communication technology has been mainly used by the Supervisory Control and Data Acquisition (SCADA) systems for the purpose of sensing, monitoring, and control of the power systems. The increase in the use of information technology is expected to enhance reliability, efficiency, and sustainability of

---

A. Gusrialdi · Z. Qu (✉)
University of Central Florida, Orlando 32816, USA
e-mail: qu@ucf.edu

A. Gusrialdi
e-mail: azwirman.gusrialdi@ucf.edu

© Springer Nature Switzerland AG 2019
J. Stoustrup et al. (eds.), *Smart Grid Control*, Power Electronics
and Power Systems, https://doi.org/10.1007/978-3-319-98310-3_17

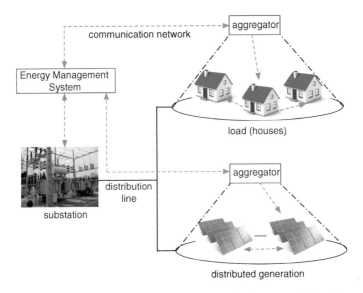

Fig. 1 Two-way communications (denoted by dash arrows) in the smart grid. An adversary could intercept the communications in order to destabilize the systems or to gain financial benefit

the future electric grid through the implementation of sophisticated monitoring and control strategies. On the other hand, information and communication technology of the power grids have started evolving from isolated structures into more open and networked environments via TCP/IP and Ethernet. Since the information and communication technology is known to be vulnerable to cyber-intrusions, potential network intrusion may cause physical damage to the power network due to the tight coupling between the power system and the cyber layer.

As an illustration, consider the demand-side management system shown in Fig. 1. The objective of such a system is to minimize peak demand and shift this load (i.e., demand response) to off-peak hours in order to improve system operation. To this end, the load control systems are equipped with two-way communications such as the Internet. Demand response can be achieved by either controlling some of the loads such as water heating, air conditioning directly or indirectly via incentives such as real-time pricing sent through the communication channel. Instead of physically accessing the power system device such as substations and gain complete control of the device, the adversary can remotely compromise a portion of the communication channels between the energy management system and aggregator. For example, the adversary could send a falsified price signal resulting in a large spike in the total load demand and destabilize the frequency of the power system [3, 22]. A real-world example of the most recent cyber-attack on power systems is the attack on the Ukraine power grid in December 2015, which is a synchronized and coordinated cyber-attack, causing a 6-hour blackout and affecting hundreds of thousands of customers [23].

Tools from network security alone are not sufficient to address the above challenges in smart grid security, since they do not take into account the physical attacks through direct interaction with the components in the grid, including the stability and control performance of the physical systems. Furthermore, the integrity of a meter can be violated without the need of breaking the cyber-security countermeasure implemented to protect the data sent through the network by placing a shunt around a meter.

This chapter presents strategies to detect false data injections on a smart grid. Specifically, we are interested in detecting a stealthy attack, that is the one which could not be detected by the standard bad data detection test. Moreover, since attacks cannot be foreseen in advance, it is desirable to design control algorithms so that the overall system becomes resilient against unknown attacks. An important issue that also needs to be taken into account in designing resilient control algorithm for a smart grid is the heterogeneity of the individual system and the openness of the networked system. To this end, we first discuss the concept of passivity-short and demonstrate its potential to deal with heterogeneous dynamics and enable plug-and-play operation of the networked system. Two distributed control strategies to make network-level cooperative control resilient against unknown attacks are then discussed.

2   Smart Grid and Potential Cyber-Attacks

Smart grid, in general, consists of four integrated sets of components:

(i) Heterogeneous and individual dynamic systems (e.g., distributed generations) that can be modeled as, for $i = 1, \ldots, n$,

$$\dot{x}_i = \mathscr{F}_i(x_i, v_i), \quad y_i = H_i(x_i), \tag{1}$$

where $x_i \in \mathfrak{R}^{n_i}$ is the state, v_i is the control input, and $y_i \in \mathfrak{R}^p$ is the output.

(ii) A physical network whose characteristics could be characterized as

$$\kappa_l(y_1, y_2, \ldots, y_n, \theta_1, \ldots, \theta_n) = 0, \quad l = 1, \ldots, n, \tag{2}$$

where $\theta_i \in \mathfrak{R}^{n_i}$ are the state variables of the physical network.

(iii) Sensory and local communication networks are represented by observation equation (estimator)

$$z = H\theta + w, \tag{3}$$

and communication matrix

$$S(t) = \left[S_{ij} \right] \in \mathfrak{R}^{n \times n}, \quad S_{ii}(t) \equiv 1, \quad S_{ij}(t) = \begin{cases} 1 \ \{j \to i\} \in \mathscr{E} \\ 0 \ else \end{cases},$$

where $z \in \mathfrak{R}^M$ is the vector of measurements, $H \in \mathfrak{R}^{M \times N}$ is the measurement matrix, w is the measurement noise, and digraph $(\mathscr{V}, \mathscr{E}(t))$ represents

the instantaneous communication topology. Moreover, the control center could
also implement dynamic estimators such as Kalman filter.
(iv) A system operator (e.g., energy management system as depicted in Fig. 1) who
     attempts to optimize the overall system.

In this chapter, we consider a class of attacks known as deception attacks which
violate data integrity and whose goal is to destabilize the system by injecting exogenous input to the system. To this end, it is assumed that the adversary is able to
intercept the communication network and corrupt the measurement and input signals of the system. Without any knowledge of the system dynamics, the adversary
can disrupt the operation of the system by launching a replay attack [31]. That is,
first he/she observes and records the readings of the sensors and then repeats them
while injecting exogenous signal to the system.

If the adversary has some knowledge of the system model, he/she can then perform
a much more powerful attack and possibly without being detected (i.e., *stealthy*) by
the standard statistical test. This class of deception attacks is also known as false
data injection attack (FDIA). One of the well-known FDIA attacks on power grid is
the attack on state estimation [16]. State estimation is a key function in smart grid
due to its wide applications such as for contingency analysis, load forecasting, and
calculating locational marginal pricing for power markets. State estimation aims at
estimating the states $\theta \in \mathfrak{R}^N$ of the power systems (e.g., bus voltage magnitudes and
phase angles) denoted by from a limited set of measurements $z \in \mathfrak{R}^M$ (with $M > N$)
by solving (3). It is known that if the adversary has sufficient information of the
physical network, he/she can then launch FDIA by modifying measurement data
from the true measurement z in (3) to $(z + \delta_z)$ for some $\delta_z = Hc$, where $c \in \mathfrak{R}^N$.
In other words, the attack vector δ_z lie in the null space of matrix $(P - I)$ where P
is projection of matrix H, namely $P = H(H^T H)^{-1} H^T$. It is shown that the injection $\delta_z = Hc$ keeps the measurement residual unchanged compared to the case when
the measurement is not corrupted. Hence, the attack is stealthy (unobservable) since
its presence cannot be detected by bad data detection test which utilizes the measurement residual. Since, in practice, the adversary has a limited budget to compromise
the sensors, the smallest set of sensors that need to be compromised which results in
network unobservability is discussed in [15]. While the above discussions focus on
the attack by a single adversary, the case where the attack is launched by multiple
adversaries is analyzed in [28].

In the following, we discuss existing and potential strategies to detect both replay
and false data injection attacks. While detecting an attack is an important first step
toward mitigating the effect of the attack, the ultimate objective in securing the smart
grid is to ensure the operation of the grid (that is the power network can still provide
essential services) under unknown attacks. Therefore, in addition to attack detection
methods, promising strategies to make smart grid resilient against unknown attacks
will also be discussed.

3   Potential Defense Strategies

In the following, we discuss potential strategies together with their challenges to detect (stealthy) deception attacks and make the smart grid resilient against unknown cyber-attacks.

3.1   Attack Detection

When the injection is not stealthy, a method to detect the attack is proposed in [15], which utilizes a generalized likelihood ratio test and incorporates historical data. Moreover, the work [12] develops a real-time detection scheme whose goal is to detect the attack as early as possible while satisfying certain error constraints. In comparison to the traditional detection test and the generalized likelihood ratio test, the algorithm has low complexity and leads to a better accuracy. For the case of stealthy attacks, the system operator can strategically secure certain critical sensors to either increase the number of sensors needed to attacks needed or make attack vector c infeasible, see e.g., [14, 20]. Moreover, a game theoretic approach to analyze FDIA involving a defender (whose goal is to reduce the impact of the attack by securing a set of measurements) and multiple adversaries are studied in [28]. However, from a practical point of view securing meters works well for transmission but not for expansive distribution network. A strategy based on reconfiguring the information structure used for state estimation is proposed in [30]. To this end, it is assumed that the power grid can be partitioned into a group of subsystems as depicted in Fig. 2. Moreover, each subsystem is assumed to have the capability of reconfiguring its information structure, performing state estimation, and reporting the result to the higher level energy management system. Assume that the matrix H

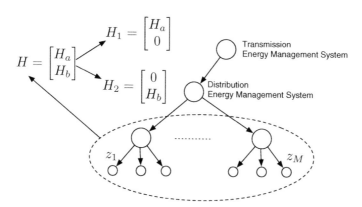

Fig. 2 Protection of power grid against false data injection attack

can be reconfigured by $\begin{bmatrix} H_a \\ H_b \end{bmatrix}$ and partitioned into $H_1 = \begin{bmatrix} H_a \\ 0 \end{bmatrix}$ and $H_2 = \begin{bmatrix} 0 \\ H_b \end{bmatrix}$ such that

$$rank \begin{bmatrix} P_1 - I \\ P_2 - I \end{bmatrix} = M$$

where P_1, P_2 denote the projection matrix of H_1 and H_2 respectively. The only admissible solution of stealthy attack vector is then given by $\delta_z = 0$. Hence, the power grid can be secured from coordinated data attacks by reconfiguring the information structure and without the need for any physical effort. Moreover, if physical constraints (2) could be adjusted real time, detectability of false data injection would be substantially increased by using dynamically changing physical topology.

As a future direction, the designer (system operator) can also take advantage of the physical system dynamics and use it to detect stealthy attacks. In particular, dynamics of local systems in (1) can be utilized to enhance injection detection, and Kalman filters can be designed to correlate local system output y_i to estimated physical network state $(\theta + c)$. In addition, since in order to remain stealthy, the adversary needs to obey to a certain structure related to the physical system, the system operator can then introduce a type of randomization (not known to the attacker) to increase the probability of detecting the stealthy attacks. For example, a concept of physical watermark is introduced in [21] to detect *replay attack*, where a disturbance is added to the control signal which acts as a time-stamped authentication signal. The attacks can then be detected with high probability by checking the correlation between disturbance and measurement. Another potential strategy to detect stealthy FDIA is by utilizing tools from machine learning to distinguish normal operation from the case under attacks. Learning-based techniques, namely supervised learning (requiring data training) and unsupervised learning are proposed in [1] to detect stealthy attacks. In both strategies, a principal component analysis which projects the data to a low-dimensional observation is used to reduce dimensionality of the processed data. An interesting key observation is that normal and compromised data tend to be separated in the projected space since the normal data are governed by physical laws while the compromised data are not. Finally, distributed detection algorithms should be developed to take advantage of better resolution and granularity of sensory data at the local level.

3.2 Resilient Control Design

Resilient control refers to those controls that are capable of maintaining or restoring systems performance under unexpected events. Given that attacks or faults cannot be foreseen in advance, it is desirable to design control algorithms so that the overall system becomes resilient against unknown attacks, which will be discussed in this section.

3.2.1 Modularized Design for Cooperative Control

As described previously, a smart grid involves a number of heterogeneous physical systems that are networked not only by their physical network but also their sensory, communication, and control networks. Moreover, the networked operation of these heterogeneous dynamics needs to be maintained even when some of the components get upgraded or exchanged. Hence, before designing a resilient control system, we first need to answer the fundamental question of how a networked of heterogeneous systems can be operated in a stable and cooperative manner. The concept of dissipativity offers a way to analyze input–output relationship of nonlinear systems and hence has potential to deal with the heterogeneous dynamics and their networked operation. System (1) is said to be dissipative if a positive semi-definite storage function V_i and a supply rate $\Phi_i(\cdot)$ exist such that

$$V_i(x_i) - V_i(x_i(0)) \le \int_0^t \Phi_i(x_i, v_i)ds.$$

The input–output pair $\{v_i, y_i\}$ of the system (1) is said to be input feedforward passive if, for some positive semi-definite function η_i,

$$\Phi_i(x_i, v_i) = -\eta_i(x_i) + v_i^T y_i + \frac{\varepsilon_i}{2}\|v_i\|^2$$

where the constant ε_i is called impact coefficient. The most common form of dissipativity is passivity [13], that is when $\varepsilon_i \le 0$. Even though control design for passive systems is well known, they are known to have limited applications. On the other hand, if $\varepsilon_i \ge 0$ the system is said to be passivity-short [29]. While passive systems are always decreasing in energy with respect to input energy, passivity-short systems may increase or remain the same in energy from input to output during the transience. This behavior is also related to oscillating systems with small or nonexistent damping, in which the output energy of the system may be similar to the input energy. An example of this is a generator that is not decreasing in energy at all times since it is producing some amount of energy. While most systems are passivity-short (e.g., it is shown in [11] that each component in the power system is passivity- short), analysis tools and control design for passivity-short systems have not been investigated until recently [25]. Passivity-short systems enjoy the nice property that its input feedforward passivity is invariant under all interconnections including parallel, series, positive feedback loop, and negative feedback loop [10]. Cooperative control theory [24] aims to design network-enabled distributed controls by explicitly considering both dynamics and control of individual physical systems as well as local communication networks and information-structured controls. Specifically, the goal of cooperative control is to achieve nontrivial consensus using only local information, that is for all individual system i in the network we have

$$\lim_{t\to\infty} \|y_i - y_j\| = 0, \forall i, j \quad \text{or} \quad \lim_{t\to\infty} y_i(t) = c_n. \tag{4}$$

Cooperative controls have been successfully applied to distributed dispatch of active and reactive power [32], distributed voltage control and loss minimization [19], charging of electric vehicles [8], load control for ancillary services [27], and optimal power flow and frequency synchronization [17]. The concept of passivity-short allows us to develop a fully modular design methodology where a self-feedback control is first designed individually for each heterogeneous system and network-level cooperative control can then be designed separately. To this end, the control input in (1) is chosen as

$$v_i = v_{s_i}(x_i) + u_i$$

where v_{s_i} is the self-feedback control designed so that the individual system becomes passivity-short with impact coefficient $\varepsilon_i \in (0, \bar{\varepsilon})$ and $u_i \in \Re^{m_i}$ is the network-enable control. It is shown in [26] that, for any connected network of passivity-short systems, their network-level control can always be chosen to be

$$u_i = k_{y_i} \sum_{j=1}^{n} S_{ij}(y_j - y_i)$$

to guarantee non-trivial consensus (4), provided that $k_{y_i} \in (0, \bar{k}_y)$ with $\bar{k}_y$ is given by

$$\bar{k}_y = \frac{\lambda_2(\Gamma L + L^T \Gamma)}{\bar{\varepsilon}\lambda_{\max}(L^T \Gamma L)}$$

in which $L = D - S$ with $D = diag(S1)$, λ_2 denotes the smallest nonzero eigenvalue, $\lambda_{\max}$ denotes the largest eigenvalue, matrix $\Gamma = diag(\gamma)$ and eigenvector γ satisfies $\gamma^T L = 0$. It is worth noting that the constant $\bar{k}_y$ can be computed in a distributed manner using the approach developed in [4, 5]. It can also be observed that the network-level control design does not require any information about individual systems except that their impact coefficient are no larger than a threshold $\bar{\varepsilon}$. This feature enables plug-and-play feature of networked operations, that is any individual heterogeneous systems in the network can be switched into and out of service. As has been demonstrated above, the concept of passivity-short and impact equivalence principle enable the separation of control design of individual heterogeneous systems and network-level cooperative control design. Hence, for the remaining of the section, we focus on potential strategies to make the network-level cooperative control resilient against unknown attacks, that is for a small constant δ:

$$\lim_{t \to \infty} \|y_i - y_j\| \le \delta, \forall i, j \quad \text{or} \quad \lim_{t \to \infty} \|y_i(t) - c_n\| \le \delta. \tag{5}$$

In order to illustrate resilient control design, we consider a cooperative power control strategy of multiple distributed generators in distribution network. Distributed generators (DGs) have received significant attention since they could reduce power transmission loss and maintain power supply even in an emergency situation by operating in an island mode. Conventional centralized control scheme which requires each

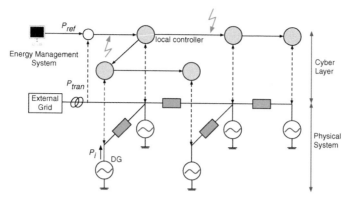

Fig. 3 Cooperative control of DGs. Solid and dash arrows denote communication between local controller of DGs and between physical and cyber layers respectively. Red arrows represent potential cyber-attack

DGs to communicate directly with a central control is not desirable in controlling the geographically dispersed DGs since it requires high-bandwidth communication and more importantly, it does not enable plug-and-play operation. Hence, distributed control scheme which relies only on local communication between geographic neighboring DGs is more suitable due to its low operating cost, less system requirement, scalability and robustness to communication failures. A cooperative control scheme is proposed in [32] to design a virtual power plant (VPP) consisting of multiple DGs connected to transmission grid via a step-up transformer as depicted in Fig. 3. The objective is to regulate the active power flow across the line to a preferred value while at the same time ensuring that the active power of each DG reaches consensus on its utilization (defined as the ratio between the dispatched active power P_i and the maximal available power $P_{i,max}$). To be more precise, the goal is to regulate the power dispatched by each DG so that the total active power transferred to the transmission network P_{tran} (which depends on P_i) tracks a reference value P_{ref} and the utilization ratio of all DGs, that is $\frac{P_i}{P_{i,max}}$ converge to the same value.

It is demonstrated in [32] that the objectives mentioned above can be achieved by designing cooperative control strategy in the form of (4) (that is by taking $y_i = \frac{P_i}{P_{i,max}}$) which relies on local communications between the DGs as shown in Fig. 3. However, it is shown in [18] that an adversary could compromise local controller of DGs by launching a replay attack to prevent DGs in the network reach a consensus on the utilization ratio and thus result in unfair split of the profit among the DG owners. As an example, when the i-th DG is compromised such that it intentionally increases its dispatched power (consequently its utilization ratio), this will yield other DGs converge to a smaller utilization ratios (i.e., generating less power) to maintain P_{tran} to track P_{ref}. As a result, DG i has more economic benefit since it generates more electricity. Furthermore, this compromised DG could also deliberately set its utilization ratio to weaken the controllability of the VPP by narrowing the adjustable range of P_{tran}.

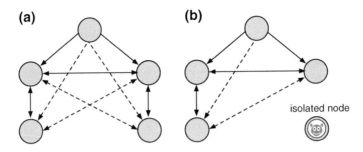

Fig. 4 **a** Communication network consisting of control graph (solid arrow) and observation graph (dash arrow). An observation link is denoted by a solid arrow if it coincides with control link; **b** network after the isolation of misbehaving node

3.2.2 Confidence-Based Resilient Control Design

An advanced cooperative control strategy is developed in [18] to make the VPP resilient against the above- mentioned attacks, that is P_{tran} tracks the value P_{ref} and all DGs reach a uniform utilization. First, given a control graph (that is a communication topology used for cooperative control (4)), an observation graph is constructed whose responsibility is mainly for surveillance, that is to estimate both the upper and lower bounds of utilization ratio of every in-neighboring DGs. The topology of the observation graph is designed such that for each DG k, any of its in-neighbors can communicate to any of its out-neighbors as illustrated in Fig. 4a. For each out-neighbors of the k-th DG, it can then estimate the feasible ratio range of DG k using the information obtained from the observation links and given that it knows the in-neighbors of the k-th DG. A distributed confidence level manager (DCLM) is then established for each DG which maintains a confidence level (a nonnegative value) for each of its in-neighbors and is updated at each iteration when executing the cooperative control. That is the confidence level of a certain in-neighboring DG decreases when the probability that the corresponding node is misbehaving increases. Finally, using the confidence level the misbehaving DG is gradually marginalized and isolated (that is all communication links relevant to it are removed) so that it cannot influence the consensus of the remaining DGs in the network as illustrated in Fig. 4b. The strategy has a limitation when detecting colluding attack which requires multiple DGs to be compromised simultaneously. For example, when a DG modifies its information to others and its output neighbors do not marginalize and isolate it on purpose.

3.2.3 Competitive Interaction Based Resilient Control Design

An alternative strategy to make the networked operation of smart grid resilient against unknown attacks is by introducing an additional information flow, namely a virtual

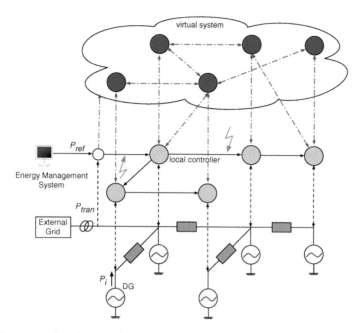

Fig. 5 Interconnection of cooperative system and a virtual system denoted by blue dash-dot arrows

system (with the so-called hidden layer and synthetic anchors) [6, 9], shielded from the adversary (thus it is called hidden layer) and interconnected with the local controller of the DGs as illustrated in Fig. 5. As discussed in [6], dynamics of the virtual system and its interconnection with the local controller of the cooperative system need to be designed such that its presence does not interfere with the operation of the original networked system. Moreover, it is designed to maintain stability of the overall system by competitively interacting with the original networked system [2]. The virtual system in principle increases the networked system's inertia against unknown but bounded attacks [6] and thus resilient operation of the networked system is guaranteed, that is $\lim_{t \to \infty} \| y_i(t) - c_n \| \leq \delta$.

It is worth noting that the design of the virtual system together with its interconnection does not require any information about the nature of the attacks and is independent of the number of attacks. The virtual system with hidden network could be implemented as an internal signal component at every node of the networked system whose security is guaranteed such as within SCADA system which makes it difficult (i.e., requires high cost) for the attacker to compromise. Moreover, additional information flow in the hidden layer can be achieved using different communication network/channel (for better security) such as internet technology (including software-defined networks). In contrast to the state of the networked system, the state of the hidden layer has no physical meaning (e.g., can be transferred by the internet) and thus is less observable to the attacker (i.e., unlikely to be attacked) or it would be difficult for the attacker to associate the information flow in the internet with the measurements of physical variables used by the networked systems. In other words, the networked systems could be made resilient if the defender has a superior information structure, i.e., has access to more layers of information than those accessible to the attacker [7]. The addition of a hidden layer also comes at a price of increased communication expenses. However, the communication expense could be small as well since the hidden network is not necessarily physical (as the primary network of cooperative systems) and hence could be done using wireless or internet. Furthermore, the interconnection between the virtual system and the networked system could also be designed to be distributed and sparse. It is also demonstrated in [6] that when less than half of the nodes being attacked, the strategy could also help the system to identify the attacks. Hence, it can be seen that such cyber information layers would substantially increase the resilience of the overall cyber-physical system and maintain open access to the system while thwarting all potential attacks. As a future direction, the strategy can be further combined with Kalman filters in order to detect attacks without any restriction on the number of nodes being attacks. Moreover, further analysis needs to be done for the case of time-varying topologies.

4 Conclusion

This chapter first briefly discusses cyber-attacks on a smart grid and their potential impacts. Potential strategies to detect stealthy attacks in a smart grid are then presented and discussed. Since attacks or faults cannot be foreseen in advance, it is desirable to design control algorithms so that the overall system becomes resilient against unknown attacks. An important issue that needs to be taken into account in designing resilient control algorithm for a smart grid is the heterogeneity of the individual system and the openness of the networked system. To this end, we first discuss the concept of passivity-short and demonstrate its potential to deal with heterogeneous dynamics and enable plug-and-play operation of the system. More importantly, it enables the separation of control design of individual heterogeneous systems and network-level cooperative control design which simplifies the design and analysis. Two distributed control strategies to make network-level cooperative control resilient against unknown attacks are then discussed. The results serve as initial steps towards development of a comprehensive analysis and design framework for resilient operation of smart grid.

Acknowledgements This work is supported in part by U.S. Department of Transportation (award DTRT13GUTC51), by U.S. National Science Foundation (grant ECCS-1308928), by US Department of Energy (awards DE-EE0006340 and DE-EE0007327), by L-3 Communication Coleman Aerospace (contract 1101312034), by Texas Instruments' awards, and by Leidos (contract P010161530).

References

1. M. Esmalifalak, L. Liu, N. Nguyen, R. Zheng, Z. Han, Detecting stealthy false data injection using machine learning in smart grid. IEEE Syst. J. **11**(3), 1644–1652 (2017)
2. B. Gharesifard, T. Basar, Resilience in consensus dynamics via competitive interconnections, in *3rd IFAC Workshop on Estimation and Control of Networked Systems* (2012), pp. 234–239
3. J. Giraldo, A. Cárdenas, N. Quijano, Integrity attacks on real-time pricing in smart grids: impact and countermeasures. IEEE Trans. Smart Grid **8**(5), 2249–2257 (2017)
4. A. Gusrialdi, Z. Qu, Growing connected networks under privacy constraint: achieving trade-off between performance and security, in *The 54th IEEE Conference on Decision and Control* (2015), pp. 312–317
5. A. Gusrialdi, Z. Qu, Distributed estimation of all the eigenvalues and eigenvectors of matrices associated with strongly connected digraphs. IEEE Control Syst. Lett. **1**(2), 328–333 (2017)
6. A. Gusrialdi, Z. Qu, M.A. Simaan, Robust design of cooperative systems against attacks, in *American Control Conference* (2014), pp. 1456–1462
7. A. Gusrialdi, Z. Qu, M.A. Simaan, Game theoretical designs of resilient cooperative systems, in *European Control Conference* (2015), pp. 1705–1711
8. A. Gusrialdi, Z. Qu, M.A. Simaan, Distributed scheduling and cooperative control for charging of electric vehicles at highway service stations. IEEE Trans. Intel. Transp. Syst. **18**(10), 2713–2727 (2017)
9. A. Gusrialdi, Z. Qu, M.A. Simaan, Competitive interaction design of cooperative systems against attacks. IEEE Trans. Auto. Control (accepted and to appear) (2018)
10. R. Harvey, Z. Qu, Cooperative control and networked operation of passivity-short systems, in *Control of Complex Systems*, ed. by K.G. Vamvoudakis, S. Jagannathan (Butterworth-Heinemann, 2016), pp. 499–518
11. R. Harvey, X. Ying, Z. Qu, T. Namerikawa, Dissipativity-based design of local and wide-area der controls for large-scale power systems with high penetration of renewables, in *IEEE Conference on Control Technology* (2017), pp. 2180–2187
12. Y. Huang, J. Tang, Y. Cheng, H. Li, K.A. Campbell, Z. Han, Real-time detection of false data injection in smart grid networks: an adaptive cusum method and analysis. IEEE Syst. J. **10**(2), 532–543 (2016)
13. H.K. Khalil, *Noninear Systems* (Prentice-Hall, New Jersey, 2003)
14. K. Khanna, B.K. Panigrahi, A. Joshi, Feasibility and mitigation of false data injection attacks in smart grid, in *IEEE 6th International Conference on Power Systems (ICPS)* (2016), pp. 1–6
15. O. Kosut, L. Jia, R.J. Thomas, L. Tong, Malicious data attacks on the smart grid. IEEE Trans. Smart Grid **2**(4), 645–658 (2011)
16. G. Liang, J. Zhao, F. Luo, S. Weller, Z.Y. Dong, A review of false data injection attacks against modern power systems. IEEE Trans. Smart Grid **8**(4), 1630–1638 (2017)
17. Y. Liu, Z. Qu, H. Xin, D. Gan, Distributed real-time optimal power flow control in smart grid. IEEE Trans. Power Syst. **32**(5), 3403–3414 (2017)
18. Y. Liu, H. Xin, Z. Qu, D. Gan, An attack-resilient cooperative control strategy of multiple distributed generators in distribution networks. IEEE Trans. Smart Grid **7**(6), 2923–2932 (2016)
19. A. Maknouninejad, Z. Qu, Realizing unified microgrid voltage profile and loss minimization: a cooperative distributed optimization and control approach. IEEE Trans. Smart Grid **5**(4), 1621–1630 (2014)
20. Y. Mo, T. Hyun-Jin Kim, K. Brancik, D. Dickinson, H. Lee, A. Perrig, B. Sinopoli, Cyber-physical security of a smart grid infrastructure. Proc. IEEE **100**(1), 195–209 (2012)
21. Y. Mo, S. Weerakkody, B. Sinopoli, Physical authentication of control systems: designing watermarked control inputs to detect counterfeit sensor outputs. IEEE Control Syst. **35**(1), 93–109 (2015)
22. A.H. Mohsenian-Rad, A. Leon-Garcia, Distributed internet-based load altering attacks against smart power grids. IEEE Trans. Smart Grid **2**(4), 667–674 (2011)
23. T. Pultarova, Cyber security—ukraine grid hack is wake-up call for network operators [news briefing]. Eng. Technol. **11**(1), 12–13 (2016)

24. Z. Qu, Cooperative control of dynamical systems: applications to autonomous vehicles (Springer Science & Business Media, 2009)
25. Z. Qu, An impact equivalence principle of separating control designs for networked heterogeneous affine systems. IFAC Proc. Vols. **45**(26), 210–215 (2012)
26. Z. Qu, M.A. Simaan, Modularized design for cooperative control and plug-and-play operation of networked heterogeneous systems. Automatica **50**(9), 2405–2414 (2014)
27. T. Rahman, R. Harvey, Z. Qu, M.A. Simaan, A distributed cooperative load control approach for ancillary services in smart grid, in *American Control Conference* (2017), pp. 1401–1406
28. A. Sanjab, W. Saad, Data injection attacks on smart grids with multiple adversaries: a game-theoretic perspective. IEEE Trans. Smart Grid **7**(4), 2038–2049 (2016)
29. R. Sepulchre, M. Jankovic, P.V. Kokotovic, Constructive nonlinear control (Springer Science & Business Media, 2012)
30. M. Talebi, C. Li, Z. Qu, Enhanced protection against false data injection by dynamically changing information structure of microgrids, in *2012 IEEE 7th Sensor Array and Multichannel Signal Processing Workshop (SAM)* (2012), pp. 393–396
31. A. Teixeira, I. Shames, H. Sandberg, K.H. Johansson, A secure control framework for resource-limited adversaries. Automatica **51**, 135–148 (2015)
32. H. Xin, Z. Qu, J. Seuss, A. Maknouninejad, A self-organizing strategy for power flow control of photovoltaic generators in a distribution network. IEEE Trans. Power Syst. **26**(3), 1462–1473 (2011)

.